Why Mining?

by

David L McKay

© Copyright 2002 David L. McKay. All rights reserved.

No part of this publication may be reproduced, stored in a retrieval system, or transmitted, in any form or by any means, electronic, mechanical, photocopying, recording, or otherwise, without the written prior permission of the author.

Printed in Victoria, Canada

National Library of Canada Cataloguing in Publication Data

```
McKay, David L. 1927-
     Why mining? / David L. McKay.
Includes index.
ISBN 1-55369-688-3
     1. McKay, David L, 1927-  2. Mining engineers--Canada-
-Biography.  I. Title.
TN140.M23A3 2002          622'.092         C2002-902958-9
```

TRAFFORD

This book was published *on-demand* in cooperation with Trafford Publishing.
On-demand publishing is a unique process and service of making a book available for retail sale to the public taking advantage of on-demand manufacturing and Internet marketing.
On-demand publishing includes promotions, retail sales, manufacturing, order fulfilment, accounting and collecting royalties on behalf of the author.

Suite 6E, 2333 Government St., Victoria, B.C. V8T 4P4, CANADA
Phone	250-383-6864	Toll-free	1-888-232-4444 (Canada & US)
Fax	250-383-6804	E-mail	sales@trafford.com
Web site	www.trafford.com	TRAFFORD PUBLISHING IS A DIVISION OF TRAFFORD HOLDINGS LTD.	
Trafford Catalogue #02-0501			www.trafford.com/robots/02-0501.html

10   9

Dedication

To Irene

Table of Contents

Part 1

Prologue — Page 1

Chapter 1 — The Start — Page 1
■Early Days 1 ■UBC 10 ■Sullivan Mine 19 ■Tadanac Concentrator 25

Chapter 2 — Sheep Creek — Page 27
■Paradise Mine 27 ■Lucky Jim 36 ■Mineral King 38 ■Lucky Jim Second Time 42

Chapter 3 — Je Me Souviens — Page 49
■Trip East 49 ■Malartic Goldfields 52 ■Malartic 63 ■Partir 69

Chapter 4 — Red Mud — Page 71
■Atikokan 71 ■Steep Rock Iron Mines 73 ■Errington Mine 78 ■Business Trips 86 ■Mine Projects 91 ■Mining Research 92 ■Activities 94

Chapter 5 — Rock Bolts & Yielding Arches — Page 97
■Rockiron 97 ■Travels 98 ■Philosophy 108

Chapter 6 — Grow — Page 112
■IMC 112 ■Engineering 116 ■Operations 121

Chapter 7 — Cominco — Page 133
■House and Home 133 ■Commuting 135 ■Travel 139 ■Construction 143 ■Startup 146 ■Flooding 151

Chapter 8 — Ontario Again — Page 157
■Ecstall 157 ■Texasgulf 163 ■No 2 Mine 170 ■Scandinavia 184 ■Kidd Creek 190 ■Head Office 195

Part 2

Chapter 9 — On My Own — Page 203
■Consulting 203

Chapter 10 — Potash & Sulphur — Page 207
■Canterra 207 ■Sulchem 213 ■SulFer Works 219

Chapter 11 — Belmoral — Page 225
■Broulan Reef 225 ■Vedron 229 ■Armistice 231 ■Ferderber & Dumont 231 ■Wrightbar 236 ■Louvem 236 ■Yorbeau 237 ■Ketza River 239

Table of Contents

Chapter 12 Sweden — Page 242
- World Mininig Congress 242

Chapter 13 Bond Gold — Page 244
- Northern B.C. 244 ■ East Amphi 245 ■ Golden Patricia 246
- Shoal Lake 249 ■ Eskay Creek 250

Chapter 14 Dynatec — Page 255
- Montanore 255 ■ General Chemicals 258

Chapter 15 Newfoundland — Page 262
- Hope Brook 262

Chapter 16 Revenue Canada — Page 266
- Drillex 266 ■ Mintronics 267 ■ Meta-Probe 267 ■ Phoenix 268
- Auburn 268 ■ Drillnamic 269 ■ Tomco 270

Chapter 17 New Jersey — Page 271
- Mount Hope Hydro Project 271

Chapter 18 Simons & Peru — Page 274
- Coroccohuayco 274

Chapter 19 Taku & Tulsequah — Page 276
- Polaris Taku 276

Chapter 20 Davy McKee — Page 282
- Kennsington 282 ■ Bakyrchik 284

Chapter 21 Wright Engineering — Page 299
- Sunshine 299 ■ Aquas Tenidas 300 ■ McCreedy East 301
- Changma 304 ■ XiYu 314 ■ HuiXiam 316

Chapter 22 Kilborn Engineering — Page 325
- Potash 325 ■ Mount Polley 327 ■ Huckleberry 331

Epilogue — Page 336

Why Mining?

Part 1

Prologue

Why mining? Why mining indeed! This question was first asked in October 1948 by Professor Leslie Crouch when I interviewed with him at the beginning of the third year of engineering at the University of British Columbia. Professor Crouch interviewed all the new mining engineering students individually during the fall term as to why they chose mining engineering as their course. Giving some sort of an answer saying something like, "... having always lived in or near mining towns, I enjoyed the people." It was a pretty lame answer but it was the best that could be given at the time. This biography, hopefully, will shed some light on why this person chose mining engineering as a career.

Chapter 1
The Start

Early Days

How does one start the story of their career other than to say, it all started in Invermere B.C. on August 4, 1927 the day I was born. I was the first born of three boys to Judith and Bill McKay. We were not what you would say very closely spaced in our birth dates as I was three and a half years older than my brother Jock and thirteen years older than my youngest brother Billy. As a consequence, as brothers, we did not have a very close relationship. Our dear mother tried to get Jock and I together when we were younger obligating me (or rather forcing me) to take him everywhere I went. Perhaps that is why to this day, there appears to be this wall or barrier in our relationship. In any case, ours was a very happy childhood. My mother and father could not have been more lovable and in fact in my eyes, even to this day, they were exemplary.

Our father started out in Athalmer in 1925 as a game warden but transferred to the British Columbia Provincial Police shortly afterward and ended up in Trail in 1961 on retirement as a member of the Royal Canadian Mounted Police (RCMP). In the 36-year interval, the McKay family moved from Athalmer to Grand Forks, then to Kaslo, to Rossland, on to Maillardville, to Penticton, to Cranbrook and finally to Trail. That's a lot of moving and the brunt of the moving was borne by our mother, although father did help out

Why Mining?

the best he could. In those days, all the furniture had to go by rail which required that the furniture had to be crated and the dishes packed in wooden barrels. I remember as a kid during these moves a carpenter would come to the house to work for a couple of days crating all the furniture. In contrast today, all one has to do is call a moving company and have them pack and load their truck and it's all done in a matter of a day or two at the most. When my parents moved, father often times was required to proceed immediately to his new posting to become familiar with his new assignment. Also, he had to seek rental accommodation for the family as the furniture would arrive later and often or not there were no storage facilities in the new town. With rail transportation for the furniture, it meant that there would sometimes be a wait of a week or two for the furniture to arrive. That meant we would be obliged to stay in hotels until the furniture caught up with the family. The hotels were not always the most comfortable either. Sometimes it would be necessary for the family to share bathrooms and that sort of thing. *(Our mother must have been very strong to endure that.)*

Although there is some dim recollection of events, when living in Grand Forks and Kaslo *(I was only three years old at the most)*, it wasn't until our move to Rossland that I had my first vivid memory of events in my life. My first memory of mining was a trip to Invermere where my grandfather, "Big" Jim McKay ran a placer operation on Toby or Horse Thief Creeks, I can't remember which. What is remembered best were the two black bears we saw on the road while travelling to the placer operation. As far as the placer operation goes, all I can remember was seeing Uncle Gordon there as he worked for my grandfather. My first trip underground occurred a year or so later in Rossland, when I was only five years old. It was during the depression and there was a lot of mining activity on Red Mountain above Rossland. It was said *(I don't know how true this is)* that Cominco, who owned the claims around Rossland, threw these open to various "leasers" in order to alleviate the unemployment in the area. These so-called "leasers" would solicit funds from various individuals through grub-staking. My father was talked into assisting in the grub-staking of one such claim by a friend of his, Bill Blair, who was a brother-in-law to Roy Stevens a mining engineer in that area.

I remember driving up the mountain in Roy Stevens' car. *(He and Bill Blair were the only two in our neighbourhood who owned cars. It was through knowing Mr Roy Stevens that I got the idea that all mining engineers were rich.)* We stopped first at a surface lease where three men were double-jacking holes in preparation to blasting a bench in an open cut. In double-jacking, two men swing sledge hammers at a drill steel held in place by a third man who rotates the steel slightly after each blow. The third man has to have a lot of faith in the men swinging the sledge hammers. After Roy Stevens had conducted his business with the men in the open cut, he drove us out past what was then called "The Black Bear" to an adit that was worked by a small crew of men. The adit was probably only into the hill about a couple of hundred feet, but in any case, the men (my Dad, Roy Stevens and Bill Blair) were given carbide lamps to carry while I was given my Dad's hand to hold to be led into the adit. There was a bit of water coming out of the sides of the adit as I remember being carried over some areas of the track that were flooded. On reaching the face there were a couple of men finishing drilling a round by single jacking. *(Those old timers must have been tough. It's a good day's work to drill-off a round with a jackleg, let alone drilling it by sledge hammers and hand-held drill steel.)* It is doubtful if Dad made anything from his mining venture as our life style didn't change. Although we were not flush with money by any means, my Dad did have a job with a steady pay cheque and that was something in the middle of the depression.

Even though the kids in Rossland didn't know much about mining they did do a lot of talking about the business. In winter we would make tunnels (we called them mines) out the snow piles that accumulated in front of our homes. In the spring of the year, there would be placer mines with the melted snow run-off. In one way or another, some mining terms would be incorporated into our games. Speaking of spring run-off, my friend, Len Camozzi and I started a small excavation (our mine) using the water that flowed off the street to down under the wooden side-walk on Columbia Avenue (the main street in Rossland) where we sawed all the cross-bracing supporting the wooden side walk above allowing us to walk from one section under the side-walk to another without having to duck under these cross-braces. There would have been more braces sawn if we hadn't been spotted by Len's older brother, Walter, who caught and scolded us for doing damage to public property. I often wonder if that section of the side walk on Columbia Avenue ever sagged from the lack of any cross-bracing (ground support). I'll probably never know as that wooden sidewalk has since been replaced with a concrete

Why Mining?

one. Another mining stunt Len Camozzi and I pulled off was to pretend to enter an old mine shaft just outside the city limits of Rossland. The collar of this particular mine shaft was timbered over with old rotten planks which could be easily lifted. After lifting some of the planks out of the way, we would throw rocks down the shaft to hear the water splash down below. The water was probably up within a couple of hundred feet of the collar. At one particular time we pretended to actually enter the shaft. A lady whose home was located a few hundred feet from that abandoned shaft saw this and had a fit with worry and then sent one of her older sons to get those kids out of there and tell them never to come back. *(A shaft collar would never be left that way in present times. The authorities require abandoned shafts secured by constructing a reinforced concrete bulkhead at the collar so that no one could accidentally or otherwise fall into the shaft.)*

When the McKay kids were growing up our mother would always drum into us that education was the key to getting a good job or position. She never said it was the key to happiness or security. She would say in a matter of fact way that... "after high school, you will either take a business course or go to college". We never questioned her statement. Our father was always supportive in what mother said so there was never any question about education. When the time came, all three McKay boys went on to college; my brother Jock, graduated as a metallurgical engineer from U.B.C.; my youngest brother Bill, graduated as a lieutenant from the Royal Military College; and I graduated from U.B.C. in mining engineering.

While growing up, I, like all kids, probably wanted to be various things but most of all I would settle for something dynamic like building roads, bridges or houses. It was never said I wanted to be a policeman or fireman even though these careers are very dynamic. I thought it would be wonderful to work in a store but I never thought of being a storekeeper. I liked to build things even though I wasn't very good with tools. To demonstrate how poor I was with tools, one time in high school woodworking class, when after completing making a desk, my friend Doug Williston and I picked up the desk to carry it home and a couple of the short legs fell off. What carried me through the industrial arts courses in high school was my drafting. I was pretty good with a T-square and triangle. My printing was quite acceptable and I could visualize images in different views. At one time, the thought occurred that it would be nice to be an architect.

It was not until grade ten in high school that the word "engineer" took on meaning. When living in Cranbrook the district superintendent for the B.C. Highways and Public Works was a neighbour. My Dad suggested I should go and ask Mr Johnstone (district superintendent) for a summer job with the road gang, perhaps shovelling gravel. I was not very tall nor very heavy; in fact, I was downright skinny but because it was wartime and labour was scarce (also I was a neighbour) I was given a job loading gravel by hand into a five-ton truck. What this amounted to was for three of us to position ourselves about the sides and back of the truck box and shovel gravel into the box until it was full. The driver would then drive the load away and dump it and return. Generally there was a 15 to 20 minute waiting period between loads so there was a chance to rest up before shovelling again.

After a couple weeks of shovelling gravel, the neighbour Mr Johnstone, got me onto a survey gang at Jaffray B.C. about 25 miles east of Cranbrook where the Southern Provincial Highway was being surveyed for a new location. This first position with the survey gang was as a cook's helper peeling potatoes, doing dishes and sweeping floors. The work wasn't too bad and there was a lot of free time during the day. The survey crew would go out in the morning and not return until supper time so there was little to do after washing the breakfast dishes. Generally, during the day I would go down and meet the morning train at the whistle stop station at Jaffray and collect the mail at the combination general store and post office. I got to know most of the ladies who would congregate at the store waiting for the mail to be sorted. It was a good place to pick up on the local gossip.

I was the cook's helper. Our cook was a little bit of a man by the name of Alex MacNeil. The first summer on the survey, showed he was an excellent cook but by the following summer his standards had slipped considerably. Everything was fried. *(What could have happened to Alex in the period of one year for his cooking ability to go definitely south?)* The afternoons as a cook's helper (cookie) were a drag. Some fishing was tried in the creek that flowed near the camp but it had been fished out years before so there wasn't much to do. One day I asked the draftsman, Charlie Wilkins (who was the only crew member besides myself and the cook who stayed in camp all day)

Why Mining?

if I could go into the mapping tent and observe what he was doing. He said he would be delighted as he would welcome some company. That was the start of my engineering career. At first I just observed what he was doing and then later he gave me a job of plotting cross-sections. Cross-sections were plotted for the rest of the summer until I went back to school in the fall. Also while working with the draftsman, he taught me how to calculate "lats and deps" even though I didn't know anything about trigonometry. I also learned how to use a slide rule that summer. I had just turned 16 years old and entered grade eleven when I decided on becoming an engineer and it would be civil engineering.

The next summer was spent with the survey crew, again on the Southern Provincial Highway but this time we were camped at Elko B.C. about 20 miles west of Fernie. The job that particular summer was to act as a rodman taking topography. This job is similar to rodding taking cross-sections except one looks for the location of the next contour instead of "breaks in ground". I found contour work interesting as a picture is drawn as you go along. Later that summer, I was given the job of "stake artist". That's the guy who writes the numbers on the stakes and pounds them into the ground. I learned how to calculate curves and spirals. *(I was really getting into this engineering thing.)* At the Elko survey camp, Geoff Montfort our "levelman" claimed to be a descendent of the famous Simon d'Montfort who was instrumental in establishing Britain's first legislative body known as parliament. Geoff took an interest in me and taught me how to operate a surveyor's level. Often in the evenings, I would practise my skill with the level by traversing from one bench mark to another and checking the accuracy. An elderly Scotsman by the name of Hughie Anderson would act as the rodman. *(He was the oldest man in camp but he was the youngest in spirit.)* During one of these levelling exercises, another fellow from camp, Ross Battistoni, who shared the same tent as myself came along with a bullwhip. Ross was snapping this whip without much care and accidentally caught the level's tripod and tipped the instrument onto the road blacktop. Of course the instrument was damaged and had to be sent away for repairs. That was the end of my levelling career and a chance to be an instrument man that summer.

It was not all work and no play at the Elko survey camp or the previous year's Jaffray camp. The Jaffray and Elko camps were located fairly close to

MacBaine's Lake (sometimes called Rosen Lake) where many young ladies spent their summer swimming and sunning themselves at their parents' summer cottages. We boys from the survey camp in turn spent most of our weekends and sometimes evenings during the week courting these young lonely ladies who seemed to welcome our attention. During the summer there was always a dance with a live band playing at the community hall at MacBaine's Lake which added to the enjoyment of the area.

In Cranbrook during my year in grade twelve, I palled around with a couple of guys who were a year ahead of me in school. One of these was Doug Bertoia, known as "Alphonse" in those days, who got turned off from senior matric (grade 13) and got a full-time job at the Sullivan mine in Kimberley which is 20 miles from Cranbrook. We would see him on the weekends and all Doug would talk about was what a wonderful place the mine was to work. Finally he asked Jack Frey and I to come to Kimberley as he could get a pass for us to tour underground. I wasn't too keen about the idea but went along with it as Jack Frey seemed to be interested. On the following Saturday (the mine worked 6 days a week then), we were escorted through the gatehouse into the visitors' dry and then were issued lamps. I found the lamps heavy and the hat and boots cumbersome. Anyway, we rode into the mine with the crew at seven o'clock sharp in the morning. There were four man trains that entered the 3900 level portal and travelled about 2 miles into the mine before branching off to disembark passengers at various locations in the mine. Our particular train stopped at the south-end shops where we left our lunch buckets *(This was an all day tour.)* in the lunch room there. Our official guide was an engineer but Doug (Alphonse) Bertoia came along with us as well. Our friend, Doug, who was so enthusiastic about the mine, was very keen to hear our feelings about what we had seen underground when we got to surface and had changed our clothes. Jack Frey offered the opinion that..."it wouldn't be too bad a place to work ...at least the money is very good". Although I found the mine interesting and was surprised at finding such facilities underground as the repair shops, steel sharpening shop, first-aid room, hoist rooms etc., I stated "I wouldn't want to work underground". Of the three of us, Doug Bertoia went on to become Deputy General Manager of KLM in Canada. He only worked in the mine for about six months. Jack Frey, who graduated in mining engineering, had to quit working underground because of his asthmatic condition that he developed in Lamaque Mine in Val

Why Mining?

d'Or, ended his career as Dean of Haileybury School of Mines. *(Guess who continued working underground until his retirement?)*

The following spring of 1945, I was eligible to join the RCNVR (navy) at seventeen and a half as I had been a sea cadet while living in Penticton. I wanted to be a part of the action that all my older friends were experiencing. This meant begging my parents to sign the papers giving their permission to join as I was not yet eighteen years old. They signed the papers on the condition that I would complete my grade twelve year of school. In April that year, the Navy called me down to Vancouver to HMCS Discovery for a medical. This I passed quite easily even though I was six feet tall and only weighed 150 pounds. A month later, I got my call to report even though I stated on the application that three months would be required to complete my personal affairs, being grade twelve. Fortunately, "VE Day" came before having to report to the Navy which saved me from having to leave school before completing the grade twelve (junior matriculation) year.

Hope/Princeton Highway Surveying August 1945

That summer was another summer on a survey crew. This time, the job was at Tashme on the Hope Princeton Highway. Tashme was an internment camp for 2200 Canadians of Japanese origin. At the time, because it was wartime and these folks looked like the enemy, we therefore thought of them as the enemy. However, it didn't prevent the survey camp fellows from enjoying the facilities that the residents of Tashme had made. For instance, movies were enjoyed once a week at their community hall which was once a barn and had been converted into a meeting site and social hall. Their hospital and medical facilities were used when we suffered an accident. My friend Alan Beesley (Law of the Sea fame) while cutting survey line injured himself with an axe and had to be taken into the hospital at Tashme for treatment. They had some very nice looking nurses at that hospital which didn't go unnoticed. We also would watch the fastball games that were played nearly every evening at the Tashme Ball Grounds. There was some very good softball played at Tashme. I got to know some of the ballplayers and would cheer for their team when it played.

My summer at Tashme was spent between being stake artist for the line crew and rodman for my friend Alan Beesley. Even though Alan and I were the same age, and he being my boss, I don't remember being resentful of him perhaps because we were good friends. We had known each other in Penticton and were very close chums. On more than one occasion when we were in public school and were attacked by a bully, we could always count on one another to come to each other's aid in a fight. To this day I can't remember him ever not coming to my aid in a ruckus. Anyway, the summer passed quickly, "VJ Day" came and the world was at peace.

Leaving Tashme for home at the end of August, I returned to school as I wanted to continue into senior matric which was equivalent to first year arts at university. My ambition still was to become an engineer. "Engineer" meant or translated as, "Civil Engineer". Designing roads and bridges and then building them, was the dream. Later there were doubts about civil engineering because once these projects were completed one had to move on. What would it be like looking for a new job all the time? The "location engineers" that I had worked for during the summers with the road survey crew all had steady jobs with the B.C. Public Works Department but they were sent off to the bush away from their families each summer. I began to wonder if I wanted that kind of life. At that time, mining engineering was an unknown.

Senior matric was very interesting. I enjoyed my first taste of physics even though "Pappy" Irving, the physics teacher said I would never make it as I didn't think logically. Did he say that to challenge me or did he mean it? One will never know. I really knuckled down after he made that remark and physics became my best subject. *(My mother had the highest mark in physics for all of B.C. the year she graduated from high school.)*

After the year in senior matric, I thought I would change my career path and try the life of a postal clerk. Having worked as a mail sorter after school and on holidays while going to high school, it was natural that postmaster Harris would hire me full-time when school ended in June 1946. I worked all summer and was earning fair money for the type of work I was doing. Also, it appeared there was a future for me in the postal service as Mr Harris was more than pleased with my work as he congratulated me about it on a number of occasions. I probably would have stayed with the postal service had it not been for my father. About the end of July 1946, he suggested that I should be thinking of going on to university. He said that he and mother were prepared to make the financial sacrifice for me to continue my education for at least one year. I thought about it for a day or two and agreed that maybe there were better things than sorting mail. The thought occurred at the time, that I would take one year at a time and give U.B.C. a try if accepted. After submitting my application in August 1946, I was accepted into first year of Applied Science. If I continued there, I would be one of the graduates of the Class '50.

UBC

Registration into first year at U.B.C. went smoothly as I had my friend Jack Frey (ApSc49) with me who had attended university the previous year and knew the ropes. Had it not been for Jack Frey who helped me through registration as well as the first few weeks of classes, I probably would have been intimidated and overwhelmed by the size of the university and gone home. Coming from a small high school in Cranbrook where we had only 200 students, I felt like a very small fish in a big pond at UBC where I considered myself to be a big wheel at Cranbrook High as I had been the president of the students council in my junior matric year. Because first and second year students were required to register two weeks early, we had to travel from Cranbrook to Vancouver in order to register, then hang around for two weeks. With nothing to do during that time but wait for classes to start, Jack and I decided to go out and look for casual labour jobs in Vancouver. We both landed a job in a tar and shingle plant in the Kitsilano Area. This plant manufactured asphalt shingles and Jack's job was to truck barrels of tar from storage to the melting pots. For some reason I was given an easier job of putting wire handles on one-gallon pails of tar that the company marketed. Anyway, the job did keep us busy and the extra money more than offset the extra room and board we were obliged to pay because of early registration.

The years 1945, '46, and '47 were heavy registration years at the university as there were many returned servicemen continuing their education that had been interrupted by the war. It must have been very difficult for some of these returned servicemen as many of them had been away from studying and the academic routine for a number of years. Also, some of these men had wives and families who demanded a certain amount of their time when home. It was so much easier for the students who were single as they could devote all their spare time to studying if they so wished.

Engineering was not easy as we were required to take 14 subjects the first year. Because I was not an ex-serviceman, I was required to take physical education training twice a week. That was considered a bit onerous. Regarding the course, engineering was a five-year course which required one year of arts prior to entering the four years of applied science. The courses in applied science for all engineering disciplines were the same for the first two years with the last two specializing in the discipline of choice. All engineering students (besides the year of arts where one carried such courses as mathematics, physics, chemistry, English and a foreign language) were obliged in the first and second year applied science courses to carry such subjects as mathematics (geometry, trigonometry, calculus and algebra), physics, chemistry, engineering subjects (descriptive geometry, hydraulics, strength of materials, graphic statics, drafting, engineering problems, surveying, etc), economics and report writing. I found all these courses interesting and at times very trying, especially some of the strength of materials problems.

Up until near the end of second year engineering, I still intended to become a civil engineer, perhaps because I had worked three summers on survey parties for the British Columbia Department of Public Works. At the end of second year, it started to occur to me that maybe civil engineering wasn't what I really wanted. It appeared to me that a civil engineer was always on the move; that is, when a project was completed, one had to move on to another one. In other words, one is always looking for a new job at the completion of a project. I wanted a job situation that was more stable. Yet, I also wanted a course that was very general like that of civil engineering. As the McKay Family had always lived in or near mining towns, the thought occurred to me that perhaps mining engineering should be tried. At least

Why Mining?

working at a mine had a certain degree of continuity and the mining engineering course was very general much like that of the civil engineering course. It was this reasoning that stirred me into the mining engineering course at U.B.C.

The third and fourth year courses for mining engineering were heavily weighted with geology (mineralogy, crystallography, economic geology, coal geology, palaeontology and structural geology) and metallurgy (chemical, physical and mineral processing). Besides these courses, a mining course contained a number of mechanical engineering courses (thermodynamics being one) and civil engineering courses such as structural design and materials testing. The mining courses were mostly in mine management, principles of mining, mining problems and ethics. The metallurgical courses were given by Professor Frank Forward who was the head of the mining and metallurgical department. Professor Forward, although a very fine man, was not a warm person and for some reason he couldn't inspire me as I failed both his subjects in my third year and had to write supplemental examinations in these before being accepted into fourth year. Fortunately these supplementals were passed handily the second time. Professor Henry Howard was in charge of mineral processing which was called mineral dressing in those days. Professor Howard was a quiet man who offered a lot of home-spun philosophy to the class. We all enjoyed his stories. Leslie Crouch was professor of mining. He had a way with words and made one think about the many alternatives that one could meet in a mining situation in order to determine a method for solving a problem. He was the first professor who actually called us into his office for a chat. His first question when you went into his office for your little chat was, "Why mining?". In my case, I said, that I had lived in a number of mining towns in the interior of B.C., I had worked one summer in a mine and enjoyed it, and I thought mining engineering was the most general course offered in engineering and therefore I would be able jump into any engineering field with little difficulty if I were inclined. I felt good talking to a professor at last. Prior to third year, I felt I was just a number with no contact at all with the professors, perhaps because the first and second year classes were so large.

Speaking of large classes, in the first year, we had an economics course from the dean of applied science, Dean Finlayson, who would lecture to 500 students at a time in the auditorium. There were 2000 in the whole faculty of applied science during that year (1946), of which 1000 were in first year. It was necessary therefore, that Dean Finlayson deliver his economics lecture twice to cover all the first year engineering students. On one particular day, the dean said, "...look to the right of you and then to the left, because those fellows*will not be here in four years time". Dean Finlayson was right because in 1950 there were only 500 in the graduating class in engineering.

* He used the word "fellows" because at the time there was only one young lady, Nelda Ozol, studying engineering and she was in the class ahead of me. However, we did have young ladies taking the courses in "nursing" who were included in the Faculty of Applied Science.

First year went well. The boarding houses were quite satisfactory. My first place was away up Cambie Street which meant taking the Cambie Street Bus to Broadway, then transferring to the Broadway Street Car. If one was lucky, the Broadway Car would go to 10th & Sasamat; if not, it was necessary to transfer at Granville to a 10th & Sasamat Car. At 10th & Sasamat was the transfer point where one caught the University Bus into the university campus. The amount of transferring from one streetcar to another and then to a bus was very time consuming especially when there were hundreds of other students trying to do the same thing. After Christmas I was able to relocate to another boarding house, the home of the Taylor family, which was a bit closer at 14th and Burrard requiring only one streetcar and bus change. This move proved beneficial because my room mate was their son, Stan, who was in first year engineering also. Stan and I were able to assist each other and between the two of us were able to get through our assignments without too much difficulty. As a result we both got through first year with a high second class average. Also at this boarding house were three other students who added some interest. We often would have a friendly game of poker before dinner where everything seemed to be legal tender. Cigarettes *(everyone smoked in those days)* counted for a penny each, streetcar tickets were seven cents and university bus tickets were valued at three cents. Of course, pennies, nickels and dimes were accepted at face value. A pot would in all likelihood consist of two or three cigarettes, a bus ticket or two, some streetcar tickets, a flock of pennies and a few nickels and dimes. *(We were high stake gamblers.)*

Why Mining?

The following year I boarded at the home of my Uncle Harold and Aunt Dorothy on Third Avenue near Alma Street. At this location, I generally hitch-hiked to U.B.C. because many of the cars going to the campus travelled Marine Drive along Jericho Beach. Second year was very difficult compared to first. The authorities were undoubtedly trying to weed out the weaker ones in that year as there was very limited space for those proceeding on to third and fourth year. All during first and second year, it was still my intention to be a civil engineer even though there were some doubts about the living component of that discipline.

Although I enjoyed boarding with my uncle and aunt, as a change for the next year, I boarded at the home of Mrs George Sainas *(the aunt to the Thunderbird Football players, Bill and George Sainas who lived next door to her on Dunbar Street)*. Board and room (two meals only as we had to find our own lunch) was $45 per month sharing a room. My roommate was Alf Dumont whose father owned a lumber mill in the Interior of B.C. Alf's father was wise to the ways of young university students as he had sent both Alf's two older brothers to college before him. His father's method to keep Alf from getting into trouble and to keep him at his studies was to make sure Alf was broke. Alf's monthly cheque was always a week late and when it did arrive it was post-dated a week. Each month Alf would have to go hat in hand and explain this to Mrs Sainas. I am sure Mrs Sainas, who was Greek and didn't always understand everything said in English thought Alf was trying to pull a fast one.

Alf and I decided we would make our own lunches instead of eating at the university cafeteria which was expensive. We put up a dollar or two each and Alf, because he had more time than I did, would shop for the bread, butter, meat, etc. Everything went well for the first couple of weeks. Alf would collect a dollar from me and bring in the necessary food supplies. Then for a couple of weeks, Alf said he had money left over and wouldn't have to collect anything from me. Then for a couple of weeks, he would collect a dollar from me again. Finally one week he asked if I could anti-up another dollar as we needed supplies. After a couple of double anti's I asked Alf what was going on. Apparently Alf had joined a campus noon-hour poker club hidden out of sight at the U.B.C. stadium and had been losing heavily after winning for a number of weeks. I'm not sure if Alf graduated or not as I have since lost all contact with him but somehow I am sure he made his mark one way or another. *(I have since found out that Alf did graduate and made his mark as a sawmill owner and operator.)*

The best place to stay during my years at U.B.C. was at Acadia Camp. Acadia Camp was a cluster of "H-huts", remnants of one of two army camps that were built during the Second World War on the U.B.C. Campus. The advantage was that one could walk to classes as it was less than a half mile from the then center of campus. The price was right too as room and board was only $48 per month for three meals a day with a shared room. Another advantage was that when one ran into difficulty with a course problem one could always find someone to help. It was also a good party place every second weekend. Speaking of parties, a number of the residents at Acadia Camp had stills for distilling rubbing alcohol and these stills were very active prior to the end of exam time. At the end of my fourth year, there was a continuous party for about four days at one of the huts at Acadia Camp. When walking by this one particular hut, its chief occupant hollered after me to come in and he would show me some real drinking liquor. When entering his room, I saw there were half a dozen others in various states of inebriation. He then poured some of his alcohol on the floor, lit a match to it and when the large blue flame emitted from the floor, he said, "...now there's real drinkin' liquor". I don't remember this student's name but I know he was taking a graduate course in engineering physics as he was considered to be scholarship material (?).

On the more respectable side, there was the Engineers' Ball held during the Spring Term. The Engineers' Ball was attended only once and that was in my second year. How I was able to afford the Ball was by squirrelling away $25 in a bank account at a branch of Imperial Bank *(CIBC now)* located inconveniently across town at the beginning of the Fall Term when I had extra cash. On the big day of the Ball, like many others, I went downtown Vancouver and rented a tuxedo, picked up my date's corsage and bought a bottle of rye, as well as a pack of tailor-made cigarettes. *(One would not be civilized smoking roll-your-owns at a fashionable ball.)* On this particular occasion, my date was Rona Raynor (Arts'49) from Kimberley who was studying Commerce. On the night of the big dance, I travelled in my rented tuxedo to Rona's boarding house by streetcar but we rode to the

Why Mining?

"Commodore" on Granville in a taxi. A table was prearranged with my friends (All ApSc'50), Jim Twaddle and his wife, Al Horner and his date as well as a few others whose names I've since forgotten. It was a grand evening. I'm glad I at least attended one Engineers' Ball while at university. Al Horner (who must have been really flush with the doe-ray-me that evening to have been able to rent a car) gave Rhona and me, as well as the Twaddles, a ride home saving us taxi fare. The whole evening cost $25.

During the last year of college, a field trip to Britannia mine was organized in which all fourth year mining students participated. This was before there was a highway to Britannia Beach. Prior to extending Highway 99, one had to travel to Britannia Beach by boat. After catching the boat at the C.P.R. Docks at the bottom of Granville Street in Vancouver, it was necessary to travel for about three hours by water to Brittania Beach. From there one had to walk a quarter mile to the lower skip station where an inclined skipway was caught which travelled up the side of the mountain for a few thousand feet. Half the skip was covered while the remainder was in the open air. Our field trip group was fortunate in getting into the covered section as this particular trip was not over-loaded. At the top, the group disembarked from the skip and were loaded into a narrow-gauge railway car which took us to Brittania Townsite about three miles from the skipway. The next day was spent touring the Bluff and Victoria Mines. The water from these mines was very high in copper content as it could leach through a rail or pipe in a matter of a few weeks by electro-chemical replacement; that is, the iron in the pipe would go into solution and the copper would precipitate out. The mine even collected the mine water and ran it through a number of wooden cells containing scrap iron where the copper would precipitate making a rough concentrate.

The 1950 Graduating Class was the big year in mining engineering as there were twenty-seven in the graduating class. One or more of the outstanding ones that comes to mind was Peter Boyko, who spent a number of years with Sunshine Mining in Wallace Idaho. Pete had a bit of a stutter but it did not hold him back from being very comical. He could take a very simple fact and make it into a very funny story. Bob Bray, hailing from Manitoba, had a very successful career with Hudson Bay Mining and Smelting. Bob was one of the guys at Acadia Camp who could be counted on when there were problems with assignments. Tommy Cochrane, was one of the oldest in the class. He had a very interesting career in mining research with the Mines Branch in Ottawa. Our paths crossed many times throughout the years. Bill Gibney and Stan (the man) Hodgson both went to Kimberley with Cominco and from there, they moved about the various Cominco operations. They would often be seen when I joined Cominco 15 years later. Peter Jack was probably one of the smartest students in the mining class, if not the smartest. Peter spent most of his career in potash with the Potash Company of America both in the States and Canada. I got to know him fairly well when we both worked in Saskatoon in the potash industry. Douglas Little, another mature student, had a very successful career with Placer Development. Bob Bray and I went to Doug's wedding after graduation and had a ball. Mel Lorimer was the very distinguished looking chap in the class with a British air of respectability. Big Angus McDougall was very successful in mining contracting and in the latter part of his career he was part owner of American Mine Contractors in Colorado. George Pinsky, alias "de Pinsk", who hailed from the Northwest Territories, went into the Oil Patch and has since retired in Calgary. Dan Poitevin, (and who hasn't heard about Dan?) Dan came out of the R.A.F. with a nervus affliction. Every time he felt pressure, especially at exam time, he would slap his chin repeatedly. Dan joined Ingersoll Rand upon graduation and had a very full career with them. Our paths crossed numerous times and to this day, my wife Irene and I always stop in to see Dan and Joan Poitevin when in Sudbury. Jim Thomson would always say, "...that's Thomson without the P". Jim managed the Braylorne mine for a while but ended his career with Teck Corp. Carlos Castro, our Cuban colleague, graduated but was never heard from since. Ned Fillip went to California and stayed in the sun. The last time I saw him was in Saskatoon when we were trying to arrange the development of a new potash mining machine through Smith Tool Company of California, the company where Ned worked. Tommy Kneen headed east and worked in the uranium field. Tom was always a quiet guy. We were both on the advisory committee for mining at Cambrian College in Sudbury. Bal Shannon had a very exiting career which seemed to carry him all over the U.S. and Canada. Bal visited me in Timmins one summer when he was on vacation (at the time, he was working in Missouri). Bill Shindle worked all his career in heavy construction at the Lakehead Area (Thunder Bay) and later ran his own construction company. It was Bill who gave George Pinsky his nickname. There were other guys in our class, including such persons as Don

Why Mining?

Codville, Bill Cook, Jim Eastman, John Leask, Jim Millar, Edgar Moore, Cliff Newton and George Trowsdale, who for some reason or another, I never ran into them after leaving U.B.C. in the spring of 1950.

The 50th Year Reunion in 2000 of the U.B.C. Mining Class of '50

Front Row Left to Right:

Cliff Newton, Dan Poitevin, Al Stickney, Don Codville & Doug Little

Back Row from the Left:

Professor Malcolm Scobie (present Professor of Mining)

Jack McGuirk, Dave McKay (author), Jim Eastman

Tom Cochrane Peter Jack & George Pinsky

Travelling back home to Cranbrook from Vancouver at the end of each fall term on the Christmas Break was via the C.P.R. Kettle Valley Line. The train would leave Vancouver at about seven o'clock in the evening after the students had a round of beers at the Castle Hotel prior to the train ride home. It was one continuous sing-song and party for most of the way. Generally because the C.P.R. authorities knew there would be a bunch of university kids on-board at a reduced-fare excursion, they shunted us to the rear of the train where they had placed their old "straw-back" coaches *(Railcars with seats of woven straw)*. No one minded this pampering (?) as we were all in a party mood having written exams and anxious to get home. On one particular occasion, there were three or four pipers *(including myself)* as well as girls performing highland dances right in the narrow aisles. The parties, as I remember, never got out of hand even with the fair amount of alcohol consumed perhaps because the drinking had to be done discreetly or else the conductor would put you off the train. The first batch of students would disembark in the early morning hours at Penticton, providing of course the train travelled through the Coquihalla Pass; otherwise, if the pass was closed the train would be routed through Spences Bridge adding four extra hours to the trip. At Grand Forks in early afternoon another batch of students would disembark. Those people still in a party mood would purchase more beer from the liquor store near the train station at this stop. The train load of students would thin down considerably upon reaching Castlegar as the Rossland and Trail passengers would leave the train there. Nelson, in the evening was always an interesting time as the train had a four-hour stop-over here in order to make connections with the Great Northern Railway coming up from Spokane. Most of the students would leave the train and head up-town. If it was a Saturday night, there would be a dance which one could go to for a few hours before the train departed. By the time we got to Cranbrook the next morning, having spent 36 hours on the train except for the four hours waiting around Nelson, the train was nearly empty. Two weeks later the same procedure was repeated in reverse, except there was no four-hour stop along the way at Nelson.

Sullivan Mine

The first summer home from university, I got a job as a "grademan" with the road engineering crew that gave line and grade to the road contractor (General Construction) who had the contract to rebuild a section of the

Why Mining?

Southern Provincial Highway. I had been part of the survey party that surveyed that section of the highway three summers prior. For some reason, the head of the engineering crew didn't take kindly to students as he called them "summer complaint". Anyway I bided my time and started to make enquiries around Cranbrook for other employment. I landed a job at the Sullivan mine through some contacts made with members of the Kimberley Pipe Band. *(In my high school days, I was a member of the Scottish Band as I played the bagpipes.)* Johnnie McDonald was the president of the band and was also general foreman at the north-end section of the Sullivan mine. At band practise one evening, after travelling to Kimberley with Alan Graham *(my former pipe teacher who was also a member of the Kimberley Pipe Band)*, I approached Johnnie McDonald about a job. John must have put in a real good word because the following week, the authorities told me to report to the Sullivan mine doctor for a medical. Incidently, the medical did include a chest X-ray but the physical part of it consisted of the doctor asking, "Can you touch your toes and are you generally healthy?"

Cominco, known then as "The Consolidated Mining and Smelting Company of Canada Limited or C M & S", was thought to be very advanced in safety for those days. Everyone starting in the underground would have to spend one week in Joe Shaw's underground safety school. "Safety Joe", as Joe Shaw was affectionately named, would bring all new recruits to a classroom at the south-end section of the Sullivan mine. There the recruits (or green horns) were given a book of safety rules and other little tidbits of information that everyone in the mine should know. The information in this manual covered bell signals for the skips (cages), stench gas procedures, blasting warnings, etc. After a full morning of safety lectures, the new recruits were issued picks and shovels then were led away by an instructor, who addressed the recruits as "engineers" and then showed everyone where they would spend the balance of the first shift laying track and cleaning ditches. I had an old pair of leather boots left over from my survey crew days that I had worn this first day instead of buying new boots that could be waterproofed. As a consequence, they were soaked right through at shift end. To my good fortune *(I thought)*, I found a pair of shaft boots in my locker *(the last occupant had left them instead of throwing them into the garbage)* and wore these during the next shift instead of buying a new pair of mining boots at the company store. The next shift, wearing these large shaft boots with the shins reinforced as well as the toes, I reported to Safety Joe's school already for the ditches. But instead of the ditches, we were given pack-boards and told to carry 50-pound cases of Forcite up four sets of ladders to the sublevel above for a large blasthole stope charge. At shift end, my shins were raw. For the next shift, I charged a decent pair of underground mine boots at the company store (Mark Creek Store).

A number of old miners would congregate early outside the dry before each shift waiting to board their coaches. The younger guys called these old timers, "the happy hour club". It goes to show even the young people then just didn't have much respect for tradition.

While waiting on the coaches that took the crews into the mine on day and afternoon shifts, there would be a lecture (safety talk) over the loud-speaker from Joe Shaw. Joe would always end his talk saying, "...be careful boys, be careful". Then a spirited march would be played over the loud-speaker as the mine whistle blew and each train's siren sounded, signalling the start of each train into the mine.

After a week of safety school, each person was assigned to their respective places in the mine. Because of the good friends in the Kimberley Pipe Band, I was given a very easy assignment at the north-end up 30 Raise on the 4200 Level pulling chutes. In one week, I had gone from "mucker" the lowest rank to "chuteman" (two ranks above) leapfrogging the motorman position altogether. There was only one train crew on the 4200 level north and I was a member of this three-man crew. I should have been assigned the job as motorman (locomotive driver) but the incumbent had a bad leg from the war and couldn't jump out of the way very quickly *(which one had to do when pulling chutes)* and therefore preferred to stay driving the locomotive. This suited me fine because a chute puller earned five cents more an hour.

With regard to being able to jump fast, one day our crew was assigned to pull backfill from a chute on 4200 level. Unknown to us, someone had turned a water hose into the backfill on 4600 level above. After pulling a few cars of fill, a mass of water, mud, sand and gravel came rushing out of the chute nearly burying both my partner *(the other chuteman)* and myself. We both dove off the chute and high-tailed up the drift to safety. After that we were

Why Mining?

very careful to inspect any new situation before getting too involved.

Dumping Granby Cars Sullivan Mine 1947

Another fortunate thing about working on the 4200 level was the fact that there were only two shifts, day and afternoon shifts. Had I been assigned to the 3900 level, it would have been as a motorman earning less pay working all three shifts; day, afternoon and graveyard. I did work one shift on the 3900 level when this level was short of crew. They kept me busy all shift where I had only 20 minutes for lunch and then I had to take my lunch at the convenience of the mucker boss. After the experience on 3900 level, I realized how fortunate it was to be on the 4200 level. *(It pays to have friends)*

On one shift when there was a shortage of muck on the 4200 level, our mucker boss sent our crew to the 4600 level to pull muck there. With a shovel each over our shoulders and with me about 100 feet in the lead, I came across a gentleman when rounding a bend in the drift, who carried his lamp in his hand just like a shift boss so naturally I thought he was one. He stopped me and said, "Laddie do you see that water overflowing ditch, get to it and shovel". He then left humming a little tune. When my partners caught up to me, I was busy cleaning the ditch very industriously. They asked, "what are ya doing"? I said, "The shift boss said to clean up the ditch". They both started to laugh and said, "...that was no shift boss that was 'shit house Jones' who nails-up the shit boxes and supplies the shit paper. Yah don't have to do what he tells yah".

Mentioning friends in high places, my shift boss was Jimmy MacLaren and his cross shift was a Scot as well by name of Jimmy MacFarland. As referred to before, Johnnie McDonald was the general foreman in charge of our section, Doogie Smith (the bass drummer in the pipe band) was the mucker boss at

4200 level south end, the Scott brothers (Angus and Hammish were pipers) were shift bosses, Roddie MacQuarrie was the general foreman down the shaft, Jock McConneky was the main dispatcher and so it went. *(There were Scots all through the mine and they looked after their own.)*

After working a good part of the summer on the 4200 level, our mucker boss, whose name I have forgotten, was about to take his summer vacation. He took me aside this particular day and said that I would be taking his place for the next two weeks. *(Here was a boy age 19 years, working for only three months underground and they give him a chance to be a mucker boss, such nepotism!)* He showed me the proper way to charge out powder from the powder magazine to the contract miners, how to check the inventory in the magazine and to order powder, make out the time sheets and write up the log. But most important of all was, it was necessary to drive one of the man-trains in and out of the mine at the beginning and end of shift. My first effort at this didn't go that well as it was my first time driving a double-tandum locomotive. After the first effort and receiving a little coaching from a friend, my locomotive operating didn't go too badly.

A while later, I was called to the engineering office for a chat with Johnnie Stewart (senior engineer) who asked me to go into the engineering office as a mine surveyor's helper for a few weeks. I said, "OK" thinking it might be a good experience. Anyway it was steady day shift. For the next few weeks, as a surveyor's helper, I applied my knowledge of surveying learned through the compliments of the B.C. Highways Department. My first boss was Andy Stirling who was in charge of development. Andy was a very patient man. He demonstrated how to open a "H-I" stick (folding rule) without breaking it. After carefully warning me about how delicate these rules were, I, during my first attempt to open the rule, broke it. Later I worked for Welchy James who had a regular beat at the north-end above 3900 level. I didn't particularly like the engineering office as even though it was steady day shift, one had to work until 3:30 PM each day instead of 3:00 PM when working in operations on the 4200 level. The reason as explained by Kenny Davies (section engineer) was that in operations, one ate on company time while in the engineering office, one ate on their own time. Getting home one-half hour earlier was important to me then.

Why Mining?

One day while underground with Welchy James, I bumped into my supporter Johnnie McDonald, general foreman, who asked, "How are you doing"? I answered, "Not so good. I would like to go back to operations underground". He said, "Leave it with me, maybe I can get you into a good stope where you can do a little mining". True to his word, a week later after returning from a tour to Idaho in the States with the pipe band, I began to do a bit of mining in a blasthole stope. Cominco had a system where all new "miners" were relegated to the spare-board. That is, you took the place of someone who was absent. One could be sent to a different location each shift. There was some moving around during the first week or two but afterward, undoubtedly because of Johnie McDonald, I was sent to a particular stope that had a good chance of making a lot of bonus. Normally in a slusher drift, Cominco paid nine cents a ton for every ton scraped into the millhole but because this particular stope was on its last leg, the rate given was 19 cents a ton because it was their feeling that the muck was more difficult to obtain. This was a fixed rate and would not be changed for the period which spanned a full month. Fortunately for us, the back of the stope caved in around the second or third day which gave a lot of free muck. We just bailed it out for the balance of the month and made a grand sum of $22 a day bonus each, which was away above normal for stoping in those days. When one considers the base wage was $1.27 per hour, a bonus of $22 was indeed a tremendous reward.

A sad note to the engineering office days was the fatality of one of the survey helpers. While I was away with the pipe band on the trip to Idaho, the chap (David Brown) who took my place was caught by a run-of-backfill and taken down a raise and covered with fill. The instrument-man, Hap Richardson, who was with David got caught in the air and water piping along the sublevel and therefore the running backfill passed under him. Hap was badly injured. The last time I saw him, years later, he was still confined to a wheel-chair.

After working from May to September in 1947 underground at the Sullivan mine, it still didn't convince me that mining engineering was the chosen discipline. I was still going to be a civil engineer even though I found the methods of mining (blasthole and open stoping) and the materials handling systems very interesting. As mentioned previously, it was not until the next summer after working on road construction and doing a bit of thinking about how and where I wanted to live and the continuity of work that the decision was made to try mining engineering.

As mentioned, the following summer, I went back to road construction with General Construction of Vancouver, the firm that had the contract for the upgrading and constructing the B.C. Southern Provincial Highway between Cranbrook and Fernie. The job that summer was drilling and blasting the rock-cuts along the right-of-way. There were two of us (Ed Soles and myself) who did all the drilling and blasting that summer. We were given new Ingersol Rand J-40 jackhammer drills using hexagonal steel with Timkin threaded detachable bits. *(Carbide bits were just being introduced into the industry.)* Ed and I would drill from Monday to Thursday and then load and blast on Fridays. By Monday we would be moved and would repeat the whole procedure again. I can't think of anything out of the ordinary happening that summer, except that our road crew did have a good softball team where we would challenge the local Wardner B.C. team and generally win. After a summer of riding a jackhammer, I decided that there was indeed something about mining that was appealing. Perhaps it was the people in mining or perhaps it was the variety of problems that one could encounter. On returning to university after the end of summer, I registered in mining engineering.

Tadanac Concentrator

I discussed previously those people who were associated with the mining industry and the mining course. But there was one more work summer (1949) prior to graduation that had a lot to do with influencing my choice of a career. During the 1948-49 semesters, my father was transferred from Cranbrook to Trail. This transfer affected my earning power as far as free room and board went. If I wanted to work at the Sullivan mine, it would mean being obliged to find a boarding house in Kimberley and paying room and board. My father suggested trying a job at the Trail smelter instead of the mine as there would be free room and board at home and therefore enough money for my year at college. I thought I needed more mining experience as it would affect my job chances on graduation and was therefore reluctant to miss what I thought to be crucial and essential. It didn't happen that way as the summer at the smelter was quite rewarding as I was assigned to the Tadanac concentrator where it was possible to gain a little knowledge about milling. This knowledge helped in my first job and the experience gained that

Why Mining?

summer in milling helped in other job situations throughout my career.

The Tadanac concentrator was one of the more desirable places to work during the hot Trail summers because the plant was fairly cool *(no furnaces)*. Unfortunately this plant was not making any money for the C M & S reprocessing old oxidized tailings from previous milling operations. The plant operators *(meaning us workmen)* did every trick in the book to keep the plant viable. One such trick was to take the tailings samples *(all sampling was by hand)* after an extra shot of Xanthate was introduced into the scavenger cells. Also, we noticed that if one took a moisture sample of the concentrate in the railcar and then added a shot of water into the car prior to it being weighed at the smelter bin, the plant would get credit for extra concentrate which was really water. Fortunately for me, I was able to maintain this choice job in the cool concentrator all summer as that plant limped along perhaps through the ingenuity and deception of the plant operators.

The Author's First Encounter With Mining

The picture, taken in 1931 at Big Jim McKay's placer operation, shows from the left: Gordon and Elsie (Thorton) McKay (uncle & aunt), the Author (in front of his father) Bill McKay with Elin (Larson) Kadey (aunt) and Judy McKay (mother) holding brother Jock .

Chapter 2

Sheep Creek

Paradise Mine

During the last year of college, the Canadian Institute of Mining and Metallurgy (CIM) held its Western Meeting in Vancouver. All mining, metallurgical and geological students, as members of the Dawson Club, were given assignments at the convention, performing such functions as manning the registration desk, operating the slide projectors, stacking chairs and other chores necessary for the success of the meeting. This entitled those who worked, free admittance into the convention. Later a small honorarium was given as well. During the convention a few of us, like college boys will do, made our way to the assorted hospitality rooms sponsored by the various manufacturers and suppliers. It was in one of these hospitality rooms that I met Henry Doelle who was the managing director of Sheep Creek Gold Mines. Hearing from others that Sheep Creek was an aggressive small mining company and appeared to maintain a very stable group of staff members, it seemed to be the natural thing to investigate this company further. At the time, this company operated two or three small mines and were actively involved conducting exploration in order to develop other operations. As it was my ambition, at least for the first job, to work at a small mine in order to obtain as much practical experience as possible in a short time. I approached Mr Doelle for a job and told him that I would be happy to start either as a mucker in one of his mines or even as a helper in one of his mills. To my surprise, he asked me to write him a letter stating when I would be available and he would consider the letter as my application. True to his word, my application was accepted and I was told to report at the Paradise mine in Invermere.

The instructions were that I would be hired as a "mucker" and to find my way to Athalmer which is situated at the head waters of the Columbia River in B.C. At that location, if I waited at the Coronation Hotel *(the only hotel in town)* "Slump's Delivery" *(who had a delivery contract with the mine)* would take me to the mine. The actual mine was 20 miles or more from Athalmer and Invermere at an elevation of 8000 feet. The mill was located only 10 miles from these towns at a somewhat lower elevation of about 4500 feet. Sam Slump, who owned the delivery firm also operated a hamburger restaurant called "Slump's Dump" across from the Coronation Hotel. The

Why Mining?

hotel housed the only beer parlour in town and therefore Slump's Dump was nicely located for those who craved a hamburger after a session of beer drinking. Sam Slump was an aggressive entrepreneur who dabbled in a variety of enterprises some legal and some not. Sam drove me and a couple of others in his delivery van along with the groceries and mail for the mine to Jackpine where the mill and office were located. The trip to Jackpine was uneventful except that the road was narrow, unpaved, rutted and very winding. It wound along Toby Creek for the most part. Not being one of the lucky ones to get a front seat with the driver, I had to sit in the back along with the groceries and other supplies. This made the ride that much more trying as there wasn't a decent seat in the back of the van on which to sit. Anyway we got there in spite of the rough ride.

Upon arrival at Jackpine, one could see that it consisted only of a small 50-ton mill, a combination bunkhouse, cookery and office, as well as residences for the manager, Jack Crowhurst and family; also, the mill foreman and accountant, Roy and Ray Gould and their mother; a duplex shared by the assayer, Ed Barrington and family, and the cat-skinner for road maintenance, Vic Pritchard and his wife Dodie. The cook, Mary Kennedy, and her live-in *(the live-in acted as bull cook)* occupied a portion of the cookery as their living quarters. Besides these, there was a separate crusher house, generator building (as all power was generated by diesel) and a tailings pond. The untreated overflow from the pond drained into Toby Creek, an action that would be forbidden today.

The mine operated five days a week producing 60 tons daily while the mill operated at a rate of 50 tons per day which required it to operate a sixth day or until the ore bins were emptied. The distance between the mine and mill was about 10 miles over a very steep narrow winding road. The run-of-mine ore was trucked from the mine to the mill by a contractor, Jack Jones and his driver Sid Thornton, using two 12-ton White trucks. Generally, the two trucks would haul three loads of ore each from the mine to the mill bins during the day and then take a load or two of concentrate to railhead in Athalmer at the day's end. Sometimes it would be a long and difficult day for these drivers as the road was narrow and winding and in poor repair due to the spring run-off. To make matters worse, the road was rutted from the snow run-off because there were very few culverts to take the water away

Sheep Creek

from the road surface. It's quite unlikely that the contractor made much money as the condition of the road was very hard on the trucks and it was necessary for him and his driver to work long hours just to keep up with the ore and concentrate haul.

After signing on, the manager, Jack Crowhurst said that I would not be going to the mine for a while as I was to learn how to assay and to perform mill test work in the lab. This was OK as I was interested in milling as well; however, the kicker was, there was no room in the bunk house. The bull cook did find a bed for me but he had to set it up at the end of the hallway. That would be my living quarters for the next couple of weeks. There were two large rooms in the bunk house to house the mill crew and even these rooms had no doors. There was not much privacy for anyone in the bunkhouse due to overcrowding; in fact, often times the trucking contractor and driver would be too tired after a long day hauling ore from the mine to the mill and would bunk into an empty bed belonging to one of the mill operators instead of driving home to Invermere with a load of concentrate to railhead. It seemed that any empty bed at night was fair game for the taking. It's funny no one complained. *(Can you imagine today's young graduates or anyone for that matter tolerating conditions like that?)* As an aside and for everyone's interest: this old log building, used as a combination bunkhouse and cookery, was formerly a bawdy house in the old days catering to miners and loggers that first worked in those parts. *(The place didn't have class but it did have history.)*

My first job consisted of assaying and mill test work. Ed Barrington (assayer) would be my teacher for the assaying part and Roy Gould (mill foreman) for the mill testing. Ed was a tall friendly guy but didn't think too highly of young engineers. He probably had his reasons. For that matter neither did Sheep Creek as I was only receiving muckers' wages of less than a dollar an hour. Ed wrote out, recipe style, the method for lead and zinc determinations by wet-assaying. These were recopied meticulously in my finest printing in a note book given to me by our employer: then a flow chart of the method of determination was drawn, again in my finest printing which was then hung on the wall of the lab. Everything went fine in the lab except that I was very slow. Instead of having all the determinations completed by noon, these were still being performed well after four o'clock in the afternoon when the next shift had arrived. Ed Barrington (being assigned to relieving the mill foreman Roy Gould who was on vacation) in exasperation, came into the lab and lit a fire under me to get going and to forget about all my fancy charts on the wall and to get on with the assaying. After his tirade, the fancy charts were forgotten and I got on with the assaying. After all these years in school and college, I was acting like an academic and not like an engineer or workman. Ed had taught me a good lesson. Within a couple of weeks, I was able to get the assays out before noon and to have all kinds of time for preparing standard solutions and conducting test work under Roy Gould's supervision. Roy had returned from his vacation by then. Roy was easy to please and showed me a lot about test work as he had conducted many of these tests when the mill was first started.

The Paradise mine ore (being treated at the Jackpine mill) was heavily laden with pyrite which at times would cause the flotation cells to overflow thus losing much material onto the floor of the mill. Other times, the ore would be highly oxidized which showed up in poor recovery. As a consequence, a laboratory test programme was set up in order to determine the best combination of grinding and flotation times as well as the pH and chemicals necessary to optimize the recovery of the lead and zinc in the ore. The test work consisted of taking a large sample of crushed run-of-mine ore, cutting and riffling it and then drawing a 2000 gram sample for grinding. Sheep Creek, prior to my arrival on site, had run a number of grinding and flotation tests against a "base test" where they would compare with each series of tests. The testing included grinding and flotation. Only one variable (grinding time, pH, quantity of flotation chemicals, flotation time, etc) was changed at a time in each series of tests. Basically, what I was doing was filling in the blank spaces in the testing programme where some of the series of tests lacked completeness. At the end of each testing, the heads, concentrates, middlings and tailings would be assayed and the results would be recorded with those of the previous tests. There were at least five test batches in a series with each having at least ten variables. This meant that there were five times ten, or fifty of these tedious test batches to complete in order to get a full picture. Even as the fiftieth test was completed, Roy Gould would come up with other components to vary. This testing went on and on, even after I was given another assignment.

Sheep Creek

One day Jack Crowhurst, the manager, approached me and said that he was sending me up to the mine to work for Johnnie Parker who was the mine foreman at the Paradise mine. After spending a month or more at the Jackpine mill assaying and mill testing, it was a happy day to move on and get underground and do some mining. *(This is what the four years at UBC was all about.)* Anyway, Jack Crowhurst said to take the flatdeck truck (three-ton) with some supplies up to the mine. Jack didn't ask if I possessed a driver's licence or not, he just assumed that I did... but I didn't. The drive up to the mine was on a rutted gravel single-lane road with at least ten switchbacks to gain 3500 feet in ten miles, that's 6.5% grade all the way. Some of the switchbacks could not be navigated in one turn and one would have to back up and move ahead and back a couple of times in order to make the turn. On approaching the mine at elevation 8000 feet, I found there was more and more water running across the road due to the spring run-off. It was early June by then and the spring run-off was in full swing. This too added to the difficulty in navigating that road. For an inexperienced driver, I found the road frightening but the trip to the mine was made safely never-the-less.

Paradise mine surface facilities consisted of a two-story office building which acted as office downstairs and living quarters upstairs for the staff, made up of Johnnie Parker, the mine foreman and Olie Brekie the shifter. There was a fairly large cookery and dining hall with the bunkhouse rooms situated above the dining hall. A wash-house (dry) was separate next to the dining hall-bunkhouse building. Outside the mine portal, was a blacksmith shop. Except for a generator building near the portal, these were all the buildings at the Paradise mine itself. The mine camp had some very spectacular scenery as it was situated in a bowl surrounded on three sides by mountain peaks about five hundred feet above the camp *(we were at the top of the mountain)* and open on the fourth side to the valley where one could see 15 miles or more in the distance to the Columbia Valley 5000 feet below. There was still a bit of snow on the ground in June; if fact, that year there was a snowfall on the 8th of July. All of this was above tree line.

Johnnie Parker, who had graduated from U.B.C a few years before me, was the mine foreman. Johnnie was a jolly sort of a guy and made jokes about most of the hardships that people had to put up with at this mine. In one

Why Mining?

instance, Johnnie joked about the time the mine was short of shovels requiring him to take the only spare shovel from the cook's coal shed to satisfy the need of the shovel shortage in the mine. I had met Johnnie before at Jackpine so we were not strangers. On arriving at the mine, I went in to see Johnnie to hear eagerly about my new assignment. To my horror, Johnnie said that I looked a bit sallow and that a bit of sunshine would do me a lot of good. He assigned me digging and installing culverts on the road from the mill to the mine. *(Remember, the roadway needed culverts...somehow I knew I would be selected to remedy that shortfall.)*

The culverts consisted of two rough 4 x 12's placed parallel to one another with short 4 x 12's cut and spiked to hold the assembly together. I was given a partner, Pete, a doukabour, who was very strong and a good steady worker. While working with Pete, I learned a few phrases of Russian, such as "doeborha" (spelling?), "dieah me cloutch", etc as well as others that are not very respectable. The objective was to dig and install two culverts a day. This meant that Pete and I would dig a trench about two feet wide and two feet deep the full extent of the road width about 15 feet from the ditch to opposite shoulder. The location for the culvert was picked where the greatest flow of water from the spring run-off seemed to be streaming across the roadway. Also, it was necessary to schedule the digging so that this would not hold up the trucks hauling ore from the mine. It was wet and messy work. After two weeks of digging in the June sun, I was no longer sallow looking as Johnnie Parker had said, and also was pretty hardened up from the heavy shovelling in the ditch, and lifting and sawing of the 4 x 12 timbers. We must have installed about two dozen culverts making the roadway to the mine much more traversable. *(I was now ready for underground work.)*

The Paradise mine was entered via an adit on the 7900 level (that is, 7900 feet elevation above sea level) which was the main or only haulage level. A two-man train crew consisting of a motorman who operated a one-ton Mancha motor on 18-inch gauge track and a chute-puller (who loaded one-ton V-dump cars by hand-chutes) did the haulage on this level. There were two other active levels, the 8000 level which was the top level then and a newly created 7800 level which was not at this time connected to the other two levels. In fact the 7800 level was just an adit into the hill for a distance of about 500 feet. There were a couple of other levels in the top mine operated

Sheep Creek

in the early 1920's which were at this time abandoned. It should be mentioned here that one of the earlier operators in the 1920s built a mill at the minesite and processed the ore there but later had to abandon this site because of lack of water year-round. There was ample water in late spring, summer and early fall, but nothing at other times because of freezing. Remnants of the old 1920's mill were still standing in the flat just below the cookery. These old buildings were a source of lumber and spikes as well as corrugated metal roofing for the mine when required. The old-timers around Invermere said that the very early operators (1910's), instead of milling the ore, hand-sorted the high-grade and raw-hided it to railhead in winter.

There were three active square-set stopes in the mine, two on the 7900 and the other on the 8000 level. One entered the mine via the 7900 portal. The 8000 level was reached by climbing through a two-compartment raise which acted as an ore-pass and manway. All the stopes (in June 1950) had completed the third lift and were excavating the fourth. No fill had yet been placed at that time. The mine was ventilated by natural ventilation with the exhaust air finding its way out from the mine via the old abandoned top-mine workings. *(As I remember, the ventilation was pretty good.)* Each stope had a crew of three, a miner, timberman and mucker. These titles were really for pay designations as everyone performed the same duties. I was the low man on the totem pole and therefore was designated a mucker and paid accordingly. There was no bonus system.

At the start of each shift, the three-man crew in the stope had to hand muck the previous round into the slide to the chute below. The chutes were spaced every fourth set so that there was little double mucking. As soon as space was cleared to install a post, the miner and timberman would quickly install it and a temporary girt, while the mucker *(meaning me)* would continue mucking. At one time in the procedure, I found myself working with a miner, whom we all called "Eric the Finn". Eric was built like you would imagine a Finn to look: short, stocky and powerful *(like a brick shit house)*. This particular day, while Eric drove a wedge to secure the post to back, my job was to hold the post upright for him. Eric told me, "Hold like a man." Then he took a powerful swing to drive the wedge home secure. I didn't realize he could pound so hard. The force of that blow on the wedge drove me and the post backward to the muck pile and onto my rear. After that experience, when Eric

Why Mining?

said "Hold", I would hang on for dear life.

After finishing mucking, and the timber set was in place, the miner and mucker would begin drilling *(Eric would be drilling, I would do the chucking)* with a leyner on a bar. Other times if we were drilling uppers, a stoper would be used. The timberman, in the meantime, would cut and install lagging over the set to support and secure the back. After drilling we would set and secure the steel mucking sheets prior to loading and blasting. This was all very labour intensive but those in the stopes were producing about seven tons per man shift which was not bad.

Carbide lamps were used in the Paradise mine instead of the electric lead-acid or alkaline batteries. The carbide cap lamps themselves were heavier than the modern-day battery lamp worn on the miner's hat except one didn't have to carry the heavy battery on a belt which was somewhat of a plus. The first thing each shift, one would have to fill the water and carbide bowls on the lamp as well as to clean the flame orifice. Each miner was equipped with a carbide container which was filled with carbide at the beginning of each shift. One was always fiddling with these lamps either shaking them or recharging them with water or carbide and even cleaning the orifice that would plug-up with residue. A sudden draft at times would blow the lamp out. One warning for anyone who has not experienced in the use of carbide lamps was... **do not use the lamp as a spitter in lighting a round** as the newly lit charge would often blow-back on the lamp and snuff it out leaving the person in the dark with a burning fuse to worry about. Don Pretty, a fellow mucker friend, while working in the adjacent stope related his experience where the miner and timberman in his stope both had their lights blown out while lighting a round. They thought they would reach for Don's lamp to complete the lighting but he refused and led them out of the stope to safety with his lamp, the only lamp that remained lit. Some of the miners at the Paradise, even though they were not on bonus, were more concerned about making the round instead of the safety of their lives and those of their crew.

One day while working at the Paradise, Mr Doelle, the managing director, came to the mine on one of his inspections. Nick Starric*, a miner who had worked for Mr Doelle before, had the courage to ask him as to when would

* Six months later, Nike Starric was killed in a cave-in at the Paradise Mine.

we be getting electric cap lamps at the Paradise mine. Mr Doelle answered, "Nick, you'll get electric lamps when this mine makes a profit".

Still on the subject of carbide lamps, I should at this time relate my experience with these lamps while surveying. I was asked while working at the Paradise mine to do some surveying in the recently driven 7800 level adit. This particular adit had been driven into the hill about 500 feet where the company was proposing to do some diamond drilling in the area. They asked me to lay out some foresight and backsight lines for the drill-holes with a transit. Well, I was very elated to be assigned this responsibility and began to set up the transit under a station in the back as all underground stations are established. While the driller, Bob Mitchell *(more about him later)* was giving the backsight, I looked through the eye-piece of the transit. Then all of a sudden, the plumb-bob string caught fire from the flame of my carbide lamp causing the plumb-bob to come crashing down onto the glass of the transit's compass, breaking it of course. *(Here for the second time in my career, without really getting started, I am responsible for two broken survey instruments.)*

About the only recreation available to the crew at the Paradise mine was reading which the men supplied on their own by swapping books and magazines amongst themselves. There was no reading room or lounge for them to sit and relax after a day's work and therefore they had to read alone in their rooms seated on their bunks. The electric generator plant was turned off after 10 o'clock at night so reading even in the rooms was impossible after that time. In the summer, there was twilight even after the generator plant was shut down but not enough for reading. The only other activity was gambling. A table and a few powder boxes was set up in the dry (change house) each evening for the "high-stakes" poker games. These were serious affairs as some of the experienced players (professional gamblers) would travel around from mining camp to mining camp just to clean up on unsuspecting and naive participants. One young kid from New Brunswick, during all the time I was at the Paradise, did not once travel to town because he was cleaned out of his money from one pay period to another. When he received his cheque at the end of a pay period, he would just endorse it and hand it over to his creditor. The latter had a number of fine weekends in Calgary as a result of his winnings. *(It was pitiful.)*

```
Why Mining?
```

Speaking of the dry (change house), it wasn't too fancy either. It had the usual hook hangup area, wash-basins, showers and toilets. The showers were something to behold as they were constructed of wood with wooden tongue and grove for flooring with plywood sides. The sides and floor were covered with green algae which discouraged many from taking too many showers in that facility; instead, they would wash in the basins the best they could. A couple of us, when going to town at the end of a pay period, would go up to Radium Hot Springs (which was near Invermere) and have a good bath there. We would bring our soap into the hot pool at the Springs and soap ourselves there, much to the horror and consternation of the tourists who were bathing at the same time. Today if someone pulled that stunt, they would be turfed out on their ear by the Mounties and Park Officials.

The crew at the Paradise mine was not very stable as compared to the mill crew at Jackpine perhaps because the mine crew was made up, for the most part of "tramp miners". The mill crew, on the other hand, seemed to consist mostly of local people where some even commuted each day from town as it was only 10 miles from Athalmer and Invermere and even closer to the Village of Wilmer. The tramp miners in the mines in B.C., during this period of time, would move from one small mine to another with almost planned regularity. For instance, in the short stay at the Paradise mine, I saw miners quit and leave and then six months later when I was assigned to the Lucky Jim mine, these same guys were seen again. These men would try to find a good place to spend the winter (generally at Britannia mine on the coast) and then travel around for bonus or just a change of scenery in the summer. For the most part, from talking to them, these guys were either escaping from a bad marriage, from alcohol, from the law or maybe even from themselves. Drugs, in those days, were not a factor.

The cook at the Paradise mine was an alcoholic and would always have a small brew of fermented oranges and potatoes brewing, provided Johnnie Parker didn't find the concoction first. One day the cook didn't show up after his weekend off. The bull cook did the best he could for breakfast which amounted to cooking a few eggs and bacon and making coffee. The lunches, we made ourselves. The problem would be dinner at night. This posed a problem for Johnnie Parker as the crew needed to be fed. The problem was solved by conscripting Don Pretty from the stope next to ours to be the temporary cook. We couldn't believe it when we went into the cookery at dinner time to see Don wearing a cook's apron slinging hash. Actually Don made out quite well as he did have some previous experience at cooking. The cooking situation was finally solved by hiring Bob Mitchell's wife, Anne, to do the cooking. That worked out very conveniently for the Mitchells with both of them working at the mine.

Lucky Jim

At the end of July, I was directed to report to the Lucky Jim mine at Zincton B.C. which was located halfway between New Denver and Kaslo in the Slocan District. This mine, owned by Sheep Creek, operated at a rate of 500 tons for six days each week producing both lead and zinc concentrates. Because the mill was located at the mine portal and a branch line of the CPR crossed the north edge of the property, there was no requirement for expensive trucking as was the case at the Paradise mine. The property was composed of a half dozen or more houses for staff, two 2-storey bunkhouses, a school (which also was used as a community centre), cookery, mill building, powerhouse and shop, assay office, dry building, office and guest house. This was a compact little operation except that the ore was not that continuous.

As I remember, the ore (a sulphide replacement) was hosted in a limestone bed along one leg of a syncline fold. The hanging and footwalls were mostly black carbonaceous slate. It was the diamond drill that made this mine. The sulphide ore occurred like raisins in a loaf of bread as there was no continuity; therefore, it was necessary to locate these pods by diamond drilling and then drift out to their location and mine them. We would lay out a hole pattern of horizontal, +/-30° and +/-60° holes and others if need be. The holes would terminate once the slate in the hanging or footwall was reached. As long as the hole was in the limestone, there was a chance of finding ore. Calculating ore reserves in a situation like this was at best, guess work.

Jack McIntosh was manager at the Lucky Jim mine at that time. He and his wife Dorothy became my friends as we exchanged Christmas Cards for many years but have since lost track of each other when they moved to California some years back. Jack was a hands on manager. He could overhaul the large diesel generators better than most of the mechanics he had on site. He knew the mine well and often acted as the mine general foreman when Mike

```
Why Mining?
```

Pleacash was away from the property. Mike Pleacash, the mine general foreman, was a fatherly type and took me under his tutorage while I was there. Mike was indirectly responsible for me going to Quebec a year later when he supplied addresses where to write enquiring about jobs in the east. Mike was formerly mine captain at Duparquet in Quebec and it was his stories of that part of Quebec that were fascinating. Following Mike around the mine was a lesson in itself as to how to handle people. *(Mike was a real good guy.)*

My first job at the Lucky Jim mine was to relieve the assayer, Walt Butler, who was laid-up occasionally with arthritis. The assay office at the Lucky Jim mine was more like what an assay laboratory should be. The grinding and sample prep and the weigh-scale rooms were all separate from the determination room. It was a pretty good setup. Fortunately Butler recovered quickly so the term in the assay office wasn't much more than two weeks. However, in that time as assayer, I was able to check-in correctly with the umpire assays on determinations for concentrates sent to the Trail Smelter. Checking in with the umpire's assays raised one's confidence as an assayer.

The next assignment underground was to work timbering a newly driven inclined winze with two staff employees from the Sheep Creek's Queen mine at Salmo B.C. One of these was Chuck Treet, who was a shifter there and was assigned to the Lucky Jim mine until the company could sort out some of the problems at the Queen mine which had closed down temporarily at that time. The other person was Bill (?) Johnson, who was the carpenter foreman at the Queen mine. This new winze would serve the lower part of the mine where a new 11th level would be developed later that year. The winze measured roughly eight feet high and twelve feet wide and was inclined at 35 degrees. Our job was to install 12 x 12 stringers across the winze at eight-foot centers and then lay two runner planks roughly 4 x 10's from stringer to stringer on which 40-lb rails would be spiked. A chicken board walkway was installed on one side of the skip-car's track. Fortunately the winze was in limestone for its entire length thus it required no ground support whatsoever. This type of ground also made the cutting of hitches for the stringers easy.

Mineral King

Around the middle part of September after completing the timbering in the winze, I was requested to report back to Paradise mine to work on

Sheep Creek

diamond drilling exploration of the Mineral King property which was located near Jackpine. My partner would be Bob Mitchell, whom I had met as a result of my previous plumb-bob and survey fiasco. Our job was to make our way up Toby Creek as far as the old road would take us by truck and then with horses, pack supplies to the Mineral King property, establishing a camp there. It took about a week to cut and slash the willows and trees that grew across the old road from Jackpine up Toby Creek. Bob and I would start out early after breakfast and wouldn't get back to camp until suppertime each night. Swinging a machete all day soon got one's arms in pretty good shape. After clearing our way to Jumbo Creek, about 10 miles from Jackpine, and finding the bridge washed out, we decided to use pack horses from that point onward. We felled a few trees on each side across Jumbo Creek, down some 100 feet from where the bridge had been located, so as to scramble across the creek. This allowed us to climb across from one side of the creek to the other without getting wet as the creek was knee-high deep. Once across Jumbo Creek, it was necessary to continue along the trail for another couple of miles until one trail branched off up the hill to the Mineral King property. The outcrop on the Mineral King property was located about 1000 feet up the mountainside above Toby Creek. To get to the outcrop it was necessary to navigate a very steep trail which traversed back and forth up the mountainside. The trail must have been at least a mile or more long with a dozen or more switch-backs.

Because this was grizzly bear country, carrying a gun was a must. Although it was unknown how I would have reacted on meeting a bear, in any case, the gun did give a feeling of security. There was a grizzly that came into the camp at the Mineral King property late one afternoon which was scared off with a couple of warning shots. I was too afraid to shoot and hit it, fearing if, wounded, it may have turned and charged.

My uncle Gordon McKay ran a hunting outfitters and guiding business in the district and therefore it was natural that we rent his services to pack our supplies into the Mineral King property. My cousin Jimmy McKay was seconded to us with a string of pack horses to pack our supplies to the property. Bob Mitchell left Jim and I to do the packing while he went back to Jackpine to gather drill bits and other requirements for the X-ray diamond drill. The experience of packing horses afforded me the opportunity to learn

Why Mining?

from my cousin how to tie (throw) a basket and diamond hitch. *(I felt like a real Westerner: here I was back in the bush riding a horse, carrying a rifle, packing horses and tying the packs with the bonafide diamond hitch.)* The first day, Jim and I made two trips up to the property but on the way back, it was getting dark, requiring us to stay over night at *(well known in these parts)* the old Earl Grey Cabin that was located below the Mineral King property. During the night after being awakened by a noise at one end of the cabin, I got up and took a shot with a 303 British Ross rifle at a pack rat that was interested in our supplies. I probably missed it but we were not bothered any further that night. After completing the packing of supplies to the property, my cousin left us one pack horse to pack any further supplies that we would need. Jim's instructions regarding the horse he was leaving were, "...this horse will carry any dead load but he won't carry a live one so don't try to ride him".

At the outcrop, one could see where others *(the old timers)* had blasted an open cut to better evaluate the ore. Without much imagination one could see that there would be a mine here as there was a lot of high grade mineralization showing in the outcrop that the previous owners* had excavated. About 100 feet below the outcrop, the old timers had collared an adit and tunnelled about 200 feet into the hill to intersect the showing. Unfortunately for them they were short by a dozen feet or more as the mineralization dipped away into the hill. The tunnel or cross-cut was rather interesting as the rails were wooden 2 x 4's with a strip of galvanized metal as a runner on top. There was an old half-ton front-end dump railcar, complete with springs, abandoned on the property.

Our living quarters for the next two months was a three-sided log cabin with the fourth side built against the rock wall of the hill. It had a dirt floor and the roof was made from corrugated iron angled so the avalanches or small snow slides slithering down from the hill would pass over the cabin. The builders did their work well. Although the cabin was in the path of snow slides, the cabin remained there sturdy and untouched over the years.

The task at hand was to drill the outcrop to determine its continuity with an

    * My grandmother, Anna Elizabeth McKay, at one time, was the registered owner of the Mineral King claims.

Sheep Creek

X-ray diamond drill. If the short holes made with this drill showed encouragement, the company planned then to move a larger drill onto the property the following year to prove-up further reserves. It was possible to drill five or six holes averaging about 50 feet each (in barite formation) which was considered pretty good work. Sheep Creek did move a drill onto the property the following year and later operated the Mineral King mine for a number of years until the 70s. At the present time, the Panorama Ski Resort has been built on the opposite side of the valley where we had cleared trail and camped.

Our daily routine was for Bob Mitchell to start drilling each morning alone while I would lead the pack-horse down to Toby Creek for fresh water.

Fortunately there was a flow of water (somewhat rusty) coming out from the portal which was used for the drill. Every second day, it was necessary for me to go all the way across Jumbo Creek to where the truck was parked for gasoline and other supplies that were left there for us on a regular basis by personnel from Jackpine. One day, once down the hill and on the level trail in the valley below, I tried to get up and ride the horse instead of leading him all the time. My cousin was right, the horse threw me and galloped down the trail leaving me wondering if he would ever be seen or caught again. The good news was that the horse stopped at Jumbo Creek so it was possible to get a hold of his halter and gain control of him. Getting across Jumbo Creek took good coordination and timing. This was done by letting out about 30 feet of rope attached to the halter and then direct the horse with a slap on his rump into the creek. While the horse waded across the creek, it would mean running down-stream 100 feet, where we had felled the trees across, and quickly scramble to the other side on these felled trees. If one reached the opposite side of the creek too soon, the horse would turn around in mid-stream and wade back to the other side. If the horse beat me to the opposite side, he would move out of the creek at a gallop and it would mean having to search down the trail a mile or so before catching him again. The reason for the 30 feet of rope on the halter was to make it easier to grab the horse on the fly because he would be moving when coming out of the creek. The trick was to meet the horse just as it got up the side of the creek bank and began to gallop. In this way, by grabbing the rope, one could steady the horse. It took some pulling though to make him stop.

Why Mining?

Getting up the hill to the property wasn't too bad once you caught on to "tailing" the horse. By tailing, it meant holding onto the horse's tail and letting him pull you up the hill. Fortunately the horse would just follow the path up the hill carrying the load...(*and me holding his tail*) dragging me behind. Generally I would be back at camp with the horse and pack about noon each day. In the afternoon, it meant acting as diamond drill helper until an hour before dinner at night when I would break-off to prepare dinner. We had it worked out where Bob would cook breakfast and I would prepare the dinner. Lunch was the responsibility of whoever was the hungriest. Later another driller by the name of Laddie Syronge was added to the crew. He took over the responsibility for lunch and the drill helper's duties. Johnnie Parker paid us a visit and said that with the addition of Laddie on the drilling, he wanted me to prepare a topographical map of the site and then handed me a brunton compass, tape and hand level. With the help of Laddie, I was able to make a baseline, do some levelling and prepare what looked like a pretty good map. The drilling went on until the first week of November when the weather got pretty cold and the trails were covered in deep snow. At that point, we packed everything into the adit, barricaded the portal and moved out to Jackpine.

Lucky Jim - Second Time

On returning to the Lucky Jim mine at Zincton that November, I was placed on staff payroll as a junior engineer at a salary of $225 per month. The job consisted of assisting Bob Robertson, the mine engineer. Bob had graduated the previous year, and had much more experience than me. *(Bob's wife Joan had just delivered twins that autumn so it goes without saying Bob was very experienced.)* Bob showed me around the mine where we would survey the stopes, lay out diamond drill holes, grab and log core as well as calculate the bonus. The advantages of working for a small mine were the variety of experiences one gained. For instance, the mine engineer was responsible for repair of the drills *(I still have a split nail on one of my fingers on account of a drill falling out of the vice and slitting my finger and nail)*, cutting and capping fuse, packing pumps, pipe-fitting, operating the hoist and supervision. Another highlight was looking over the mine plan and listening to Mr Doelle when he visited the mine to hear what he thought should be done and what his engineers should accomplish. It was an education in itself.

The only drawback was that we were obliged to work six days a week, that is, until Bob Robertson and I began campaigning for a 44 hour week instead of the 48 hour week that was then the custom. Jack McIntosh finally gave us his "OK" whereby after Christmas of 1950, we were able to take every second Saturday off.

Sheep Creek Mines Ltd Zincton B.C. 1950

This picture was taken from the No 3 level portal. The mill buildings are in the center foreground along with the two two-story bunkhouses to the right above. The building at the extreme left are the office and staff quarters with the white manager's house to the right of the office. Between the manager's house and mill buildings are the cookery and assay office. Other buildings are residences for a few of the employees who had families living on site. The CP Rail spur curves around the top of the picture with the Denver-Kaslo highway just below it. The snow slide path is the lightly-treed area at the extreme top right.

```
Why Mining?
```

A special duty for the newest engineer on the property was to run the projector at the weekly movie held in the school building which doubled as a community centre. Bob Robertson gave me a couple of lessons as projectionist and then I was on my own. It also was the job of the projectionist to help with the collection of tickets and with the stacking of chairs. Bob was a pleasure to work with for the short time we had together as he and his family left Zincton for the Silver Standard mine in early March of 1951. Unfortunately, Bob had an accident a year later at the Silver Standard mine where he lost an eye* taking a chip sample. He left mining shortly afterward and began a career in teaching mathematics and science at a high school in Vancouver.

The Lucky Jim mine had exceptionally good ground where it could employ open stoping. Large caverns were excavated by benching using Gardner-Denver D-88 drills mounted on tripods. The work in setting up these drills on a tripod was a two-man job. The whole setup was very cumbersome to handle and very slow to move about. The Swedes (Atlas Copco), at that time, were very active in introducing the "jackleg drill and integral steel" into the mines. They scheduled a week's drilling demonstration and I was selected to run the time tests to check the penetration and so on. The miners were very sceptical as some of them had seen the jackleg drills before but not with the integral steel. They were non-believers until I got to try the jackleg one morning and drilled over 40 feet in an hour which was unheard of with the D-88 on the tripod. After that, there was no problem in introducing the jackleg underground. Regarding the Swedes, they were not Swedes at all. They were Canadians from Eastern Canada with Scandinavian names such as Dick Dell and Henry Jensen, the drill demonstrators and Larry Thiessen, the salesman. It was Dell and Jensen who reinforced my interest in going to Eastern Canada to further my career as they said that the mining there was a good five-years ahead of what they had found in B.C.

The Lucky Jim mine was more or less divided into three sections of levels. The top levels (1st, 2nd & 3rd) were accessed by driving a jeep truck up the switch-back mountain road and entering at the 2nd level adit. The ore from the top mine was trammed to surface on the 3rd level and hauled down to the mill bins by a jig-back tram-line. In a jig-back system, there are two buckets

* Safety glasses had not come into general practice at that time.

on two separate cable lines where the loaded bucket (in this case a 1-ton bucket) while descending, pulls a live-line connected to the empty bucket up to the loading station. Braking is accomplished on the live-line. It is a pretty good system requiring no power except when a bucket jumps the cable then it takes manpower to get it back up onto the cable again. I did a bit of spare supervision in the top mine and had a chance to operate the jig-back tramline.

The middle section of the mine contained the 8th level which was the main level for haulage to the mill. All ore below the 4th level (which was inactive at the time) was transferred by orepass to the 8th level. The ore from below the 8th level down to the 11th level was hoisted using the newly commissioned winze that I had helped in timbering the summer before. In order to speed up the development of the lower (11th) level, Jack McIntosh suggested that I work steady night shift (8 pm to 4 am) supervising the development crews consisting of two driftmen, two raisemen and a trammer. On day shift there was adequate supervision (two shifters and a general foreman) but for night shift there was only one other shifter *(besides myself)* for the whole mine. My other duty on night shift was to operate the hoist in the winze. This required emptying the pocket about twice each shift with the trammer acting as skip-tender.

There were a few experiences in the winter at Zincton that should be mentioned. One was the snow slides in late winter. The mine was located in a very narrow valley which experienced very heavy snowfalls. In late February when the temperature became a bit warmer, often times the property would experience snow slides on either side or sometimes both sides at the same time. That particular winter spent at the Lucky Jim, they experienced snow slides on both sides of the townsite locking everyone in for a week until the snow plows (D-6s) could establish a roadway. One could, if necessary, climb over the slide and walk to town about 10 miles distant. However, those that did climb over the slides were not going to town but to the bootlegger's place at Three Forks about a mile or two down the road. John Henry the bootlegger, ran a pretty honest house, at least I found it so, when the only time going there was with Jack McIntosh, the mine manager. Maybe it was out of respect for Jack that John Henry poured good sized drinks.

Another character around the Zincton area was Christine Larsen the local

Why Mining?

madam who held forth in Sandon about five miles from Zincton. Sandon was going through a revival with a lot of diamond drilling going on in the area and with the rich Viola Mac mine in operation. One evening while coming back from a weekend in Trail (where my folks lived) I got off the bus at New Denver hoping to find a ride up to the mine about 10 miles away. Luckily I came across Pete, a doukabour who worked at the mine and owned a car. He kindly offered to give me a ride to the mine. On our way to Zincton, instead of staying on the road to Zincton, Pete turned off toward Sandon saying, "I've got to pick up Frenchie". Frenchie, who was at least 70 years old, was our blacksmith. Anyway he said no more until we got to Frenchie's house, and then said, "There's probably a party going on". Thinking no more of it, we went up to the house and knocked at the door enquiring if Frenchie was ready. The old lady *(which was assumed to be Frenchie's wife)* was very friendly and asked us to come in and wait. She sat us in the parlour saying that Frenchie was packing and would join us shortly but in the meantime ..."would you boys like a drink as we have beer and whisky, whatever you want". I thought, gee this old grandmother is very broad minded to offer us a drink. We settled for a beer each. Then noticing in the next room that there were people who were dancing and some of them dancing very inappropriately, I began wondering at the time, if the old girl was blind letting her daughters dance so scandalously close. Another party came to the door when we were having our beer and the old girl sat him down in the parlour as well. This new chap ordered a beer and gave the old girl money for it. It then dawned on me that I was in a bawdyhouse for the last half hour, not realizing where in hell I was. *(Talk about being naive!)*

The Robertsons when they lived at Zincton were fun people and their twins (a boy and girl) were getting to the interesting stage. Some very enjoyable evenings were spent at their home on the property. After Bob and Joan Robertson left Zincton, I found my spare time at the Lucky Jim a bit dull. However, all was not lost. By springtime, I had enough money saved to buy a secondhand car for the princely sum of $600 which would give me some mobility. *(A sum of $600 saved doesn't sound like very much, but when one thinks that, when earning only $225 a month, paying $60 room and board to the mine, returning $25 for income tax, and sending $50 home each month to help my parents send my brother Jock to college, I think this was exceptional.)* This first car was a '38 Ford Coupe with poor brakes. It was a

Sheep Creek

bit hair-raising to go down the hill from Zincton to New Denver as it was necessary to put the car into low gear and hope the brakes would do the rest. In any case, the vehicle was my transportation out of camp at least once during the week and on weekends. It also was the means to become reacquainted with a couple of young nurses at the New Denver Hospital. *(I say reacquainted because the previous summer I spent two days in the hospital fighting a virus.)*

During the first year while being stuck in the backwoods of B.C. in a mining camp, my buddy Jack Frey, who had moved out to Val d'Or Quebec, was enjoying such things as French Canadian hospitality which included pretty girls and parties, city living, night clubs, Sunday movies, etc. His letters would include a snap shot or two of one of his many dates where they would be enjoying themselves at a party. I was very envious. Finally, after talking to Mike Pleacash and the Swedes (when they were at Zincton), I began a writing campaign, sending letters of application to points east where one could live a little and at the same time advance one's career. Letters were sent to the Beattie, Horne, Canadian Malartic mines without much success. Finally I received a reply from an unlikely source, Malartic Goldfields. Apparently, my letter to Herman C Herz (manager of Canadian Malartic) was passed on to Freddie Cook (chief engineer) who phoned Andy Anderson (assistant manager of Malartic Goldfields) enquiring on my behalf. *(The Malartic camp in those days was closely knit with close cooperation between the mines.)*

A job offer was made by Malartic Goldfields where I would be given a job as a "machine runner" in a stope with guaranteed earnings of $250 per month. That is, if the wages and bonus didn't equal $250, the company would make good the rest so that the gross income would be $250 monthly. This was a good deal so naturally I jumped at it. I gave my month's notice to Sheep Creek following the Malartic Goldfields' offer. Jack McIntosh received this resignation with a lot of class and wished me well. In fact the night before I left, he drove me down to John Henry the bootleggers' place at Three Forks for a couple of "rum and cokes".

When closing the bank account in New Denver, I ran across Dr Robbie Robinson whom I had met socially through Jack McIntosh and professionally during my stay in the New Denver Hospital the previous summer. After

Why Mining?

telling Robbie that I would be leaving for Malartic shortly and said good bye, he said, "I hope you enjoy winter because they get lots of it there." Robbie had previously moved from Noranda Quebec to New Denver and knew what he was talking about. I wasn't aware that the weather in Northwestern Quebec would be any different from that of the B.C. Interior. *(Having never been out of B.C., how would I?)* Looking at the map, Malartic seemed to be the same latitude as New Denver, so why would it have a long winter? *(I would find out about the weather the following winter after moving to Malartic.)*

After making final arrangements at the mine, I left Zincton and drove via way of Nelson where I sold my car after owning it for a total of four months. It was fun owning it but I wasn't about to drive it across Canada with the condition of the roads in the early 50's, although at one point it was considered. The way east would be by air plane. Now with one last visit home to Trail, I was ready for a new adventure east.

Paradise Mine Remnants 1970

In the forefront are the truck bins which received ore from the 7900 level and at the back left are what remains of the changehouse and dry.

Chapter 3

Je Me Souviens (au Je veux oublier)

Trip East

The journey to Quebec would be the first trip east as well as my first flight on a commercial aircraft. Although having flown in a Cessna air plane, flying around Cranbrook, this would be the first in a very large plane; that is, if you call a DC 3 plane large. Canadian Pacific Airlines, in those days, operated the flights from Vancouver to Calgary, stopping at such intermediate points as Penticton, Castlegar and Cranbrook. The flight east started from the Castlegar Airport which is nestled in a valley at the confluence of the Columbia and Kootenay Rivers. It's not the best place to land an aircraft even at the best of times, but in this case, the flight took off in good weather in June 1951 so it wasn't too intimidating. The DC 3 aircraft had a limitation that prevented it from flying high enough when loaded to clear the Rocky Mountains east of Cranbrook, and therefore had to fly through the Crowsnest Pass in order to make its way to Calgary. Whether or not the plane would fly the Crowsnest Pass, depended on the weather or cloud conditions in the pass. The pilots received the weather condition reports at Cranbrook. On our particular flight, the weather conditions were favourable and clear through the pass so that there was no delay in reaching Calgary that day.

People told me that for the best view of the country side, try to be located in the last row of seats in the plane or at least those seats behind the wing. Fortunately, one of the window seats in the last row was unoccupied. Also, another fortunate thing occurred as the other seat in the last row was reserved for the pretty air stewardess, one whom I had met at university. *(In those days all air stewardesses were nurses and were generally very pretty.)*

From Calgary, I was scheduled to fly Trans Canada Airlines (TCA) to Montreal and from there switch back to Canadian Pacific Airlines (CPA) for the Montreal to Val d'Or leg of the journey. As planned, the plane arrived in Calgary on time at noon but there was no connecting flight to Montreal and therefore, it was necessary to wait until eight o'clock that night for the next plane for the east. To make matters worse, TCA found my baggage to be overweight by twenty-five pounds. My baggage included a suitcase of clothes and personal things, a duffle bag of underground diggers and a set of bagpipes. TCA wanted extra for the bagpipes which were to be sent air cargo

Why Mining?

on separate flight the next day. *(What choice did I have?)* The flight was in a North Star aircraft which stopped at various cities (Regina, Winnipeg and Toronto) between Calgary and Montreal. Unfortunately, the flight arrived in Montreal late by thirty minutes arriving at eight o'clock the next morning while my flight via CPA to Val d'Or had taken off at the same time that the TCA flight was landing. What this meant was that I would have to spend that day and night in Montreal waiting for the next CPA flight to Val d'Or which was at eight o'clock the next morning.

Fortune shone on me again. Who do I meet at the Dorval Airport in Montreal but an old buddy, Doug (Alphonse) Bertoia, who introduced me to underground mining at the Sullivan mine in Kimberley while attending high school in Cranbrook. Doug, at the time, was the head passenger agent for TCA at the Dorval Airport so I was in luck. With Doug's help, he arranged a change of flight plans to Val d'Or for the next day and then invited me home to stay overnight at his place. Doug was able to get off early from work that day. We went to his home where I was introduced to his much pregnant wife "Gaby". *(Gaby has since become a favourite in the McKay household. She is a gem.)* Gaby made dinner and then told Doug to take his friend out and show him the town. *(What a good sport!)*

The first stop in doing the town was to see the Old Montreal Royals Baseball Team play at Delormier Stadium. What a thrill that was, as this was "Triple A" ball. The Royals were playing the Syracuse Chiefs but I don't remember the score or who won. Our next stop was generally to tour the downtown by foot and by streetcar. Montreal was then the principal city of Canada. Not only was it the financial capital, it was the cultural capital and the sports capital as well. Toronto, in those days was dowdy in comparison. *(Oh, how things have changed! Montreal is now very rundown, its streets are full of potholes and there are hardly any building cranes in sight on the skyline.)* We finally ended up at the Bellevue Casino. The Casino was reported to have the best floor show with the largest cast in all Canada. One could believe it as there were no floor shows to speak of outside of Quebec and this one was the biggest in Montreal. The price for a beer was two dollars a bottle which in those days was steep. But when one considers there was no cover charge for a tremendous show, the price was about right. An incident which was not part of the floor show, which happened to everyone's delight anyway, was

Je Me Souviens

when one of the high stepping girls in the chorus line bent over, her brassiere clasp broke and her brassiere slipped down showing her very nicely shaped feminine bosom. *(All the things I had heard about Montreal were reinforced with that occurrence.)*

The next morning was uneventful other than Doug Bertoia getting me squared around and onto the CPA flight to Val d'Or. The flight was also in a DC 3 aircraft where again I was able to get seated in the last row along with a pretty Canadienne stewardess. *(I just can't and won't say "Quebecoise". "Canadienne" sounds more refined.)* The flight to Val d'Or was over bush country. One could see a road along the plane's flight path and with the clouds of dust coming from each vehicle as it proceeded along that route, indicated that the road below was unpaved. We landed in Val d'Or after a heavy rain fall there. *(What else is new?)* After a taxi ride to the Bourlamaque Hotel, I checked in and waited for my buddy Jack Frey who had arranged to meet me on his noon-hour break.

The first thing I noticed after checking into the hotel room was the flickering of the lights because of 25 cycle power. The north country didn't change over from 25 to 60 cycle power until much later. It's funny but one does get used to the flickering lights.

After meeting Jack Frey and having lunch together, I accompanied him to the survey office at the Lamaque mine where he worked. Jack Toivanen was the chief engineer at the time; also, there was Jack Shaver who was underground superintendent then but became manager some years later. After shift, a few of us went back to the bar in the basement of the Lamaque Hotel which the boys dubbed "the sump ". The popular beer amongst Jack's friends was India Pale Ale or IPA for short. I, for my part, thought Molson's Ale was great. *(In those days Molson's still had the apostrophe.)* That night the gang went to the Morroco Club and took in the floor show. *(This was living.)*

By the next day, after a night out on the town of Val d'Or and touring Montreal two nights before, I was beginning to get a bit short of cash so thought it best to get to work and earn some money. But first it was a necessary wait around Val d'Or for the morning flight from Montreal in order to retrieve my bagpipes. I was fortunate in having a friend like Doug Bertoia

Why Mining?

at the Dorval Airport on the lookout for my pipes. These arrived from Montreal in "OK" condition, so now I could leave Val d'Or and therefore, took the bus (l'Autobus Abitibi) to Halet Quebec where the Marlartic Goldfields mine (MGF) was located. Halet was located about five miles east of Malartic and about a mile south of Dubuisson. I noticed the telephone poles, when riding the bus, were not the tall straight poles as those in B.C. These poles were short and twisted and were often leaning on account of being installed in muskeg. This was my first introduction to muskeg. In fact the whole countryside around Val d'Or and Malartic was nothing but "stunted trees and muskeg swamp with a bit of sand esker thrown in" and Halet was right smack in the middle of this swamp environment.

Malartic Goldfields

The bus driver let me off where the mine road intersected the highway. The highway was unpaved as was the mine road. As it was a good mile hike into the property from the highway and finding the heavy load of carrying a suitcase, a duffle bag with my diggers and a set of bagpipes a bit cumbersome, I opted for hiding my bags under a bridge that was close to the highway instead of struggling with these all the way into the mine. The first headframe in sight was that of No 2 mine of Malartic Goldfields (MGF) but instead of stopping there, I went to the store that was nearer to the road and made enquiries. The store was owned and operated by Leo Landry who had the concession of supplying mine clothes and credit to the employees of MGF. Any credit extended was deducted from the employees pay cheque plus interest on the next pay period. With six hundred employees and this being the only store in the village of Halet, the guy couldn't lose. Contact was made with Andy Anderson by phone and shortly after, Bob Dempsey, the chief engineer, arrived by car to collect me as well as my bags from under the bridge.

Bob Dempsey was a very quiet and reserved man. He kept everyone at a distance. He never got close to anyone, certainly not any of his subordinates. He was a clever man and a very able chief engineer although the only knock on him would be the fact that he very seldom went underground. Bob's main job was to take the geological and assay plans of the development and lay out the new stoping plans. Other than that, it's anyone's guess how he spent his time. He managed by leaving everyone on his own. I don't remember him ever giving anyone a direct order or for that matter ever instructing those in his charge anything about engineering or mining. He was always around and could be approached easily by just walking into his office. Other than that you were on your own. It should be mentioned that Bob was an ardent curler. MFG had two sheets of natural ice and a very comfortable club room where you would see Bob during the winter months honing his curling skill. At the club parties, Bob would be one of those who would stand at the back with a glass of rum and coke *(his favourite drink)* and watch, mostly. He was never one to push to the forefront. He was honest and sincere, and therefore, he was very likeable.

Bob took me in to meet the assistant manager, Andy Anderson, who was in his late twenties. Herb Cox was the manager at the time, but he seemed to be away most of the time. *(A number of years later, I introduced myself to Herb Cox at the Engineers' Club in Toronto when we were both seated at the common table for lunch. He didn't remember me but he was gracious enough to reminisce about some of the people we both knew from MGF.)* It was Andy who wrote and offered me the job at MGF as a machine runner. Andy was a very friendly fellow, who took an interest in his engineers and staff, a people person really. Anyway, Andy said that MGF had lost one of its surveyors to Inco and he was offering me this job instead. I told him that my interest was in production but would take the survey job in the meantime to get started. He said the same pay conditions ($250 per month) would apply so there was nothing to lose. Besides this, by joining the mine rescue team *(which I did)*, it would add an extra $5 per month to my salary. After this I was signed on and was given all the necessary financial details by Art Symons, the chief accountant. Art was a real company man and a good accountant for a gold mine that was on the Emergency Gold Assistance Fund. *(Gold was worth $35 US or $37.50 Can.)* People always said, if there was a way of figuring out a pay package, Art would be sure to figure out the one where there would be the least return to the employee. This was true when applying for my vacation the following spring. What I had thought was my vacation entitlement was much less than what Art figured. Anyway getting back to Andy Anderson, it was he who asked me to stay with MGF (two and a half years later) after I handed in my resignation in order to join Steep Rock Iron Mines. Andy offered to match what Steep Rock Mines had offered. My mind was made up. Steep Rock, at that time, was doing all kinds of engineering

Why Mining?

projects that looked interesting which could further my career. It was flattering and much appreciated, the fact that Andy was concerned enough to ask me to stay. He was another class person.

After signing on, Dempsey took me in hand and introduced me to his staff. Afterwards, our first stop was to the staff house where I was shown my private room by the lady who looked after the place. The same lady not only kept the place clean but also made the beds each morning. What a difference compared to Jackpine, Paradise or Lucky Jim. This was paradise in comparison! The place was very POSH. It had a very large common room where we often had some very good parties. There was laundry and dry cleaning service where a guy came once a week to pick up any clothes needing washing or dry cleaning. It was a good place to stay. There were only two other guys staying in the staff house at that particular time. One of these fellows was Bob Goran, the shop foreman, who had recently arrived from the Buffalo Ankerite mine in Timmins after it had closed down. The other chap was Eli Brdar, a Yugoslav, who was the machine doctor and a very good one too. Malartic Goldfields wanted to hold on to Eli so badly that they made the special concession to let him stay in the staff house even though he was not on staff. Eli never learned to pronounce Val d'Or in spite of all my coaching. He always said,"Wal d'Or".

(Years after, I had other feelings about staying at the staff house. One of these was that I had isolated myself among English speaking people where I did not require learning French. Had I boarded in town at a home of a French speaking family with small children, I would have been compelled to learn French. In those days speaking French was considered second-class. The language around the mine, especially amongst the staff was English, particularly in the main office. We didn't have one Francophone in the engineering office until the second year I was at MGF, when a young Francophone was hired as my survey helper. Many of the businesses in the Town of Malartic were run by Anglophones who spoke English exclusively. And to top it all off, the majority of the Anglophones were Protestant to boot. The businesses, however, did hire some bilingual staff to accommodate the Francophones. I often wonder now how the Canadiens (Francophones) felt when their language was down-graded to second-class status. At the time I thought nothing about it and went along with the flow.)

Je Me Souviens

Our next stop was the cookery where they had a staff dining room. *(Again I was isolating myself from the French speaking people.)* From there, Dempsey took me to Landry's store where a few incidentals (mine boots and socks) were bought on credit and from there we went over to the dry at No 2 mine.

When Malartic Goldfields first stared, it began with the sinking of No 1 shaft at the east end of the property. The office, warehouse and shops were built next to No 1 shaft as well. That particular shaft was a standard three-compartment shaft sunk to a depth of twelve hundred feet and then extended to fifteen hundred feet later. Some years later, better ore was found further to the west. A one-mile drive was made on the 1200 level out to this new ore and, as a consequence, a new five-compartment shaft (No 2) was sunk to meet the 1200 level drive from No 1 shaft. A very modern mill was also constructed near the No 2 mine to handle all the ore from both mines. The ore from No 2 mine was conveyed on surface to the mill while that from No 1 mine was trucked. The facilities at No 2 mine (both mine, mill and dry buildings) were first-class even by today's standards. These buildings were constructed of concrete foundations, with steel frames, supporting insulated corrugated steel cladding panels. The interior partitions were of hollow masonry block construction. Later, a new shop and warehouse complex was constructed in the area of No 2 shaft.

No 1 Headframe   **No 2 Headframe**
Malartic Goldfields 1952   **Malartic Goldfields 1952**

Why Mining?

The geology at MGF was fairly simple. The ore was located in diorite dykes measuring about 100 to 200 feet long, averaging about 13 feet wide, striking east-west and dipping to the north at 65 degrees with a rake to the west at a slightly steeper angle for the most part. The diorite dykes were very competent while the country rock, what was locally known as "greenstone", was in many cases very incompetent, especially adjacent to the diorite dykes. There were some porphyry dykes that contained gold but any high assay of gold in these would be the exception at Malartic Goldfields.

The mining method was for the most part flat-back cut and fill but there were a few rill cut and fill stopes and some shrinkage stopes as well. An experiment was tried at one time with blasthole mining but it was unsuccessful. The level interval was 150 feet with a 150-foot crown pillar left above the first level (one-fifty level or un cinquante niveau). Everything was track haulage (twenty-four inch gauge) with chute loading for the most part but there were a few mucking machine draw-points to make things interesting. All stopes were started by drifting along strike on the diorite dykes and slashing these to width and then taking down the backs. The backs for the most part were excavated to twenty feet above the track. Afterward 12 x 12 B.C. fir timbers at 5-foot centers covered with spruce lagging were installed horizontally from the foot to hangingwall of the stope. These timbers were hitched into the walls using bull-horns and pin steel and supported intermediately on both sides of the track by 12-inch (or more) diameter poplar timber stulls (posts). Chutes were installed every fifth set and manways were installed at both ends of the stope. The diamond drill played an important part of the stope layout. Prior to laying out a stope, the dyke was sampled by drilling angle holes parallel to the rake of the dyke at twenty-foot centers. Any low-grade zones were marked and if these were extensive, they would not be mined. This was done by box-holing through these areas and silling above where the better ore was located so as to begin flat-back cut and fill mining at this higher horizon. A twenty-foot sill pillar was left as the stope approached the level above. This sill was mined in most cases by rill cut and fill stoping on an angle of thirty-five degrees (angle of repose). Some timber support was used in the mining method either in a form of timber cribs or as props. When one considers that this was the age before rockbolting, the methods used at Malartic Goldfields were really quite safe. As mentioned, there were some shrinkage stopes in the very narrow dykes but these were very few. Most of the fill was distributed

Je Me Souviens

by slusher but there were a few stopes that depended on the hand-trammed fill-car for filling. When I first arrived at Malartic Goldfields, sand fill was used exclusively as hydraulic fill was only then being introduced into the mining industry. Early in 1952, a small group from Goldfields made a trip to Red Lake to observe hydraulic sand filling of stopes and came back sold on this method of filling. Thus all stopes, in the lower section of the mine below the 1200 level, were developed for this method of filling.

As mentioned, the job assigned to me was in the survey office instead of production. There were three survey crews at MGF, two on stope surveys and one on development. My assignment was that of the development surveyor which was fine as it got me around to all the new parts of the mine. The guy who showed me around this beat was a Brit by the name of Bob Goldsmith who later accepted a job in Timmins. Bob was a bit on the lazy side and, during his last few days before leaving, would spend all afternoon just refilling a stapler. *(Bob since went to Malaysia. I not only took over his job but I took over his girl friend. But that's another story which we'll deal with later.)* Bob showed me around the mine and the surveying system used at Goldfields. Unfortunately for me was the fact that Bob's survey helper was leaving also. I only got to meet this helper once and he was gone. Not only was I inexperienced in mine surveying and unfamiliar with the mine, there was a new fellow to break-in as helper who was unfamiliar with the mine as well. With this new helper, Tom Black, we somehow found our way around and looked after six raise crews, eight drift crews and a shaft sinking crew without too many complaints. We also were responsible for surveying the diamond drill holes which included giving fore-sights and back-sights prior to drilling and then later surveying the collars and determining the azimuth and dip.

A very interesting mine development which had been planned just prior to arriving at Malartic Goldfields was the extension of No 2 shaft another six hundred feet to the 1800 level using its own shaft crew. The plan called for a sinking hoist to be set up above the spill pocket horizon. Sinking was to be through the cage and manway compartments requiring a heavy bulkhead to be installed at the loading pocket and below the spill pocket for protection of the sinkers below. Sinking would be carried on simultaneously while the mine operated. New stations were to be collared at the 1350, 1500, 1650 and 1800 levels with a new loading pocket to be excavated between the 1650 and 1800 levels. This plan was executed during the first year after arriving at Goldfields.

It was the responsibility of the development surveyor for the surveying and extending lines in the shaft and for the shaft plumbing at the shaft stations so that survey points (stations) could be continued out into the mine. What was found very interesting, was the drive out on the 1800 level as crews set up a slusher-trench in conjunction with the lip-pocket to handle the muck from the cross-cut drive out to the north of the shaft towards the ore zone. The final plan called for conveying the run-of-mine ore from the 1800 level up to the 1650 level loading pocket. This required driving a fifteen degree incline conveyor gallery from the north-end of the 1800 cross-cut to the 1650 level station to intersect above the loading-pocket at this level. A 48-inch Meco belt conveyor was installed in the gallery which brought ore from a feeder located just below the 1800 level up to the loading-pocket below 1650 level. It was very gratifying that the gallery broke through on target as it proved my shaft plumbing and surveying for the cross-cut drive on 1800 level and the gallery drive were correct. *(I guess I would have been fired or at least lost much face if the gallery had broken through at the wrong place.)* However, I was pretty sure of the surveying as everything was double-checked before the 1800 crosscut had been excavated too far. Even the shaft plumbing was done twice on two different Sundays just to make sure this was correct. As a rough check, line was taken from the timbers to be doubly sure. *(Now-a-days, one has to only drop one line and use a azimuth seeking theodolite. So simple!)*

While at Malartic Goldfields, we did find a shaft plumbing error between 1050 and 1200 levels at No 2 shaft where each level survey differed by about eight feet on average. Apparently prior to my arrival, a previous surveyor had misread a survey angle while plumbing these levels. This discrepancy was noticed every time a raise would be driven between 1200 and 1050 levels as the break-throughs would be out about eight feet and some would be more for those raises that were further from the shaft. In any case, prior to plumbing the extension of No 2 shaft to the 1800 level horizon, we plumbed the shaft again from collar to 1200 as a check before plumbing to the 1800 level just to make sure everything was to our liking.

Je Me Souviens

Tom Black was my first survey helper (*or rather partner as I liked to call him*) for about six months and then he was promoted to stope surveyor. Tom was a small wiry man who married a big woman. (*I don't know the reason for this but it often happens that small men like big women.*) Tom was also a very optimistic man especially when it came to money. Tom was always in debt but it didn't seem to bother him. For instance, one Saturday after work when we went to the post office in Malartic together to pick up our mail, Tom's mail was all bills. He immediately shoved the bills back into the post box and said, "no use spoiling the weekend". That was it. Many pleasant weekends were spent visiting Tom and his wife Jean at their apartment in Malartic. They liked to play cards especially "hearts". They were fun people but not all together forthright. When Tom and Jean left Malartic, it was on the "QT" as they had bills to pay all over town. (*I often wonder how they made out when they left for Guelph Ontario. To do what? I never did find out what happened to them.*)

After six months with Tom Black as the helper and prior to his leaving, he was promoted to stope surveyor which meant breaking-in another helper (or partner) for development surveying. The unlucky new chap was no other than Alex Makila who ended his career as assistant manager of the American Barrick property which was situated just north of Malartic Goldfields. Alex spent all his career in the Val d'Or - Malartic area as it was home to him and his wife Aneta. To be very honest, Alex was not a very good survey helper, perhaps because he was trained as a geologist and therefore was much more interested in geology than simple mine surveying. (*And could you blame him?*) The only reason Alex took the survey job was to get his foot in the door, so to speak, and be next in line for an opening in the Geology Department. Alex was a very kind sort of man, he couldn't hurt a fly. Loyalty was another one of his traits. Fortunately for Alex, the opportunity to get into the Geology Department was soon in coming which required the training of another partner. This time the new partner was a "Canadien", Jacques Goudreau. Jacques had just been discharged from the Air Force and this was his first civilian job. Jacques was a fast learner, and because he was so quick, he was transferred to the stope surveying section to fill a void there due to a termination. My next partner was Claude Jolin, a kid from Amos whose father owned not only the saw mill located on the Malartic Goldfields property but a couple of others in the Amos area as well. Claude had half a year at McGill

Why Mining?

but quit in mid-term. He was a welcome addition as we all found him to be a comical guy with a very dry sense of humour. He was always peddling something...suits, caps, coats, you name it. *(I think most of these goods were hot but Claude always denied it. I did buy a suit from him once with the assurance that it wasn't hot. It was a good suit too.)* When leaving MGF, Claude and I corresponded for a couple of years, then it was only Christmas cards for a while, but later somehow we lost touch with one another.

Regarding Alex Makila moving to the Geology Department at Malartic Goldfields, Alex could not have had a better mentor than Clarence Wilton who was the chief geologist during that era. Clarence embodied everything that a chief mine geologist should be. He produced some of the most comprehensive maps and sections that one would ever see at any mine. These maps and sections were always up-to-date because he and his staff were always just behind the survey crew mapping the geology in the drifts and cross-cuts as soon as they were driven. The core logging was always up to date as well. Generally most mine geologists dream of the chance to get into exploration and neglect the mine geology if there is any exploration to do around the camp. Not Clarence...the mine geology came first but he was not averse to dabbling in a bit of exploration though. There was some exploration to the west and east of the two shafts both on surface and underground. There was some exploration drilling done in the Barraute area that caught Clarence's attention also. As an aside, Clarence liked to get some company gravy such as free meals and gas etc; therefore, the short excursions off the property to attend to exploration problems offered him this chance. On one particular occasion he had me travel with him to Barraute to survey diamond drill holes and to help him log core. Instead of carrying a lunch like the rest of us and eating it in the bush, Clarence insisted on having lunch at the local hotel as it was a meal he could charge to his expense account. I would follow him into the hotel, order a cup of coffee to keep him company and eat my own lunch which was brought from home. The hotel owner let me do that, or perhaps I should say, he tolerated it. Anyway on this one particular day, Clarence was served a cockroach with the stew that he had ordered. Instead of objecting and walking out, Clarence just moved the cockroach aside and kept on eating. Having no stomach for the place, I got up and left and ate the remainder of my lunch in the truck. Clarence was a good guy though. My first winter, I curled on his rink along with Wally Gibb (geologist) and Henry

Burk (purchasing agent) all from Malartic Goldfields. Clarence was a terrific shot-maker. It's too bad he didn't have a good rink behind him as we didn't win much that winter even though we curled in most of the local bonspiels at Perron (which was their last spiel), Val d'Or, Goldfields, and Noranda. Poor Clarence seemed to end up with some odd people playing for him. One winter at the Goldfields Curling Club (it was open to anyone who wished to curl), Clarence was teamed up with a "FR RANGER" whom he thought to be Frank Ranger. However, the Frank Ranger turned out to be Father Ranger (pronounced "Ronjay") complete with cassock and all. *(Can you imagine the sight of a priest curling in his cassock? Better still, how about a bunch of Scots in kilts playing against a bunch of priests in cassocks. Now that would be a sight!)* Anyway, Clarence weathered that winter too.

The curling club was the center of activity at Halet in the winter time. With two sheets of ice and a fine club room, it was always a good place to socialize. Even in the summertime, there were a number of good parties in the club house. Some would turn out to be all night parties and then into a shivaree to some unsuspecting couple's home. On one such party, we *(I was a part of the group with pipes and all)* marched through Johnnie Waugh's house (John was the mine superintendent) from the back door through to the front door very early in morning. We felt safe in doing this as we had John's brother-in-law, Jimmy Morash with us. Jim, who was the assistant mine superintendent, enjoyed a good party but unfortunately, he did not enjoy good health as he died a year later.

Jim Morash's place, as assistant mine superintendent, was taken over by Murray Kennedy who was a very clever mining supervisor. Between him and Johnnie Waugh, they ran a pretty efficient mine. A very excellent mill superintendent was Gerry Roach, who later worked for Rio Algom. *(I saw Gerry a few times after leaving Goldfields at the C.I.M. Conventions in Toronto years later.)* Gerry O'Halleron was the maintenance superintendent during the years at Malartic Goldfields. *(I caught up with Gerry when I went to work in Timmins for Texasgulf-Kidd Creek in 1971.)* A few other people worthy of mention are Gene Graham (later of Ingersol Rand and U & N Equipment) who was a shifter at MGF, and a good one too. Another guy was Elmer Gauthier, a captain. Elmer met an untimely death (which had sinister overtones) in Peru. When the authorities were transporting his body out from

Why Mining?

the mountainous region where he met his death, the transport truck, while rounding a sharp bend in the road, tipped the coffin with the body over the bank. It was nearly a week before they could reach the ledge where the coffin had come to rest. Bill Hatherly, a captain, was a very intelligent supervisor with a lot of mining common sense. Bill met an untimely death with cancer while he was employed by Angus Martin of RAM Raising in Sudbury. A couple of other interesting guys were Bill Ball and Pappy Hylands of the Bonus Department. Pappy was the comic of that group while Bill was the serious one. MGF was fortunate to have had many good people. Most of these supervisors had come from the Lake Shore mine at Kirkland Lake and all of them, to a man, were top notch mining men.

Shortly after arriving at Malartic Goldfields, Herb Cox, the manager, came in to the engineering office one afternoon and suggested to all the young engineers, especially those who were planning to make a career working in Quebec, that we should strive to learn to speak French. With this in mind, he said that the company would be making the curling club party room available for lessons provided there was enough interest shown. A number of us signed up along with quite a few supervisors' wives from the townsite of Halet. Twice a week lessons were started that fall in the curling club rooms with a teacher travelling in from the French School at Malartic.*(Herb Cox was forward thinking. In fact, I have since looked back and found a lot of my opinions have changed over the years, especially those regarding Quebec and the French language. How I wish now that I worked harder at my French lessons. When I lived in Quebec, I was so hard headed that I felt the old "Red Ensign" flag was the one and only flag for Canada. I couldn't understand why Louis Tanguay, (later with the Quebec Department of Energy & Resources) when he worked at Malartic Goldfields didn't have the same feeling as I had. It was the Francophones who pushed for a more distinctive flag for Canada while the Anglophones dragged their feet. When we finally adopted a new flag, it was a bit late, the quiet revolution had begun and the "Fleur de Lis" was being embraced by the Canadiens. The "distinct society" has caused a lot of questions in Canada, yet we unconsciously think in terms of Quebec as being distinct. On the one hand, we say all ten provinces are equal and in the next breath we will say French Canada and English Canada, with French Canada meaning Quebec. When we refer to French Canada and English Canada, are we not unconsciously saying there is a partnership between the two language groups in Canada? Maybe the Quebecois have it right and we in English Canada have to catch up just like we had to do when adopting the "Maple Leaf" flag.)*

Malartic

The first summer in Malartic was quite eventful for the town; that is, it was the summer the main street (rue Principale) was paved as well as the highway to Val d'Or. Jacques Micquolin was the newly elected local deputy and also was a cabinet minister in Duplessis's government. Malartic voted right and therefore got its paving. The town was going through a bit of a boom in spite of the fact that gold was only $35 US an ounce. East Malartic had found new and higher grade of ore at depth. Barnet Mines had built a new headframe and Sladen Malartic built some new houses to the east of Malartic. It was the year the old National mine headframe was torn down. Of course Sladen Malartic was operating as was Canadian Malartic mine. Also, Malartic Goldfields was increasing production to 2000 tons daily. To the west of town, the O'Brien mine was going full force in Cadillac. Besides all this, a new and bigger Catholic Church was being built right on rue Principale to replace the one that had burnt down. Yes indeed, Malartic was on the move. For a guy coming from mining camps of Zincton and Jackpine, Malartic looked just terrific. What excitement, especially Sundays! Coming from B.C., I had never experienced an open Sunday before.

Speaking of building the new Catholic Church, the Protestants were also having a building drive to enlarge their church at the same time. One canvasser on a dare went to Father Renaud, the "curé", requesting a donation for the Protestant Church. Father Renaud asked if it was their intention to tear down the old church and build a new one on the same property, which the canvasser answered in the affirmative. "Then, said Father Renaud, I will donate $25 to tear down the old church". It was also Father Renaud who drove the largest car in town, a big black and chrome Cadillac, with all the bells and whistles. *(Would I have loved to have been able to afford that car!)*

On first arriving in Val d'Or Jack Frey informed me that he would be leaving active mining (especially working underground) because of the asthmatic condition he had developed. Having followed Jack across the country and to renew our acquaintance from high school and university, I found myself alone

Why Mining?

in that part of the world without my old buddy. Before he left, we did have some fun around Val d'Or as I would travel into Val d'Or every weekend and bunk in at Jack's apartment which he shared with a couple of other young fellows, Bill Zawadski, later of the Mines Branch in Ottawa, being one of them. A weekend would involve going down to an afternoon "jam session" at the Chateau Louis to listen to Benny Couture and Howan Giguere and their band rehearse, drink beer and tease the whores. On one such occasion, there was this hooker who appeared to have a "live one" in toe at the table next to ours so we young bucks all clapped and cheered to get her up on stage and sing "J'attendrai" for the audience. She reluctantly got up and went through the whole song *(she sang quite nicely too)* but during her performance her hot "John" had taken up with another girl and left. *(I guess he was too hot to wait or "attendre".)* Saturday nights were spent first at the Siscoe Club (at Siscoe Island, the site of the old Siscoe mine) where we would watch the first floor show at 11 o'clock and then travel into Val d'Or and catch the second show at the Morocco Club at 1 o'clock there. There were always willing girls who liked to go out clubbing so we had lots of company. It was fun. Speaking of the Morocco Club, it should be mentioned that the evening of Jack Frey's Farewell Party, given to him by the staff at Lamaque mines, we ended up stopping the floor show. Our gang, after the party had broken up at the Bourlamaque Hotel, all went downtown Val d'Or to the Morocco Club and with me in front playing my bagpipes, marched into the club and onto the floor as the one o'clock floor show was in progress. Who can compete with a bunch of rowdies led by an inebriated piper? The floor show had no choice but to stop. *(The Canadiens do love the pipes.)* That first summer in the Val d'Or - Malartic area was probably my best during my single days. *(It was also my last as a young bachelor for I got married the following year.)*

After Jack Frey left Val d'Or I was fortunate, with all the activity in Malartic, to find a number of young engineers and geologists with whom to make friends. All of us had recently graduated and were all a long way from home. All were single, so we could come and go as we pleased. What a summer we had! I mentioned Bob Goldsmith, a graduate of London School of Mining. He was one of the chaps in the gang prior to his leaving for Timmins and later to Malaysia. Ernie Isaac, MAPEO and Ontario Ministry of Labour, also a London School of Mining graduate was another. There were a couple of

Je Me Souviens

geologists who were taking their Master's at McGill who also chummed around with the group. The McGill students, who were working during the summer for the Quebec Department of Energy and Resources, were quartered at the Chateau Malartic Inn. *(It was pretty nice of the government to house "offices and lodging" for their geologists at the Chateau.)* There were a couple of others whose names I do not remember but we did have quite a gang.

Our Sundays started out at the Chateau Malartic Inn in Mr Authier's office (Mr Authier was the Inn's owner and manager) drinking beer as the bar could not open until noon after Mass was over on Sunday. *(Being a good catholic boy, I always went to Mass and therefore had none of this beer drinking in Authier's office except when Mass got out early.)* After the bar opened at noon, we would have an eye opener and then order lunch as we had to get to the ball game at two o'clock. They had a very good baseball league composed of teams from Val d'Or, Amos, Golden Manitou Mines and Malartic to name a few. The teams were made up of mostly U.S. university students and some local talent. It was pretty good caliber ball for that part of the world. After the ball game, the group would generally stop off at one of the hotels which could be the Beauchene, Manor Boutin, Chateau Malartic or whatever. After a short beer session, we generally had dinner at the Chateau Malartic because the kitchen was clean. That says a lot. Some of the eating places in Malartic in those days left a lot to be desired. In the evenings, there were the movies as everyone went in those days as televison had only come into being in the larger centers and Malartic was not one of them. On Sundays there was always an English language movie but we would have to sit through a number of French language movie previews which were scheduled to be shown during the middle of the week. The French language previews *(These were all from France.)* would show nearly the whole movie that it was promoting. Honestly, the whole preview took at least fifteen or twenty minutes to show and would invariably start out in the first sequence with a couple embracing and caressing, and then the next few scenes would show the same couple bashing each other around and finally the closing sequence would again show them embracing in each other's arms. The same story or series of scenes would repeat itself in every movie. These movies were not popular even among the Canadiens. *(Since that time, the Canadiens have taken a hold of their culture and are no longer dependent on France for their*

Why Mining?

entertainment.)

While working at Malartic Goldfields that summer, the refinery at East Malartic mine (a few miles to the west of Goldfields) was robbed of four gold bricks. The robbers got away with the bricks by stealing acetylene equipment from the East Malartic mine shops and cutting their way into the refinery. They carried the four bricks to the fence line where they had entered the property but because the bricks were so heavy, they decided to bury the three heavy ones and carry with them the smallest of the four bricks. The authorities found the three large buried bricks the next day but the smallest brick was never recovered even though the thieves were caught, charged and convicted. Incidently, the thieves were a father and son team who worked at East Malartic so one could say it was an inside job. Apparently, they had disposed of the small brick through the "gold syndicate" that operated in the north.

Regarding the small brick, everyone knew that the head high-grader was Johnnie (I forget his last name) from Val d'Or who had purchased the brick...but no one could prove it and the two thieves were not about to tell or were just too afraid of the consequences if they did. The head high-grader, Johnnie, was also the manager of the Val d'Or baseball team. The following Sunday after the incident, it so happened that the Val d'Or ball team was playing in Malartic. Whenever there was a close call on the field that went against the Val d'Or team, Johnnie the manager, would go out and try to argue with the umpire. As he attempted to argue with the umpire, the crowd in the stands would holler, "Hey Johnnie what did you do with the brick?" I forget who won or can't even remember if it was a good game: it was just fun to be there and to hear all the comments from the crowd.

A lot of funny things happened at the ball grounds. One night an angry husband, suspecting his wife was having an affair with a Mr Bourque, a local business man, sneaked up on them as they were parked one dark evening at the ball grounds which were very secluded near the east edge of town. Upon reaching their parked car, the husband saw that Bourque's trousers were thrown over the back of the front seat, so he reached in through the open front window lifted the trousers and took off with them. *(I often wondered how Bourque explained his missing pants to his wife when he got home that night.)*

We had other characters in town worthy of mention. There was quite an infamous whore that the boys nicknamed the "Black Diamond". *(Black Diamond slickers were worn by shaftmen and were considered the best.)* As far as I know, the Black Diamond was never a "madam" ...just an independent business woman, a loner. This one particular evening, just as the patrons of first showing of the Sunday evening movie were coming out of the theatre, two policemen were dragging (literally) the Black Diamond to the police station around the corner from the theatre. While crossing rue Principale in front of the theatre, the Black Diamond's bloomers fell down to her ankles. The policemen, being gentlemen, stopped and unhanded her so she could pull them up. But instead of pulling them up, she pulled them off and then swished them into the faces of the startled policemen. All this happened in front of the movie crowd to its delight.

Rupert McEwan, an old prospector, was given credit for a number of finds around the Malartic Camp and, as a result, the mine managers in the area decided to look after him in his later years. Rupert was not a person to save his money; in fact, he was known to throw away what money he had to his friends. On one occasion, he came out from the Beauchene Hotel and threw hand fulls of his money up into the air for anyone to catch and yelled, "Who wants money?" and then would throw another handful up into the air. As a way of looking after McEwan, the mine managers would take turns at giving him employment at their mines every fall and winter. What he did in the spring and summer is anyone's guess... possibly some prospecting. During my second autumn in Malartic, it was MGF's turn to hire and look after McEwan. He was assigned to the carpenter shop which was considered a very soft touch but Rupert McEwan decided he would make things even softer by sitting on the work bench smoking and refusing to work. It was not a good example for the other workmen to see this guy getting away without doing a tap of work. Finally in desperation, Herb Cox, the manager, came down and appealed to McEwan to please get off his butt and try to put in a fair day's work. Rupert still wouldn't budge. Rumour had it that McEwan was sent home and told to stay home but was probably paid never-the-less.

The destruction of Rock d'Or was before my time but one did hear much about it from those who lived in Malartic during that period. Rock d'Or was the "Sodom and Gomorrha" of the north. Apparently, it got so bad that the provincial government of Duplessis ordered everyone out and then contracted bull-dozers to level the village. Nothing remained of the place after the dozers finished. *(My first wife Mary, who was a little girl at the time, recalled one time while she was riding home on a bus past Rock d'Or to her home in Heva River (which was the west of Rock d'Or), she saw naked people chasing each other from one house to another.)*

After a pretty full summer, the first one to leave our Malartic gang was Bob Goldsmith who left around the middle of July. His departure left a bit of a hole but there were others to take his place. Besides, there were a number of nice Canadienne girls to squire around that made Bob's departure a non-entity. I mentioned previously that on Bob's departure, I took over his girl friend. Well that's not quite true, not just then. It was rumoured that Bob's girlfriend was moving with her family to Sudbury at the end of the summer. So I didn't make any effort to move into Bob's former territory but instead dated a couple of other nice Malartic girls. In September the geology students from McGill left to continue their studies. It also put an end to our claim to the Chateau Malartic as our base because of these students leaving. It meant the government rooms at the Chateau would be unavailable to us until the next summer, or perhaps never. Then Ernie Isaac said he had applied for a job in Peru and would be leaving in the Autumn. That left me on my own to find a new group of friends. It was my good fortune to find out from Ernie Isaac that Bob Goldsmith's old girl friend had not left town after all and was quite available there in Malartic. It didn't require anyone to draw me a picture, I made my move.

The young lady in question, was Mary Sutherland who worked at Blais Telephone Ltee as their chief and bilingual operator. Between squiring Mary around most of the autumn and doing some early curling, I kept pretty busy. Finally, things got serious, maybe because of being lonesome for home or just plain love-struck. But in any case, Mary and I did become engaged to be married but set no date. In the meantime just prior to Christmas, one of her two brothers with whom she shared an apartment was killed in a car accident. Her parents of course were quite shaken up about this terrible accident and prevailed upon Mary to move to Sudbury to be with them. We, therefore, had to carry on our courtship by mail, telephone and the occasional trip for each ...for her to Malartic and for me to Sudbury. Finally, the next spring we were married in Sudbury with my old buddy Jack Frey as my best man and Doug Bertoia and his wife Gaby standing up and supporting me. All the other guests were friends and relatives of the Sutherland family.

My wife and I settled down in an apartment on rue Montcalm in Malartic for the first year. It was a typical square box of a building with four apartments. The heat was supplied by a space heater (which one had to supply themselves) set up in the middle of the house and we hoped the heat would flow evenly to all corners of the apartment which it didn't. The building had no basement and little in the way of a foundation as the floors sagged. Our apartment was on the ground floor and, therefore, it was very cold. The people who lived above us were very comfortable in their apartment as they got all our heat. We stayed there only the one winter but the next spring we moved to an apartment above a dry goods store on rue Principale where it was much more comfortable but the stay there was quite short only six months as I had accepted a job in Atikokan with Steep Rock Mines and therefore moved in the fall of 1953.

Partir

After reading an advertisement in The Northern Miner for engineering staff at Steep Rock Iron Mines and being very interested in that project, I submitted my application for employment. Ken McRorrie, who was chief engineer at the time, interviewed me over the telephone. Undoubtedly, I didn't impress Ken as the job went to Al Sumner who, at the time, was working at East Malartic. *(Al Sumner later moved from Steep Rock to Geco Mines and worked there until his retirement.)* A few months later, Walt Bannister (later of Inland Cement), who was then the mine engineer at the Steep Rock Errington underground mine, phoned and offered me a job. The job was not that great as the position was only for a surveyor but the future looked bright and the fact that Steep Rock promised me a new house, was incentive enough for us to move.

There was no highway into Atikokan in those days so everything, meaning furniture, had to go in by rail. This meant crating everything or selling it all in order to make this move. We sold our chesterfield suite and bought a new

Why Mining?

one from Sears to be delivered to us on our arrival in Atikokan. The refrigerator which operated on 25 cycle power was traded in at Sears for a new 60 cycle powered fridge that again would be delivered in Atikokan. This lessened the packing and the crating that had to be done. As I would be doing the crating, I was more than happy to unload as many of these bulky things as possible. Also we were responsible for paying all the moving expenses so it was to our advantage to decrease the expense as much as possible.

A couple of days before transporting the crates of our belongings to the railway freight shed, my wife's brother, drove up from Sudbury and took her back with him so that she could have a few extra days with her family. I would join her in Sudbury later and then would travel together by train to Atikokan which was 800 miles to the west of Sudbury. It was just as well that she left early as the table and chairs were crated the day she left and the bed and mattress the following day. I slept at the staff house in Halet after that. A couple of friends with a borrowed truck helped to transport the furniture to the railway freight shed. The freight agent was real kind in giving me a much discounted weight after weighing the load because he knew that this move was charged to my account. *(In the 50's mining companies looked upon young mining engineers as though they were a dime a dozen. If I had been a doctor, or a lawyer, or even a teacher, I would have had my way paid to Atikokan.)*

**The Author & Joe Gotch
Malartic Goldfield 1951**

Chapter 4
Red Mud

Atikokan

In 1953 the only way to reach Atikokan was by rail (except maybe by canoe or pontoon plane). The "No 11 Highway" (the continuation of Yonge Street in Toronto) which was under construction from Shebandewan to Atikokan would not be completed for another year. Anyway, the CNR passenger train travelling between Thunder Bay and Winnipeg via Fort Frances and Rainey River was quite comfortable. (*In those Steep Rock years, we always referred to Thunder Bay as "The Lakehead" which even now seems to be so much more an interesting name than "Thunder Bay".*) The journey between Thunder Bay (Port Arthur) and Atikokan took all of six hours to travel the 140 miles between these points with stops along the way at Fort William, Shebandewan, Kashabowie, Sapawe and finally Atikokan. *(Some names eh?)*

There was a fair size crowd at the station on our arrival. Not to meet my wife and me, we weren't that important. But in the years before the highway opened, it seemed the thing to do in Atikokan was to go down each evening to the train station and meet the train to see who was coming to town. Being an experienced traveller by now, *(Didn't I travel all the way from the west to the east?)* I wired ahead from Port Arthur to the Atikokan Hotel (the only hotel in town) for a room with bath. Upon arrival in Atikokan I was very confident that a room would be there all ready and waiting. That was a pious thought as the hotel was booked for weeks in advance. Of course it was the shock of my life when told that there were no rooms available. However, again luck was on our side: the kind lady behind the desk knew of someone who could possibly help out. She called Ed Jackson of Steep Rock Personnel at his home and he was able to find a place that would take us in for the night. I had visions of sleeping in the station which wouldn't have been a first for me but I was a little bit concerned for my wife. The taxi that took us to the private home was driven by none other than Bob Moffat who now operates Moffat Mine Supply that has branches in a number of mining localities across Canada. Bob, who arrived broke from Manitoba, did very well in Atikokan. The last time I saw him was at a CIM Convention in the late 80's when we had breakfast together at the Royal York Hotel in Toronto.

```
Why Mining?
```

Atikokan, in 1953, was no larger than maybe 3000 people, if that. None of the streets or roads around the town were paved nor were there any sidewalks. Everything was gravel, and red gravel at that, as it contained at least 10% hematite. It takes but as little as 5% hematite to turn clean gravel, reddish brown in colour. Once this gravel gets wet, it is nothing but red mud, and in Atikokan, it rained quite often. As a consequence, the town is generally a mess even in winter with the reddish brown snow. There was no way that one could keep a pair of shoes clean; in fact, when people came calling, they always took their shoes off at the door. Try as one might, it was impossible for anyone to wipe their shoes clean without tracking red mud onto the floor when walking into a house. As a consequence, most of the people wore boots (even the women).

The town centre was made up of two blocks of stores with empty lots in each block. At the far end of main street was a combined public/high school. There were two churches, a very small old Catholic church and a recently built First Protestant church which later became the Anglican church when the United and Presbyterian parishes completed building their own places of worship. There was only one hotel at the time across from the CN station. The residential area was made up of four sections: Don Park in the west section of town (built by Steep Rock to house its employees) was quite nice; the old section around the station and main street was composed of many nondescript buildings; a bunch of squatters' buildings along the road to the mine (the Black Diamond of Malartic fame moved into this section along with her live-in, Jimmy Tester, a miner who worked underground at Steep Rock); and a newer section of town that was in the process of being built along Birch Road where many of the senior Steep Rock staff were to be housed. My wife, after her first walk around the downtown area on our first day in Atikokan, cried uncontrollably.

The next morning after breakfast at the only restaurant in town, I called my future boss, Walter Bannister. Walt was a big man with a very large head. I mention his head because it was at least a size 8.5 or better and housed a very active brain. Of all the people that I have met in the mining industry, I would say Walt was by far the most intelligent. He not only had a good personality, he was very honest as well. Is it any wonder he rose right to the top and held a very responsible position with Inland Cement Company in Edmonton some years later. I met Walt only once after leaving Steep Rock and that was in Saskatoon where he gave the local branch of the CIM a very interesting talk on the manufacturing of cement.

Steep Rock Iron Mines

Walt drove me to the main office where I was signed on and then he gave me the grand tour of the open pit and dredging project. A tour of Steep Rock in those days was most impressive. Steep Rock Iron Mines had just completed mining in the Errington pit at the south-end of the property and had started mining in the Hogarth pit at the north-end. Two very large dredges (the "Steep Rock" and the "Marmion") were operating at the southeast side of the Hogarth pit finishing the dredging there before the great move in the winter of 1954-55. Another dredge, "the Seine", which was considerably smaller, was operating on the mud flats above the Errington pit. Besides these there were a number of monitors operating along with booster pumps and barges. It was very impressive indeed. We then drove to the Errington underground where I would be working. This mine was somewhat unique as well. The headframe was a knob of rock with a sheave wheel installed on top. The shops and part of the dry were excavated into this knob. The hoist house was partially outside the knob of rock as were the offices. It was very different.

Walt Bannister explained that Steep Rock Iron Mines was an independent supplier of iron ore; that is, it wasn't tied into a steel mill and therefore had to sell its ore on the open market. Some years there were large sales while other years sales could be very lean. The shipping season, because of ice on the Great Lakes, was between early April and mid December, about eight and a half months. This meant that stripping operations would be conducted in the open pit during the winter months, and mining of ore during the summer months or shipping season. There was some stockpiling of ore during the winter months but an attempt was made to keep stockpiling to a minimum. Walt went on to explain that, like the open pit, the intention was for the underground to do as much development work in the winter, minimizing the stockpiling of ore and to mine most of the ore in the summer. This would take very careful planning and scheduling. The big unknown was the projected sales. Sometimes the total sales commitment would not be known until sometime late in the spring. Steep Rock did pretty well at projecting its sales

Why Mining?

and doing only the required amount of stripping or development that would carry them through the following shipping season.

Pop Fotheringham was the president and general manager of Steep Rock Iron Mines when I arrived. Pop was a kindly fellow with a friendly smile but somewhat aloof (maybe shy) from the people who worked at the mine. Perhaps if he had gotten more involved with the operation at the time, Steep Rock may have had less union problems. Pop had good people working for him in Administration with the likes of Neil Edmonstone as vice president of finance, Fitz Fitzgerald as assistant secretary-treasurer and Chick Baron as chief accountant. Louis Zucchiatti was another stalwart and a joker to boot in Administration. The financial projections made by these people in Administration (which included sales and expenses) carried Steep Rock Iron Mines through some very trying times.

Walt then introduced me first to Ken Dewar, the Superintendent of Underground Mining, whom I had heard about from others who had worked for him. Ken, on first impression, didn't come across as a friendly sort. He seemed to resent young engineers who worked in the engineering office regardless of the amount of practical experience they had at other operations. With all the drilling and blasting I did at the Sullivan, Lucky Jim, and Paradise mines as well as dancing on the end of a jack-hammer all one summer with General Construction, I had more than paid my dues. One had to get to know Ken Dewar to appreciate him. On one occasion when a number of engineers were leaving the office at 3:50 pm to catch the bus to go home after work, Ken stood by the front door and said, "I see there can't be that much work as you young engineers are all going home early". I was afraid that Ken was about to discharge a few of us, thinking that there were too many engineers. After paying for the move to Atikokan, I was broke and couldn't afford to lose this job. It was not a good feeling. Boy, did we ever keep busy after that. *(Talk about motivation!)* Ken could also tear a strip off a salesman that didn't deliver. I remember one salesman who had sold Steep Rock a tugger hoist that didn't deliver the necessary pull. Ken told the poor guy "...that tugger didn't have the strength to pull the pants off a sick whore". Ken could be sadistic and at other times be a very gracious host. For instance on the last day of work before Christmas, he would tell the engineers to put away their maps and drawings and clear the tables as "...we are going to have a party".

Red Mud

All the shifters, engineers and draftsmen would be called into the big engineering room where food and drink would be served very generously by Ken. The same thing would be repeated a week later before the New Years holiday. I got to know Ken better much later (in the 1960s) when he headed Kilborn Engineering.

Bannister next showed me where I stood on the organization chart. This was the first time I had ever seen an organization chart at any mine where I had previously worked. My job was to be a transit-man in a three-man survey party headed by a party leader who was Norm Kipnis. Norm Kipnis decided he wanted some practical experience underground, so he left the engineering office shortly after my arrival. He accepted a job in Mexico a short time later but ended his career in Newfoundland as director of the mines branch in that province. The other party leader was Gerry Hamilton whose path I crossed many times later in my career. Shortly after my arrival at Steep Rock, Gerry was assigned to construction where he acted as project engineer on the underground crushing station that was being built. With Gerry moving to full time on the crusher project and Norm leaving for the underground, I was assigned to be party leader over two survey crews. What this meant was for me to go with the crew that had the most difficult task for the day leaving the other crew to fend for themselves. Yes, there was more money which came in handy. Gib Hele was the chief surveyor as well as Walt Bannister's right hand man. Gib brought his survey knowledge, gained from the war as an artillery captain, to the job at Steep Rock. With this knowledge and his experience at Kerr Addison Mines (which had perhaps the best survey standards and practices of all the mines in Canada) gave Steep Rock a very well-run survey office.

Another interesting person employed at Steep Rock was Merv Upham who was the mine captain, but later, with the arrival of Wyatt Hegler, was promoted to mine superintendent. Merv was a very dynamic fellow who buzzed around the mine like a blue-assed fly. He always wore yellow slickers in the mine and with his short quick strides you knew it could be none other than Merv when he approached you in the dark underground. He was a good organizer and was very personable. He was very keen on "time and motion" study; in fact, he started a course on "time and motion" study for supervisors and engineers of which I was one of those chosen. I have never regretted

Why Mining?

taking that course and can thank Merv Upham for that opportunity. Merv, although a good organizer, was not technically strong. For instance, there was the case of a pump down in the sump at the bottom of the B-1 Errington shaft where the total head exceeded that of the pump. Merv rushed up to Ken Dewar and said that the pump appears to be working but no water was coming out of the pipe at the station above the shaft sump. Ken, who liked to have a little fun, said, "...now Merv, how have you got the discharge connected? Has the water from the discharge got a chance for a straight run before you have it directed up the shaft?" When Merv said, no, he had an elbow at the pump and the water was directed straight up the shaft. Ken then said, "...look Merv, you got to give the water a little run first so as to give the water a bit of momentum, then it will travel to the top." Merv rushed off saying, "That's it, that's it". After he left, Ken Dewar chuckled away at his little joke which was heard by a few of us next door to Ken's office.

On another occasion, I was time-studying a miner operating an Eimco mucking machine which had very worn flanges on its wheels; in fact, the flanges were almost non-existent. Nearly every time the bucket would eject its load, the front wheels of the machine would slip off the track and the load would be directed in line with the front and back wheels (because of the centralizer) and not necessarily into the railcar behind. Merv and Wyatt Hegler (Wyatt was then the mine captain) came along and observed this fiasco. Later on surface Merv said to me that he thought that was the worst demonstration of mucking machine operating he had ever seen. I tried explaining to him that it wasn't the operator's fault, but the machine's, as it had literally no flanges on its wheels. Merv wouldn't have any of that and said "...but didn't you notice that he threw the load off to the side". Hegler said, "Well that's because of the centralizer". Obviously Merv didn't know the machine had a centralizer. He was a real good guy though. Merv died of alzhiemers in 1999 which was very sad.

Regarding the house that was promised it must be said that, in spite of all the difficulties met when a town like Atikokan mushrooms from nothing to a town, eventually, of 7000 people, Steep Rock did its best to house everyone satisfactorily. The company house that was promised me was not available when arriving in town as the present occupant had met some delays in building his own private home and therefore other arrangements had to be made. Steep Rock suggested, as a temporary measure, a move into a much larger home still occupied by a former Steep Rock employee (Clarence Nelson of Nelson Insurance) who had also met some delays in building his home. What this amounted to was sharing a home with the Nelsons. This was done for a month until the Nelson home was completed and we then had the house to ourselves for a while. As mentioned the housing situation was pretty tight and because of this, I was asked to share my house with another employee Bill Yanisiew who worked in the engineering office. *(I met up with Bill Yanisiew again at Texasgulf later in my career.)* This we did for the next five months until the Yanisiews could find suitable accommodation for themselves. However in the meantime, the first house that was promised (which was smaller and more in keeping) became available and we moved into it the Saturday after New Years 1954. Wyatt Hegler moved into the Nelson house that we had just vacated. So from October to the following spring, we had moved from Malartic, camped in with the Nelsons for a month, moved to the first house promised us after two months, and shared the Nelson house and the first house promised with the Yanisieu's for five months. By April of 1954, we finally were settled in a house all to ourselves. *(Living with other people around is not my cup of tea.)*

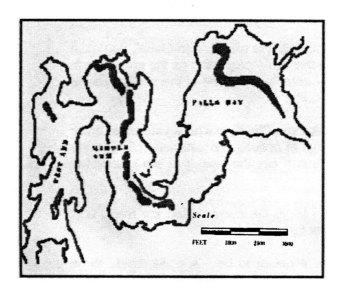

Steep Rock Lake in the Shape of a Letter "M"

Why Mining?

Errington Mine

The Steep Rock Iron orebody was confirmed by diamond drilling through the ice on Steep Rock Lake after iron ore float was found in the vicinity. It was at the persistence of Julian Cross (one of the first directors of Steep Rock Mines) who is given credit in finding the orebody in the Steep Rock Lake basin that it became a mine. Steep Rock Lake was in the form of the letter "M". The iron orebodies were under the lake and the Seine River flowed through the lake. The first mining attempted in 1938 was to sink a shaft on the north shore of Steep Rock Lake and try to mine from underground. The high water pressure encountered while cross-cutting to the ore zone quickly negated this effort. This meant that the Seine river had to be diverted and the lake drained before any mining could be attempted. This was a big engineering task and people like "Pop" Fotheringham, Ken McRorrie, Sid Hancock, Watkin Samuel and others whose names escape me showed much courage when they attempted a project of this magnitude. Much credit should go to those people who financed, or arranged financing, such as Joe Errington, Major-General Don Hogarth and Cyrus Eaton in getting the project off the ground and seeing it through to completion. The timing of the Steep Rock project was right for the time because, had it been ten years later, it would not have stood a chance of being built as the Taconite plants in the US as well as the Quebec and Labrador iron ore finds would have nosed Steep Rock Iron Ore Project out of the picture. Also because of World War II, it was thought that the Messabi range would run out of iron ore and the Allies required other sources of iron ore for the war effort. *(Timing is everything especially when you are promoting a property.)*

After the Seine River was diverted, it was necessary to pump out Steep Rock Lake and then remove the mud and silt that was two hundred feet deep in places. This was no easy task and that's where the three dredges came into the picture. If one counted the two dredges operating on the Inland Lease, one would count five dredges in total. *(Steep Rock leased one-half of the Steep Rock Range to Inland Steel of Chicago.)* The first removal of silt was by ground-sluicing; that is, by monitoring the silt into a pool and pumping the water and silt into a catchment area. The capital cost of this method is cheaper than dredging but the conditions must be suitable. Referring back to the letter "M" (Page 77), the Inland Lease was more or less the right half (east half) of the "M", while Steep Rock kept the left half (west half). It was the middle leg of the letter "M" that Steep Rock would mine. The Errington orebody, situated at the south end of the middle arm, had less silt than the other known ore zones and therefore it was the first one to be developed as an open pit mine. As mentioned the silt above the Errington orebody was sluiced with the silt and water being pumped to a catchment area, except that there wasn't a catchment area. The silt and water were pumped directly into the Seine River system, a practice you wouldn't even dream of doing today. Later, however, because of lawsuits against Steep Rock from the pulp and paper company located down stream from the Steep Rock Project, another diversion of the Seine River was necessary so that the west arm of Steep Rock Lake could be made into a catchment area for the silt. Using the Letter "M" analogy, the west arm of Steep Rock Lake would be the left leg of the letter "M" and the Seine River was diverted further west of this.

The first ore came from the Errington Pit which, was as mentioned, uncovered by monitoring and ground sluicing. The orebody was situated near the east shore against a very high bank which was the footwall, while the hangingwall was more towards the center of that part of the lake. Because of the relatively steep east bank of Steep Rock Lake, the original infrastructure was built on the hangingwall side of the orebody as were the first two mine shafts. The Errington pit operated from about 1944 to 1953 where some nine million tons of direct shipping iron ore were mined. The equipment used in this initial operation was that which was left over from the diversion project consisting of churn drills for the blast holes, $1^1/_2$ and $2^1/_2$ cubic yard diesel and electric shovels for loading and 15-ton Euclid trucks for hauling. A comical incident happened during the initial mining where the whole afternoon shift one Saturday evening departed for the Atikokan Hotel beer parlour to watch a Stanley Cup Game. When Sid Hancock, who was the general foreman at the time, (Sid later became the mayor of Atikokan) arrived at the hotel, he fired all the occupants there on the spot. Then, he was naive enough to ask them to drive the trucks to the CNR roundhouse that was used temporarily as the mine repair shop. Of course afternoon shift crew didn't oblige and it was up to the staff and the crew of the next shift to drive the trucks back.

About the same time, the construction general foreman, Cam MacDougall, was directing his crew to dismantle an old shack that was used in the early days of development. The idea was to save the lumber for another project.

Why Mining?

Instead of labouriously removing the nails one-by-one in the conventional manner, Cam calculated that, with a few pounds of stick powder hung inside the shack, and exploded, the nails would be loosened so that the task would be made easier. Cam looked in his handbook for the required powder factor per cubic-foot of building and then with his slide-rule calculated the volume of the shack and the amount of powder required. After calculating the weight of powder required, he informed his foremen of the exact weight to be used and then went back to his office. As a last minute check, he recalculated the weight of powder required and discovered he had placed the decimal point too far to the right, and, as a consequence, he had told his foreman to use ten times too much powder. Cam ran out of his office and jumped into his pickup truck and raced to the site of the shack that was to be dismantled but, alas, too late. Just as he was rounding the bend to where the shack had been, he could see the shack blown through the air. Not one piece of its lumber was salvageable.

At this time the Errington B-1 underground mine was under development where the mining method contemplated would be "block caving". *("B" was the original name of the orebody, "1" was the first shaft for that orebody, while Errington was the name of the mine. Later on, for tax reasons, there was a B-2 mine but it was accessed from the Errington B-1 shaft.)* The Errington mine was developed from the hangingwall by long crosscuts on the 700, 900 and 1100 levels from a three-compartment shaft. The ore was hoisted via a series of conveyors to surface from an underground crusher station located between the 900 and 1100 levels. The rock and mineral sequence in the Steep Rock Area from foot to hangingwall was granite, limestone, hematite-geothite, ash rock and greenstone. Within the ore zone itself, there were sequences of "paint rock" which was a very incompetent member, a mixture of very fine siliceous material and iron-oxide. Besides this, there were zones of high manganese and phosphorus, all which were contaminants. The idea was that the ore would cave (which it did) and grade control would be accomplished by pulling the various drawpoints in sequence and thereby exercising control. *(Things just didn't work that way.)*

From the main crosscuts at the orebody, hangingwall drifts were driven outside the ashrock contact in the greenstone. From the hangingwall drift, production crosscuts spaced at 250-foot centers were driven through the

Red Mud

orebody into the footwall carbonate and were connected by a footwall drift which led to the exhaust raise that was driven to surface. The ventilation flow was for the fresh-air to enter the mine via the three-compartment shaft at the hangingwall, split the flow amongst the three main crosscuts which would channel the fresh air out to the hangingwall drifts and then into the ore zone through the production crosscuts. There were a couple of main fans located on the 700 and 900 level crosscuts pushing air out into the ore zone. As I recall the ventilation was fairly good even though there were no fans located at the footwall to exhaust the air to surface.

Draining the Orebody 1100 Level Errington B-1 Mine 1953

The ore zones at Steep Rock were discovered under Steep Rock Lake which had to be drained before any mining could be started. The Errington underground which was below the original Errington pit was plagued with water both from the residual water contained within the mineralization and from the water which drained into the pit above. As a consequence, all three level crosscuts (700, 900 & 1100) had to be driven carefully by diamond drilling drain holes in advance so as not to be flooded out. The 1100 level was really the drainage level which tapped the water within the orebody to this level allowing the other two levels above to be developed. In order to advance the 1100 level, it was necessary on occasion to utilize forepoling (spiling) and numerous drain holes to advance the 1100 level drifts and crosscuts. As a result of the amount of water to be handled, it was necessary to install a first class pumping station just below the 1100 level. This pump station consisted of a number of dirty water sumps and four large clean water sumps which supplied a positive head to four Mather-Platt 1000-igpm pumps which pumped all the way to surface at a head of 1600 feet. The flow into the mine averaging 2500 igpm was controlled by valves on the drain holes; however, on occasion all four pumps would operate delivering 4000 igpm. The pumping system was the pride of the Errington mine but even so, disaster struck one evening when there was a power outage.

Why Mining?

With the power outage, the emergency procedure routine was set into place: the flood control doors in the mine were closed and the pumproom water-tight doors were also closed. One would have thought everything was "A - OK". Unfortunately, on the day of the power outage, crews had been cleaning one of the clean water sumps and had forgotten to close the clean-out valve connecting the sump to the pumproom. As a result, all clean water sumps filled but the recently cleaned sump started to drain water into the pumproom through the clean-out pipe whose valve was left open. Because the pumproom water-tight doors were closed, this acted as a dam causing the pumproom to flood which paralysed the whole pumping system thus flooding the mine. The mine was down for a couple of months while it was being pumped out with a Byron-Jackson (BJ) submersible pump. *(This is not the only time that I was in a flooded mine.)* As a result of this set-back and because of the weak iron ore market, it was decided that the Errington mine would not go into production in the 1954 Shipping Season. *(The shipping season in the Upper Great Lakes was April to mid-December.)*

The mine depended on steel arches for ground support. When first arriving at the Errington, the mine was using 6" wide-flange steel members for cap and posts spaced as close as 18" center-to-center in some cases. Later on the mine employed the German Toussaint-Heintzmann (TH) yielding set which worked very well as a passive support. At Inco they used the same TH steel section in a rigid set which, in my opinion would have worked well at Steep Rock in the active mining headings where block caving was being performed. *(It is said, hindsight is 20-20.)*

Initially at the Errington underground mine (which was developed below the exhausted Errington open pit mine), the method was to drive scram drifts at 35-foot centers and 125 feet long in both directions from the main production crosscuts meeting the scram drift driven from the adjacent production crosscut 250 feet over. In other words, a production crosscut may have as many as four scram drifts in both directions in some of the wider sections of the orebody. The block was designed to be 100 feet long by the width of the orebody and 150 to 200 feet high. The 6" wide flange rigid sets used to support the back were installed at 2-foot centers in the scrams. Drawpoints were located every 25 feet alternating from side to side along the scram drift. After driving the scrams, their walls up to the cap elevation were concreted.

Red Mud

The floors were installed with heavy 90-lb rails so that the large 60-inch scrapers pulled by 125-hp Ingersol-Rand slusher hoists would not dig into the floor. The back of the scrams were covered with 6-inch diameter lagging. Prior to drilling the undercut, the drawpoints were belled out. The undercut was drilled off from inside the bells in such a way that the back was completely undercut when the undercut holes were blasted.

It should be mentioned here that a number of mine tours to the Cleveland Cliffs mines (Mather A and B) in Ishpeming Michigan were made by underground technical and supervisory staff. Reciprocal visits were also made by the Cleveland Cliffs staff. On one such visit made by the Cleveland Cliff group prior to the Errington mine going into production, they were asked what they thought of the chances of success. Their answer was very blunt. They said, "the installation with the concrete is very elaborate but the mining of the orebody would not work". This didn't go over well but all the same, time proved the Americans right.

The initial attempt to block cave at the Erringtion was a disaster. It was thought that, in order to assist the cave, a superimposed load of rock (pit stripping from the adjacent Hogarth open pit mine) should be dumped into the old Errington pit. It was thought that this would not only help in the cave but it had the economic advantage as well as it would not be necessary to truck the waste rock out of the Steep Rock Lake Basin. This superimposed load not only caused pressure problems but it caused dilution problems as well. The walls of the scram drifts were so badly squeezed-in that the scrapers could not be pulled through from one end to the other. The lagging in the back was broken in many places and all the caps for the rigid sets were badly deformed. But worst of all were the heavy 36-in wide flange beams spanning the production crosscuts on both sides of the scram drifts which were deformed also. One had to see this to believe what force was exerted on those heavy beams. Because of deformation in the scrams, it was impossible to pull the drawpoints in an orderly fashion and, therefore, they were pulled with no sequence whatsoever, causing rat-holing where the waste from the stripping of the Hogarth open pit would end up in the drawpoints. In one incident, a Euclid truck radiator ended up in a drawpoint showing that the Errington pit was not only used as a waste dump but as a garbage dump as well.

Why Mining?

In order to salvage something from the initial mess, the mine purchased the German TH yielding sets and installed these where heading recovery was possible. The TH sets worked much better than the 6-inch wide flange sets but still there was much to be resolved but there was hope. Heavier TH sections of the yielding sets were tried which gave encouraging results. The TH set was designed to yield which gave it increased strength upon yielding but this yielding caused problems when trying to pull a scrapper from one end of the scram to the other. This latter problem was partially solved by installing chain conveyors instead of scrappers and slusher hoists. A chain conveyor is an expensive solution with its high initial cost and its high wear and tear with chains breaking and pans wearing out. Anyway, it temporarily solved the problem of the scram squeezing in on the scraper.

After the disaster on the 700 level block, it was decided to try again on the 900 level block. This time instead of driving the scrams longitudinally, it was decided to drive them across the strike in a transverse fashion. This worked out very well as one could control the pull, retreating from the footwall to the hangingwall. The heavier TH sets were installed with TH rigid dividers and chain conveyors were installed instead of slusher hoists and scrapers. No concrete was used which decreased the development time and cost.

The ground control problems were now much more under control but a problem of grade showed its ugly head. The Steep Rock orebody was not homogeneous. The ore zone was a mixture of geothite, hemitite, paint rock, high sulphur inclusions as well as high phosphorus minerals. In block caving, what reports to the drawpoints is a mixture of what the orebody contains. There is no selective mining. An attempt was made at grade control where certain drawpoints would be closed off when they looked to be running low on grade. The ore grade to be shipped to the iron plants was 52.50% iron and anything less than this, but greater than 35% iron, had to go to the concentrator. Anything less than 35% iron went to the dump. All material with a hint of sulphur went to a special stockpile on surface. Trying to segregate, ore grade, low iron, sulphur and waste was very difficult, using a conveyor to lift these materials out of the mine, especially when there was no surge capacity or storage areas where these materials could be segregated within the stoping area. It meant that the conveyor system was always stopping and starting to clear one type of material so that it could handle another. There was a lot of down time which added to the cost.

Just prior to my leaving Steep Rock, two other mining methods were being tried. One was the conventional "sublevel caving" and the other was "longwall top-slice". The sublevel caving method was quite successful in that it did allow some degree of selective mining but still, at that particular time, under the conditions that existed within the Errington underground mine, it was impossible to mine economically. In this method, scram drifts supported by yielding arches were driven transversely across the orebody from the hangingwall to footwall at 35-foot centers from one another. The same sequence of scrams were repeated on sublevels 35-foot elevation below but somewhat offset from those above. Mining started at the footwall and retreated to the hangingwall by drilling and blasting a fan of holes (after a slot raise was blasted) and mucking this broken material with an Atlas Copco T2G mucker. After each blast, the front yielding set was pulled out and retrieved and mucking would begin again until the waste prevailed over the ore. Had a small scooptram been available at the time, it is believed this method would have had much more success but, because of the limitation in productivity of the T2G, it could not deliver the production that was required to be profitable.

The longwall top-slice was an attempt to develop a new mining method. It consisted of installing a timber and wire mesh matte at the top of the orebody and lowering this under control similar to that done in longwall mining in the coal mines. The difficult part was the installation of the matte. This was done by driving a drift heading (longitudinally) along the hangingwall the full length of the block to be mined. At the far end of this drift, a crosscut was excaved from hangingwall to footwall. Lagging poles and wire mesh were placed in the excavated crosscut and then the support was removed, allowing the back to cave. Then, right beside the caved crosscut, a second crosscut was driven and more poles and wire mesh were placed. The support in this crosscut was removed also and the back was allowed to cave. The sequence was repeated until the whole stoping area to be mined was covered by this matte (or artificial roof). Mining commenced at the footwall, immediately under the matte, by driving a drift heading on strike supported by yielding friction props for the full length of the stoping area. After this heading was completed, a second drift heading was driven adjacent to the first one. The idea was that

Why Mining?

the friction props, when taking weight of the matte would yield uniformly allowing the matte (or artificial roof) to sag. Also, it was the plan to retrieve the yielding props after the longwall face had advanced two adjacent headings. Alas, none of this worked as the side pressure bent the props as they could sustain vertical pressure only. The idea was good but, again, the cost to install the matte in the first instance along with the maintenance of the props and repairing of the matte made the method uneconomical.

An attempt was made to try and maximize the mining capacity again by block caving, but this time everything that came from the drawpoints would be taken with the idea that it could be sorted out in the mill. This was right up to a point but the mill could not handle the fine material or slimes that formed. As a consequence, at times the tailings would run 35% iron or even higher. Many studies were made and many consultants looked at the problems of ground support (rock mechanics), in mining methods (block caving, sublevel caving and longwall top-slicing) and in metallurgy (cycloning, jigging and heavy media separation) to no avail. To attempt to mine iron ore which yielded $11 per long ton at the lower lakes by underground methods (where the ore itself was not homogeneous) was a lesson in futility. Underground mining at Steep Rock ended a few years later in 1964. Underground mining was again attempted in 1971 but terminated a year or two later for the same economic reason.

Business Trips

After the mine flooded in 1954 I, along with an number of other engineers and supervisors, were given training in "time and motion study" in order to keep busy as the Errington mine was placed on hold until the market conditions for iron ore improved. Harry McIntosh (an engineer from the McIntyre mine in Timmins) along with Merv Upham (who was now mine superintendent) gave this course. *(I don't ever regret taking the time and motion study course as it helped me considerably in organizing people and work.)* Anyway, we worked all June, July and August in the Hogarth open pit time-studying churn drills, trucks and shovels. In the autumn and until the next spring, we were all back in the Errington mine time-studying the mining development. Time and motion study got a bit boring after awhile but I was again lucky to get posted the following spring to the General Engineering Department as a project engineer planning a new underground mine.

Red Mud

One might say, "Why plan a new underground mine when the first mine had not a proven mining method?" The fact was, management didn't know at that time that the method chosen "block caving" would not work. This was the year 1955 and the iron ore market was still weak and the Errington was still "on hold". It wasn't until the summer of 1956 that the first attempt at actual underground mine production was pursued. The following year (1957) the Errington mine did have some success as it produced nearly half a million tons but never the 1.5 million as originally planned. Anyway, it was my job to research new methods of material handling as well as mining methods for the new Hogarth A-2 underground mine. The Hogarth underground was to be directly below the Hogarth pit which mined the "A" orebody. The numeral "2" designated the second shaft for this orebody. *(Originally in 1938, it was thought that the Steep Rock orebody could be mined by underground means without draining the lake. The A-1 shaft was sunk in the hangingwall and a crosscut was driven out to the ore zone. Water problems forced the abandonment of this scheme before reaching the ore.)*

The studies for the new mine centered on where the mine plant would be located. As there was very little space for a plant on the west or hangingwall side and because of the long crosscuts necessary to reach the orebody from the west, it was suggested that the Hogarth underground mine should be accessed from the east or footwall side which meant extending the railway and other infrastructure around the north end of the property to the east side. Studies were made on conveyor hoisting versus conventional shaft hoisting. Shaft hoisting again won out. Friction hoisting was new and this was studied as well. The first friction hoist in North America was at the Cliffs Shaft in Ishpeming Michigan. Steep Rock sent me, along with my boss Harry Mulligan and another engineer John Helliwell, to Michigan to see this installation as well as to visit other mines such as the Cleveland Cliffs and Mather A & B in Michigan; the Petersen, Canon and Sherwood mines in Wisconsin.

The Cliffs shaft, as mentioned, had the first friction hoist installation in North America. It was an ASEA hoist with a gear box mounted on springs. It was different, it was impressive and it commanded a lot of attention as the Cliffs mine received many visitors during the first few years of this hoist installation. The Mather A & B mines had impressive drum hoisting plants complete with bend and turn sheaves. These underground mines employed block cave mining

Why Mining?

which seemed to work for them. They had the advantage of fairly good ground outside the orebody, something Steep Rock did not have. The Petersen mine had an impressive hoisting plant also, but what was of interest there was its large pumping station underground. The pumping station employed large reciprocating piston pumps that generated a high head but relatively low volume compared to their size. But what really was impressive on the Wisconsin leg of this trip was the stockpiling system found at the Cannon mine. This installation utilized shuttle conveyors which required a lighter conveyor gallery structure than the tripper type stockpiling systems utilized at the Mather B and Petersen mines. The Sherwood mine was a little jewel of a mine. At this mine everything was handled by conveyors. Sublevel stoping was employed where the ore was dropped down to drawpoints equipped with feeders which loaded conveyors that transported the ore to the shaft. Very few men were required which meant the productivity for the mine, including the staff, was in the 40 tons per man range.*(That's long tons.)*

On this trip, our group stayed one night in Hurley, Wisconsin which had a bit of a reputation for being a Sodom and Gomorrah from the early days of mining in that area. Being drinkers and not lovers, we headed for a friendly restaurant and bar as Harry Mulligan liked his drink and John and I enjoyed our food and sometimes a beer as well. Perhaps we stayed too long as John and I had pretty good hangovers the next morning. Fortunately, I could hide mine but John Helliwell was very sick. Of course, Harry was used to the morning-after anxieties and showed no ill effects whatsoever. As a consequence, Harry suggested that John Helliwell not come into the Petersen or Montreal mines that morning as John would only make a bad impression. John, years later, said he was very resentful of the fact that an alcoholic told him that he was a poor ambassador because of his drinking.

Another trip was made with Harry Mulligan a year later to Quebec City to attend the Canadian Institute of Mining and Metallurgy Annual General Meeting (CIM AGM) in 1956. The reason for this trip to Quebec City was to obtain first-hand information on friction hoists as there would be several papers delivered on hoisting at the convention. Also, Harry would be delivering a paper on conveying. My purpose for being there was to offer moral support for Harry. It was terrible weather that day in late April when leaving Atikokan for the Lakehead on our way to the convention. The drive

Red Mud

from Atikokan to the Lakehead was treacherous as it was snowing a wet snow which made the road very slippery. *(I can never get used to calling the Lakehead, "Thunder Bay". I think the good people of Port Arthur and Fort William missed out when they named the amalgamated cities, Thunder Bay.)* After arriving safely in Thunder Bay, we caught the plane to Toronto and then flew on to Montreal. At Toronto, there was a delay as the plane had to be de-iced. *(It didn't give me a good feeling.)* Anyway, we got to Montreal safely but were obliged to stay overnight there at the old Laurentian Hotel before catching the early afternoon train to Quebec City.

Harry stayed up all night drinking. I tried to stay with him but had to give up at about 3 AM. The next morning was Sunday and being a good catholic boy I got up fairly early and went to mass at the Basilica across the street. *(Harry was still up but pretty inebriated.)* Upon returning to the hotel room, I found Harry was fast asleep. Anyway, Harry had good recuperative powers and was able to devour some breakfast and then catch the train to Quebec City.

Harry Mulligan and I behaved ourselves well in Quebec City as Harry had to deliver a paper Tuesday afternoon at the convention. By attending a number of papers on hoisting and shaft sinking, I was able to obtain much good information. One delegate had a working model of a Cryderman shaft mucker in his room at the Chateau Frontenac where the convention was held in those years. When Harry and I went to this particular room, the delegate who owned the exhibit of the shaft mucker had a call girl in bed with him so we had to excuse ourselves and come back later.

The delivery of Harry's paper went well and it was well received. It was after this that these Steep Rock delegates kicked up their heels. The first stop after the Chateau was a night club for dinner and to see a quartet of four bearded men. It was quite rare to see men with beards in those days. The meal was a breast of chicken washed down with numerous martinis. For some reason, Harry left but I stayed on to catch the second show. By that time, it was time to go and, feeling no pain, I caught a taxi for Lower Town and ordered a plate of spaghetti for some idiotic reason. Anyway when it was time to leave, I asked for my coat. To my horror, there was none, nor did I have one on arrival. Anyway, the lady at the counter was very kind and asked where I had been before coming to her establishment. The name of the restaurant where

Why Mining?

Harry and I had been failed me. It was known that it featured a quartet of four bearded men. The lady at the counter looked in the entertainment section of the local newspaper and found the place featuring the four bearded men. She immediately called a cab and gave instructions to the driver to drive to the restaurant and retrieve my coat. Sure enough, on arriving at the restaurant, they had my coat and Harry Mulligan's name tag as well. After retrieving these, I hired the taxi for the remainder of the night. The driver drove out to the "Coq d'Or", a drinking establishment outside of Quebec City that kept open all night. The driver spoke no English and I spoke very poor French but somehow a little liquor can loosen one's tongue where one doesn't mind how badly the words come out and we got along fine. Every once in a while I would say, "Ou est mon manteau" and the driver would hold my coat up and say "Voici est le". We got along fine and he delivered me back to the Chateau safe and sound with all my belongings. Harry was good enough to advance me some money to cover the extra expense on my expense account. Harry, who was the boss, would be approving my expense account anyway.

The following year, Steep Rock sent Bud Chesney, Monty Johnson and me on a trip to Northern Michigan for the purpose of looking at mucking machines and stockpiling methods. Bud Chesney was a dour sort of guy who kept everyone at arms length. On the other hand, Monty Johnson was completely open and at times came on a bit too strong. Bud was the new mine superintendent at the Hogarth A-2 mine and was interested in viewing mucking equipment, while Monty was a designer *(and a very good one)* who would be drawing the general arrangements for the new stockpiling system at the Hogarth A-2 mine and was interested in seeing what was used in the industry. I went along with the group as the project engineer on the planning of the Hogarth A-2 mine and it was in my interest to go on this trip even though a similar trip was made the year before.

The trip to the mines in Northern Michigan started out fine as we were flown down to Marquette Michigan in the company's Twin Beach aircraft where a car was rented in order to get around to the various mines. It was in January and on that particular day the flying conditions to Northern Michigan were good and the skies were clear. It was on the way back that the skies clouded up and it was impossible to land anywhere in Northern Michigan and, therefore, we were left on our own efforts to get back to Atikokan. This

Red Mud

meant dropping the car off and taking the bus. The bus took a very long circuitous route to Duluth, Minnesota, which took us first, south, to Ecanaba, then west to Iron Mountain, northwest to Ironwood, Michigan and Hurley, Wisconsin and finally to Duluth Minnesota. It was planned we would stay in Duluth that night but I got the idea, "...why not catch the train to Fort Francis Ontario and then the early morning train to Atikokan from there." This didn't go down well with Monty Johnson as he wanted to stay in Duluth and go out on the town. He should have had enough "tom catting" as it was, for he was out every night while in Ishpeming where we stayed. Anyway he got over his disappointment and came along with us on the train home.

Mine Projects

After much investigation on the subject of footwall versus hangingwall shaft location, size and depth of shaft, friction versus drum hoists, as well as the number of hoists, it was decided that the A-2 shaft would be located on the footwall whereas the B-1 and A-1 shafts were located on the hangingwall side of the orebody. For a start, it was decided that the shaft depth would be 1400 feet below collar with levels at 750, 1000 and 1250-foot horizons utilizing three friction hoists. I was rather proud of the shaft layout for it enabled the installation of all three hoists on one floor without the use of bend sheaves. It had one cage and two skip compartments along with their counterweight compartments, a ventilation compartment to be used during shaft sinking and during initial lateral development, as well as a manway, pipe and electrical cable compartment for a total of eight compartments. All conveyances travelled on wooden guides supported by 6-inch wide-flange steel shaft sets spaced at eight-foot centers. *(This was my first shaft design and to this day I am still very proud of it.)*

The other proud conception was the mine pumping system. On the 1250-foot horizon which was the lowest level, two Byron-Jackson submersible pumps were installed which stage pumped to four Mather-Platt plurovane pumps located on the 750-foot horizon level. The Mather-Platt pumps were installed so that they were under positive head. Cleaning sumps were located out at the orebody while the clean water sumps were located next to the pumproom near the shaft. It was a very neat setup with the Byron-Jackson pumps as these pumps removed the problem of flooding the mine as was the case at the Errington B-1 mine.

Why Mining?

Another project that was of interest was the building of the South concentrator. This project was decided in September 1958 and nine months later it was operating. By September 1959, it had treated 250,000 long tons of hematite ore. When the test work was done on the ore, it appeared that the ore could be treated utilizing gravity methods such as sink-float for the minus 3-inch plus 1-inch material, jigs for the minus 1-inch plus one-eighth material and cyclones for the minus one-eighth material. Unfortunately the material it was designed to treat was not that which was delivered as the Steep Rock ore contained some very fine material which ended up in the slimes. Some months the tailings contained as much as 35% iron. In other words, the profit went out the tailings pipe. The winter of 1959-60 was one where considerable revision work had to be done in order to make the plant work within reasonable standards. My job was to redesign the pumping and piping system as well as some of the internal conveyors. The winter was spent drawing spool sheets for the piping and designing my first conveyor, but I must admit that I had a very good detail man in the name of Stuart Johnson right behind me, making the design look good. *(Stuart worked with me later at Esterhazy Saskatchewan and again in Saskatoon. Stuart died very young (33) of cancer. I felt his loss very deeply as I considered him one of my best friends. He was the most accommodating man I have ever known. I don't think he had an enemy in the world.)*

Mining Reseach

After the South concentrator project was completed, I was promoted to mine engineer for the Errington B-1 and the Hogarth A-2 mines. This involved directing the mine engineering at the Errington and going up to the Hogarth one morning a week to observe what was needed there. The interesting part of the job was at the Errington. At this operation block cave mining was employed with yielding arches trying at the time to produce a clean ore from a non-homogeneous orebody. Steep Rock contracted all kinds of block caving and rock mechanic experts to come to the Errington mine and observe the mining problem but no one was able to solve it. While in university in the late 1940's, I was enthralled by the work of Professor Buckey of Columbia University in New York, where he wrote many papers on barodynamics and block caving. Much to my disappointment, I found Professor Buckey rather naive and impractical in his suggestions when meeting him when he visited the Errington mine. Steep Rock had the Mines Branch of

Ottawa doing rock mechanics work under the direction of Don Coates from McGill University. They did very good research but little of this work was of any use in trying to mine iron ore. The problem was really not a mining problem but a metallurgical one. Steep Rock were trying, as mentioned previously, to mine selectively by block caving. This is impossible. The ore reporting to the boxholes was a mixture of everything within the orebody. There was no way in which one could pull "X" tons of ore and then "Y" tons of waste without contaminating the former. Also, the stresses within the orebody due to the cave would build up with the delay of switching from one material to the other. These stresses would deform the steel arch sets where it was necessary to stop mining and rehabilitate the scrams. If the mine could have just mined the material as quickly as possible and then sorted it all out within a concentrator on surface, then perhaps the Errington B-1 mine would have had a chance. With the "paint rock" (highly siliceous iron oxidized material) and the slimes generated by fine-grained hematite, the recovery (within the South concentrator which was built to treat the Errington ore) was very poor where the tailings were at times never below 30% iron.

A very fine gentleman by the name of Ken McRorrie, who was general superintendent of mining started a mining research group to try to solve the underground mining problem at Steep Rock. Parallel to the mining research were on-going studies conducted on metallurgical recovery. The research group not only conducted studies in rock mechanics but in mining methods. They looked into blasthole mining as one method but this method had the same problem in that it was not a selective mining method. As the metallurgical problem had not been solved they were attempting to find a mining method where they could mine a clean ore and yet still solve the ground support problem. They came up with a combination of long wall mining and top-slicing method where a mat of timber and wire mesh was installed in the back which would be supported by yielding props. Although this method did solve the selective mining problem, it did not solve the ground support problem.

The next mining experiment to be tried was much more successful and perhaps would have been totally successful if the present day equipment would have been available then. The mine tried sub-level caving using arches for support and auto loaders (Atlas Copco T2Gs) for loading and hauling.

Why Mining?

The unfortunate part was that the T2Gs were too small and slow; whereas perhaps a 2-yard or a 5-yard LHD scooptram would have been more successful as this type of machine would have combined speed as well as versatility. In any case, the Errington B-1 mine did operate on-and-off from 1954 to 1973 producing a total of 3,706,844 tons.

Activities

Underground mining was not the only problem at Steep Rock. It did have labour problems in the form of the United Steel Workers Union of America. The local's president was a fellow by the name of Alex O'Neill who was a very abrupt individual. He probably was a very good president but he sure was a thorn in the company's side. I often look back now and wonder if the union could have been more cooperative and if the company had extended its hand in friendship, between them they could have perhaps solved the mining problem together. *(Maybe I'm being naive.)* On one such occasion, Alex took the whole mine (open pit and underground) out on a wild-cat strike all because, Jack Cole, the mine captain accused a miner of attempting to steal a piece of rope. Alex O'Neill called this episode the "Incident of the Golden Tassel." It went all the way to arbitration, a costly procedure on both sides. In another incident, (this one happened in the pit) a foreman (Svene Knutsen) on a Saturday when the pit was not operating, jumped on a bulldozer to repair a drainage ditch that was overflowing. His fast action saved the pit from flooding. Had he not done this, the open pit would have been shut down for a few days (putting the production crew out of work) while the pit was being pumped out. The union grieved over this incident saying the foreman performed work normally done by the bargaining unit employee.

Fortunately, being out of the line of fire and keeping my nose clean, I didn't have any trouble with the miners thanks to an incident which happened at the Lion's Club Annual Carnival. On this particular occasion, as a member of the Atikokan Lions Club, I was put in charge of the rock drilling contest. Being able to get a compressor donated by the mine and a number of jackleg drills, steel and hoses from Joy Manufacturing Company and being fortunate to recruit Dave Friesen to help run this event, the contest went off without a hitch... well nearly. Everything appeared to go well until a few of the miners came along and started to make a few snide remarks about Dave and myself like "...what do staff guys know about drilling?" Dave, who had worked underground at Steep Rock as a development miner, was quite confident that he could handle his own with any jackleg drill. I was not so sure as I hadn't handled a drill since the Zincton days and that was ten years back. However, I was familiar with the Joy drill as I had played around with them while Dave and I were setting up the booth for the Carnival. The miners from the Errington mine, where Ingersol Rand and Gardner-Denver drills were used, were not familiar with the Joy type of jackleg as it had an automatic retractable leg. Anyhow, after being challenged for some time, Dave and I, to save some of our dignity, had to accept the challenge and drill as much as we could in five minutes. Dave, when he drilled, posted the second best time and was beaten by a big German immigrant miner by the name of Ab Shultz. I was lucky and was able to collar my hole without incident and then the drill hit a slip in my favour or it could have been a vug or a soft section of rock. To make a long story short, I came in third. From that time on at the mine, there was never a challenge again.

Speaking of the Lions Club, the Atikokan Lions Club was very active in the community. Each year as one of the many money-making enterprises (the carnival was another), it would put on a minstrel show where members would blacken their faces and perform the show on two successive nights. Generally after the second performance there would be a great after-show party. It was on the day after one of these parties, when I was so hung-over from beer drinking and heavy smoking that I couldn't look at a cigarette. It was a perfect time to quit smoking which I did. *(At the age of 30, it was about time to grow up.)*

What better experience could a young engineer have than those eight years at Steep Rock Iron Mines. A person could get experience with nearly every conceivable piece of equipment and method used in mining. One could see five large dredges in operation in the various dredge pools and see, in the winter of 1954, the engineering feat of moving two of these very large dredges (Steep Rock weighing 850 and Marmion weighing 900 tons) intact rather than dismantling them, over-land, a distance of two miles and a vertical lift of 250 feet. Besides dredging, one saw high-powered monitors used to clean up the sides of the dredge pools.

In the three open pit mines, Steep Rock had the old-fashioned churn drills, the

Why Mining?

newer rotary drills, draglines and electric shovels as well as 15 and 35-ton Euclid trucks and Athay wagons. Steep Rock had an inclined skipway capable of hoisting 35 long tonnes (equivalent to one Euclid truck load) from the pit-bottom to the pit-lip above. Two underground mines were also projects of interest where various mining methods (block caving, sublevel caving and longwall top-slice) were employed.

A young engineer would see various types of pumps from submersible and plurovane underground pumps to the large 36-inch diameter dredge pumps. Hoists of various types (incline and vertical drum hoists as well friction hoists) were used at Steep Rock. Besides these, there were conveyors for hoisting ore and waste from the pit as well as from underground. Ore screening plants and the two concentrators were other installations that commanded much attention. Later on, Steep Rock constructed a 1,300,000 ton-a-year pellet plant to add to its establishment.

Steep Rock University was a great place where many engineers received first-class hands-on training from top-notch people.

Errington B-1 Office & Dry   **Hogarth A-2 Headframe,
                                   Shops & Office**

Steep Rock Iron Mines 1980

Chapter 5
Rockbolts & Yielding Arches

Rockiron

After eight years at Steep Rock, it was time to move on. It was plain to see that there would not be any advancement by staying and besides underground mining would never be successful under the conditions found at Steep Rock. This was a sad conclusion but the writing was on the wall. Therefore, in the summer of 1961 while vacationing in Sudbury, I visited the Rockiron plant there and was offered a job as chief engineer (*and the only engineer*) by Tom Kierans who was part owner.

Tom Kierans was a big redheaded Irishman who had formerly worked for Inco as a general foreman and later as a safety supervisor. He started Rockiron from a friend's garage making rockbolts. In those days, a rockbolt consisted of a one-inch diameter shaft of steel with a thread on one end and a slotted tongue on the other. The anchoring devise was the slotted tongue with a steel wedge rammed into the slot so as to wedge or expand the two tongues to the wall of the drill hole. It was not until sometime later that the rockbolt plug was invented. It was really a modification of a rawl plug used by civil engineers.

Anyway, Tom Kierans made the offer of $10,000 per year, a full range of fringe benefits, a chance at owning part of the business and car mileage etc. working as an engineer/salesman for Rockiron. What Tom did not tell in full when joining the company was that Rockiron was in receivership. As far as a chance to own part of the business, that was pie-in-the-sky for the creditors owned the company during that period in time. Having said all that, there were never any regrets for my decision in accepting his job offer. What it did was give me a chance to visit many mining operations all across Canada, meet many people and learn how to talk to various strangers. The fact that the company was in receivership was an added incentive to increase sales so that it could show a profit each month or otherwise the receivers would foreclose.

Regarding sales for a period during the first six months, Rockiron did not have enough sales to show a profit, so Tom Kierans would delay the month-end closing until there was a profit. During those few months, the month-end was delayed until the 2nd day of the following month, then that month's

Why Mining?

closing would be delayed until perhaps the 6th day of the next month. Finally after six months, Tom Kierans and company found itself in such a quandary that the closing date was on the 20th day of the following month. Fortunately for Rockiron, Steep Rock Iron Mines came through with a very large order of steel yielding arches where Tom was able to close the books at the end of the month as well as spread out some of this sale over the next three months instead of absorbing it all in one month. *(What some small businesses have to do in order to show their creditors that the enterprise is sound and viable!)*

Another trick in showing a profit was in inventory. Rockiron's year-end was December 31st. Therefore, the auditors would be out checking the books in January. Now, Sudbury receives a lot of snow in December and January so that fact gave Kierans the opportunity to inflate the inventory by saying everything was in the yard knowing full well that the auditors would not go out and shovel snow to check the inventory of steel or finished products. *(Tom Kierans was very cagey.)*

Travels

During the first month on the job, I travelled with Tom Kierans all the way out to the west coast and back, stopping at the various mining camps along the way. Tom drove a large 1960 Ford station-wagon with a big Thunderbird engine. His cars were always green in colour. *(He was Irish.)* Anyway, the station-wagon was loaded with all kinds of products made by Rockiron as this would be a full 30-day tour and Kierans wanted samples of all products with him. The first stop was at Wawa where we stopped at the Helen mine of Algoma Steel and later visited one of the open pits to see Hugh Smith who was the pit superintendent. Not much time was spent at either place except to let the personnel know that Rockiron was still around and open for business. That afternoon we drove onto Manitouwadge, staying over-night there.

The next day, a visit to Geco Mines of Noranda was made. At Geco there were discussions with Jack Graham, manager; Lorne Brooks, mine superintendent; Merle Marshall, chief engineer and Ed Fothergill, ventilation engineer. I was very impressed with Jack Graham, as he portrayed to me what a mine manager should be. Anyhow, both Graham and Brooks were interested in Rockiron's grouted rockbolt for the brows of drawpoints at Geco. A few

Rockbolts & Yielding Arches

months after this visit, Geco gave Rockiron a contract to install a number of grouted rockbolts in the brows of the drawpoints. As I remember, Geco was the first mine in Canada (or maybe the world) to have installed a grouted rockbolt that could be pre-stressed. The method used in this particular installation was to install and stress the bolt and then inject epoxy resin by using a 'mixing and pressure unit machine' through a valve in the bolt's washer and an exhaust tube running the full length of the bolt. It was a very delicate machine to operate as one had to be careful that the resin didn't setup within the injection valve; otherwise, one had to take the machine's injection head apart for cleaning. It was also a very cumbersome machine to move around, requiring electricity where, in some parts of the mine, only compressed air was available as a power source.

My first encounter with Merle Marshall was at Geco Mines. It wasn't until a few years later at Texasgulf-Kidd Creek in Timmins that I worked for Merle. I got to like the man later but my first impression of him at Geco was negative. Bill Fothergill, the ventilation engineer, was a gentleman who one couldn't help but like and, as a result, we were able to establish a good working relationship with him.

After conducting business at Geco we proceeded to Atikokan to visit Steep Rock. It should be mentioned at this time, these business visits generally took a full day, especially if it was possible to gather an audience; that is, if the mine operators were willing. Kierans would set up his projector and give a lecture on TH yielding arches which followed with a discussion. Then he would give a talk on his latest project, the pre-stressed and grouted rockbolt. *(Kierans was very innovative and entertaining. His audience got much interesting information from these sales talks.)*

At Steep Rock, the same procedure was followed but first a tour of the underground was made to view the action of pressure on the TH arches. After the underground tour, Tom would again set up his projector and give a lecture on TH arches even though Steep Rock had been using them for a number of years. It made for good entertainment. Also, it was Tom's theory that the more time you can spend with a customer the more orders you're going to get. *(We'll get into more of that later.)*

Why Mining?

Red Lake was an interesting camp in the early '60s. Madsen Red Lake Mines was operating under the direction of E G Crayston as manager and Al Heather as mine superintendent. Unfortunately Rockiron didn't get to first-base at this mine as they didn't acknowledge a ground support problem. Their purchasing agent said if Rockiron could beat "X" dollars for a six-foot rockbolt that he now purchases from his present supplier, he would change suppliers. Kierans declined to bid because as he said later... "when the next rockbolt salesman comes around, that same purchasing agent would give out the Rockiron price in order to try for an even lower bid". *(Probably all sales people have experienced something similar.)*

A number of calls were made in the Red Lake Area to such mines as: Campbell Red Lake Mines, Dickenson Mines, Cochenour Willans and even crossing to McKenzie Island to McKenzie Red Lake Mines. At the latter mine, we didn't know how we would be received because Mr McKenzie who ran the mine was a bit eccentric. Many a salesman was literally chased off his island and told not to come calling again. Fortunately this didn't happen to us.

The trips into Red Lake *(we made a similar trip a year later)* were not too rewarding as I do not recall ever obtaining any business from that mining camp.

The next mine visited on this trip was the K-1 Mine of International Minerals and Chemical Corporation at Esterhazy. This mine was in the late stages of shaft sinking in the summer of 1961 and would not be starting production until a year later. Merv Upham was the manager and Alex Scott was shaft superintendent in charge of shaft sinking. It was a very friendly visit with Merv Upham, who as everyone knew, was a very outgoing individual.

From Esterhazy, we travelled to Saskatoon to visit the Potash Company of America mine at Patience Lake. This mine had been in production but was shut down because of water problems in the shaft. This particular shaft was sunk without the use of tubbing. Instead, the mine operators froze the ground and poured monolith concrete rings to withstand the water and ground pressure. They got into production but soon after the freezing melted, water leakage developed in the shaft. They tried to grout the leakage to no avail. They finally had to cease production and re-freeze the shaft and attempt to

Rockbolts & Yielding Arches

grout the areas within the shaft where the leakage occurred. It was at this stage of the operation when this visit was made.

While in Saskatoon, we stayed at the Bessborough Hotel which is a fine old lady of hostelry and ate at its fine diningroom called at that time the Sunset Room. The old hotel lost something when they changed the diningroom around to Aerial's Cove, a sea coast motive. *(What that has to do with the prairies, one could never understand.)* Anyway, the next day we stopped for the night at Kindesley near the border of Saskatchewan and Alberta and registered at Chan's Motel there. After looking around the room that we were directed to, I said to Tom Kierans, "It's dirty, let's leave". Tom said, "You're too fussy and being difficult". Anyway, I pulled out and carried my things down the street to the old hotel in Kindersley and stayed there for the night. The place was old but it was clean. The next morning, Tom picked me up at the hotel and we proceeded onward to Vancouver with a stopover in Invermere B.C. where my folks were living at the time.

At Invermere, Tom wanted to proceed by plane as he had business to attend to in Vancouver. After driving him to Cranbrook to catch the plane and then driving back to Invermere, I spent another day with my folks. The following day I drove alone to Vancouver with a diversion to the Bralorne mine on the way. On reaching Lillooet and phoning the Bralorne mine to tell the people there that they could expect a visitor within the next hour, the guy at the other end of the phone gave a little chuckle and hung up. One could see why, as it took about three hours to drive the short 50 miles to Bralorne. The road was narrow, windy and full of ruts. A tire was blown as well. On reaching the mine, the mine staff were not surprised that it took me so long to arrive. They knew the road's condition and had a good laugh at this Easterner who thought 50 miles of B.C. driving was like Ontario. *(I was hoping to see Jim Thomson who was the manager at Bralorne but he was away at the time on business. Jim and I were in the same class at U.B.C. in mining.)* At the mine, the usual slide show on yielding arches was given which was of no real use to those at the mine but they were interested all the same. *(Probably when you're as isolated as they were, any diversion from the day-to-day routine is a boost.)*

At Vancouver, I was able to catch up with Kierans. From there, it was a drive

Why Mining?

to Britannia mines to go underground and see some of the workings in the No 8 orebody. There was an application for arches; however, I can't remember for sure whether or not an order was received from them. Probably eventually there was. There was also an underground tour at the Giant Nickel mine near Hope B.C. where they were using arches to support their scram drifts. Pictures were taken of the arches supporting the back in the scrams at the Giant Nickel mine. It was possible that without these arches, they could not have mined in some of their scrams.

On the way back from Vancouver, a visit was made to the Craigmont mine which was under development at the time. They had driven three adits in about 1000 feet or so each and had encountered bad ground in a couple of locations where they had installed heavy (10 x 10) timbers. This looked like a very good application for arches and so Tom Kierans decided to wait-out the Craigmont operators and get an order. Bob Hallbauer, *(later president of Cominco)* was the newly appointed mine superintendent at Craigmont. I could see no point in hanging around but Tom insisted that we stay and wait for the order. This was embarrassing, as I knew Bob from my Zincton days when he was a student at U.B.C.. Bob's father, Ted, was the master mechanic at Zincton and Bob was my helper for a month prior to my leaving for Quebec in 1951. Finally, when the Craigmont staff were calling it a day, Tom gave up and we drove on to Penticton B.C. and stayed overnight there.

The next day we drove to Nelson but made stops in Greenwood and Salmo B.C. to the mines in those vicinities. Bob Webber, an old classmate of mine from U.B.C., was the mine manager at the Emerald mine at Salmo. *(Bob was killed some years later in a plane accident, flying home to Prince George where he managed the Endako mine for Placer Development as it was called in those days.)* As the Emerald mine was in the last stages of mining, there was little that could be shown to Webber that would be profitable to him and, therefore, the stay was cut short.

The next afternoon was spent at the Sullivan mine in Kimberley. After meeting all the mine engineering staff headed by Kenny Davies and Norm Anderson *(who later was president of Cominco),* Tom showed them our slides on mining with arches. We were invited back the following morning to meet some of the others whom Kenny thought would be interested as well.

This was done the following day, which paid off with an order for steel arches.

After visiting the Sullivan mine the following morning and driving on to Fernie B.C., we had lunch with the manager of Crows Nest Collieries who arranged an underground visit at the Michel Coal mine. Later that day we drove to Michel and went underground in the afternoon at the coal mine. This was my first experience underground in a coal mine. The mine was quite unique as the coal seam was a good 40 feet thick *(that's right 40 feet)* and dipped at 35°. The operators drove (10' x 10') rooms at 25-foot centers on the footwall along strike and shot the coal in between. The coal was loaded onto a conveyor by a duckbill loader. The coal lying between the two rooms (because of the angle of repose) was scraped down to the lower room so that the duckbill could load it. It was thought that there was a good application for arches in this mine and so did the operators as well; in fact, the following spring, the colliery manager came out to Sudbury and was toured around to the various mines in Northwestern Quebec (Quimont and East Malartic) so that he could see for himself how these arches are used and what force they could withstand.

Arriving back on surface at the Michel colliery, I remember quite well being in the shower at 8 PM that evening and then driving on to the Town of Frank on the border of B.C. and Alberta where Tom Kierans and I had dinner. Frank is the location where Turtle Mountain came down and covered the whole town leaving few survivors. After the dinner that evening, we decided to drive on to Lethbridge and stay the night there. On arriving at Lethbridge, there was something going on in the town as there were no rooms available, making it necessary to drive on to the next town which was Medicine Hat. When we arrived at Medicine Hat, we were not tired, so drove on to Regina where the car was gassed up. After a coffee at Regina, we continued on to Winnipeg and had breakfast there. Along the way, Tom and I would switch places where we would drive until tired and then it would be turn about. Kenora was the freshen-up place where we went to a barber shop for a shave and then continued on to Port Arthur which is now part of Thunder Bay. Tom had relatives there and wanted to see them so we overnighted there. I was quite happy to oblige as there was a football game on TV that night which was of interest. In that last day-and-a-half, we had visited the Sullivan mine office in

Why Mining?

Kimberley, lunched with the manager of Crows Nest Collieries, went underground at Michel and drove from Kimberley B.C. to Port Arthur Ontario a distance of 1500 miles all in 39 hours. The next day was an easy drive home to Sudbury.

On another time, Tom and I travelled to Val d'Or together making various calls on the way there. At Val d'Or Quebec, Tom flew on to Chibougamau while I doubled back to Timmins Ontario and made a few calls at the mines at that location. While in Timmins, Tom called and wanted me to meet him in Val d'Or that afternoon as he had cut short his stay in Chibougamau. On racing over to Val d'Or and while driving by the cemetery west of Val d'Or, the Quebec Provincial Police (QPP), as they were called in those days, stopped me for driving too fast (56 miles per hour in the 30 mile per hour zone). The officer asked for my driver's licence, which I had, but when he asked for the car's registration, it couldn't be produced as Tom Kierans who was coming into Val d'Or later that day had it, as it was his car. That wasn't good enough for the cop; therefore, I had to follow the officer to the police station where I was booked. Fortunately, the magistrate was available (this was Friday afternoon) and therefore was able to have my case heard that afternoon. The first thought was that I would have to sit in jail over the weekend but under the circumstances it was lucky for me even if it was a kangaroo court. The officer read out the charges, all in French, speeding in a 30 mile an hour zone and not in possession of the vehicles registration. "Je suis culpable", I said, in my poor French accent. The fine was one dollar for every mile over the speed limit which amounted to $26. Also, for not producing the registration, the cost was $30. An official stamp and seal cost a further $5 and the cost of the court was $4, making the total cost of this adventure $65. That was a lot of money in 1962. What made it even worse, they allowed me to drive the car away without the registration. There was no checking up to see if the car was stolen. They were quite contented to get some money from an Ontario resident, therefore, the registration didn't matter. Tom Kierans was howling mad and threatened to write his cousin Eric Kierans who was at that time the Minister of Justice in the Quebec Government under Premier Lesage. He never did write but he probably let his cousin know of the incident the next time they talked.

Arriving late at night in Timmins without reservations at the Empire Hotel and

Rockbolts & Yielding Arches

finding that there was some sort of convention going on that day and no rooms were available, I tried the Matagami Inn *(it was new and the best place in town then)* and again to no avail. At the Matagami, they were good enough to say that there was a hotel by the Matagami River called the Riverview which may have a room. This was tried and successful *(if that's what you call it)* at the Riverview Hotel. At the Riverview, one registered for a room at the bar of the beer parlour. The bartender passed me the Register which was kept under the bar. It was a bit sopping wet from being splashed from rinsing beer glasses. Anyway, I registered and was told to pay in advance the full sum of $10, I think. On viewing the room, which was above the beer parlour, it was something to behold. The furnishings comprised a bed with poor springs and two-drawer dresser. The light fixture was a plain bulb hanging from the ceiling. There was no bathroom. The bathroom was down the hallway and was shared with all the guests in the hotel. Dreading the thought of spending even much time let alone the night there, I headed back to the Matagami Inn where they had good entertainment in their club and closed down the Matagami bar. *(The Matagami Inn was in Mountjoy Township and therefore could stay open until 12 midnight because it was in a later time-zone than Timmins. Nearly everyone in Timmins would travel out to Mountjoy Township after the Timmins bars closed at 11 o'clock at night to enjoy an hour's more drinking time.)* After closing the bar at the Matagami Inn, instead of going to the Riverview Hotel, I went to a cafe and got something to eat thereby delaying as long as possible going to bed in that horrible room. Finally, biting the bullet, I had to go to bed. It wasn't a long sleep as I arose very early in the morning. After looking high and low for an electrical outlet to plug-in an electric razor and again to no avail, I had to go down the hallway and plug the razor into a wall-plug used by the housekeepers for their vacuum cleaner.

Not all of the sales trips to Timmins or Val d'Or were as stressful as the Riverview Hotel or kangaroo court experience. There were a number of good trips to both places. When visiting the various mines, one should always (out of courtesy) first meet with the purchasing agent. It doesn't help to ignore the purchasing agent. On my first trip to the McIntyre mine, Mr Shady (purchasing agent) gave me the lowdown on the new manager. He proceeded to criticize the new mine manager, saying what a "prick" he was. *(His words.)* He was talking about none other than Pete McCroden who later retired as

Why Mining?

director for the Ontario Ministry of Northern Development and Mines. *(Apparently Shady talked about Pete McCroden to all the salesmen as others told me they had to listen to the same story.)* After the visit with purchasing, I visited the staff at the McIntyre mine and met Shorty Malcolm who was very interested in yielding arches. He introduced me to Pete McCroden who, as mentioned, was recently appointed manager of the mine. Surprise, surprise, Pete gave me an order for 50 arches on the condition that Rockiron show them how they were to be installed. This was the agreement because an order was an order and most salesmen will agree to anything to make a sale. When the arches were delivered, it was my task to instruct the crews on how they were to be installed. As I remember now, I worked along with their mine captain by the name of "Bim" who was an ex-Haileybury grad.

Shortly after this, when going out to the Pamour mine, I met Bill Marshall (manager) and York Williams (mine superintendent). They were interested in the grouted rockbolt that Rockiron had developed. This was just before Christmas in 1961 and it was getting pretty cold in Timmins. Anyway, they were told Rockiron wanted to experiment with a new method of inserting the grout into the hole for this new rockbolt and they could expect a good price. They agreed to work along with Rockiron on this experiment as long as our crews installed the bolts where they could do some good. It was agreed that Rockiron would install 100 bolts in the brows of the newly excavated drawpoints in one of the stoping sections. Rockiron agreed to supply resin kits and rebar bolts at a reduced price as well as supervision and labour to install the bolts. Pamour would supply the equipment and tools as well as drill the holes for the bolts. They also agreed to supply two miners who would learn how the bolts were to be installed.

When the time came to install the bolts at Pamour, it was mid January and it was a good 35º below. It was the kind of cold that one gets in the Timmins Area where there is an eerie haze all around. Tom Kierans took three of us (which included two servicemen from our Sudbury shop) up to Timmins, and it's good thing he did. The first day of installing bolts, using the transfer method, was not too successful. Tom insisted on installing a headed bolt, instead of a threaded one, because it was cheaper to manufacture. The heads of the rebar bolts that were brought were too large for the rockbolt dollies used at the Pamour so, as a consequence, these bolts had to be pushed into

Rockbolts & Yielding Arches

the hole by the chuck of the stoper and tightened by hand. When pushing the bolts into the hole, often times the stoper would slip off the head of the bolt and would go flying to the ground. By time the bolt had been pushed to the bottom of the hole (more often by pounding with a sledge hammer) the resin had begun to set thus preventing the stressing of the bolt. Pamour's captain, Fats Ferra, was not impressed. In order to complete a successful installation, what was needed were bolts with nuts that would fit the rockbolt dollies. The two servicemen that had been brought from the shop worked late the first night in Pamour's machine shop cutting off the nuts and plate washers from each rebar bolt, removing the rebar-deformation from the end, and threading that same end for a standard nut and washer. They did enough bolts to keep underground crew supplied for the next day. The second day went better than the first. At least now the rockbolt dolly could be used with a stoper to push the bolt into the hole and torque it. By the third day, the crews got a little system going, mixing the resin, transferring the resin, inserting the bolt and stressing the bolt by applying the proper torque. The headed-bolt would have been successful if there had been a dolly that would fit the head of the large nut that was originally used. In any case, the test was successful as we learned a lot about what should be used to make a successful grouted rockbolt installation.

It was always difficult trying to see Bill Hargrave, mine superintendent of Kerr Addison Mines. Every time when I called at Kerr, he was always too busy. Eventually by calling him long distance from Sudbury and telling him that I would be passing through Virginiatown the next day, he (after some humming and hawing) said he would see me. The next afternoon after driving up to Kerr Addison from Sudbury and making the courtesy visit to the purchasing department, I then asked if Bill Hargrave was available as I always had asked in the past. This time, the purchasing department received the affirmative answer from Bill and I was then directed over to his office which was next to the headframe. Bill sat at his desk on a chair that seemed to be higher than the one given to me. He was a very imposing figure with the meanest looking large dog I had ever seen sitting right beside him. One didn't make a move for fear of that dog. It was one of the most intimidating calls I had ever made. Bill turned out to be a pretty good guy and was quite friendly to me latter when I was with Texasgulf-Kidd Creek.

Why Mining?

Philosophy

On these long trips, Tom had me as a captive audience where he would expound on his various ideas ranging from navigation and water power projects to government. Some of these ideas were not so far fetched either. Tom's GRAND Canal (Great Replenishment And Navigational Diversion) scheme is a major project that may some day be forced upon us here in Canada by the Americans. What Tom proposed in this scheme was to construct a dam across the width of James Bay, turning this body of water into a fresh-water lake. James Bay is a very shallow body of water; in fact, it is mostly a mud flat at low tide. He also proposed to dam La Grande Rivière (just as Hydro Quebec had done later in the '70s) and generate electricity to pump the fresh-water from the James Bay up over the height of land into the Harricana River Basin. There would be a series of canal locks on the Harricana to facilitate shipping. A canal would be excavated east of Rouyn-Noranda joining the Harricana and Ottawa River systems. Also, the old ship canal between the Ottawa and French Rivers systems proposed in the 1800's would be resurrected as a barge canal. This is the old fur trader or voyageur route. Again the extra water channelling down through the Harricana and Ottawa River systems would be diverted at Mattawa to Lake Nipissing by pump stations and then down the French River to Georgian Bay and Lake Huron. This would replenish the Great Lakes and allow more water to be drawn through the Chicago Ship Canal into the Mississippi River system. All this extra water would be sold to the Americans who would pay for the whole project.

The benefits for Canada would be: 1) a barge canal from Montreal to James Bay and thereby opening the north, 2) a shorter distance by canal from Montreal to Lake Superior, 3) electrical power from La Grande Rivière system, and 4) the draining of the low-lying areas in Northwestern Quebec would thus increase the forestation of this area. The GRAND canal would connect the three great water basins (Hudson Bay, Gulf of St Lawrence and the Gulf of Mexico) of North America so that navigation could be achieved between one another. At the time that Tom expounded on this scheme in the '60s, the environmentalists were not a factor as they are today. Can you imagine the hue and cry that the environmentalists would put up at the very mention of damming James Bay and backing up the flow of water into Hudson Bay? *(As said in the beginning, this scheme may well be forced onto us by the Americans as they do need water and they somehow get their way when dealing with their neighbours.)*

Another one of Tom Kieran's schemes was on direct democracy. Tom believed that the people knew best and they should have the opportunity to voice their opinion periodically by the simple matter of voting through a voting machine *(Not the type used in Florida during the 2000 Presidential Election)* at the local post office or government building. What Tom suggested (and this was before the age of personal computers) was that everyone should be issued a personal dog-tag with his social insurance number (S.I.N.) on it which could be read by a voting machine. The members of parliament would debate the issues and the people would vote on the bill every two to four weeks by going to a local government office and keying into the voting machine with their personal dog-tag and voting. The votes would be counted electronically and forwarded on to Ottawa where the results would be published. I mentioned to Tom that didn't he think the roll of the members of parliament seemed to be down-graded, relegating them into just a debating group if this scheme ever went through. He said, "No, they are just a bunch of talkers now so what's changed?" *(Backbenchers in our present system of government are really just a bunch of highly paid trained seals who vote the party line under compulsion and never, if ever, represent the wishes of their constituents.)*

His scheme on negative tax was another idea. Tom suggested that the gross national product be calculated each year and the amount it increased would be the amount that the government should print to cover expenses. There would be no tax for anyone. Those persons whose cost of living is below the national average would be given money from that which the government had newly printed. *(I won't comment on this scheme as I'm not an economist.)*

Tom Kierans and Rockiron did come up with a number of very innovative items for the mining industry. Rockiron was the first company to commercially produce rockbolts. At the time, the rockbolt had a split rod and wedge at the anchor end and a threaded nut at the collar end. This later changed to a modified rawel plug for the anchor. Rockiron was the first to introduce high tensile steel for rockbolts. They also were the first to introduce the TH yielding arch into North American mining. The invention of the quick

Why Mining?

assembly mill hole segments, used in cut and fill stoping, was another first. The prestressed and grouted rockbolt was a first. Later, instead of having to use a machine to inject the resin, Rockiron came up with the transfer method using a plastic rod and donut piston and then, much later, a plastic rod and plunger which sucked the resin into the tube before injecting it into the rockbolt hole. Tom Kierans was indeed a very imaginative person.

One time in the shop in Sudbury, I was doing some cost accounting to determine the profit on the various orders that were current. The company had a standing order for supplying all the rockbolt needs for the Rio Algom mines in Elliot Lake Ontario. This amounted to about 125,000 to 150,000 bolts annually. This order kept two or three men constantly busy cutting lengths of steel and threading the ends. These machines, the shearer chopping steel and threaders humming away threading steel made the shop sound like a real busy enterprise. My calculations showed that we just broke even on the direct costs if the overhead cost was ignored. On informing Tom of these findings, he said he knew about it, "...but, it's good for morale in the shop when the shearers are chopping and the threaders are humming". He was right, that activity was the pace setter in the shop. It gave the impression to others in the shop that the plant was really busy.

Tom Kierans was not altogether welcome at all the mines visited and for the record nor was I. Tom used to work for Inco as a general foreman at the Creighton mine and later as the safety engineer for the Sudbury Operations. The people at the Creighton mine were first-class, especially Bert Masse the underground superintendent. On one occasion, a call was received from the people at Creighton mine to come out and demonstrate to them the new grouted bolt using the transfer method of injecting the grout. I went out to Lively and gave the demonstration and it went fine. The operators at Creighton were really impressed. I was walking on cloud nine on my return back to the shop in Sudbury. A few days later, John McCreedy, who was then superintendent of mines, called requesting me to come out to Copper Cliff to the mine engineering office and demonstrate the grouted bolt. I was really on cloud nine now being called by the "big man". Normally no one got to see McCreedy unless they had something very important. Upon arriving at McCreedy's office, he took me out to the large engineering office where he had many of his men (ex-hockey players such as Jim Dewey and Gar Greene

Rockbolts & Yielding Arches

to name two) assembled there. The grouted bolt demonstration went very well. Everyone seemed impressed. Then McCreedy called me into his private office where I thought he would be issuing an order for these bolts. Instead, he reamed me out in no uncertain terms. He told me never to go out to Creighton mine again or any other mine unless given his permission. If there was anything new to be shown, it must be shown first at Copper Cliff and then he, McCreedy, would determine if it was worthy enough to show to the operators at the mines. To say the least, I was not then, *(and even though the man is now dead)* and never will be, an admirer of McCreedy. It was the opinion of many sales persons who called at Inco at that time, that McCreedy was considered a bully who surrounded himself with "yes men" of the likes of Dewey and Greene who were afraid of him.

The Toussaint-Heintzmann Yeilding Mine Support Arch

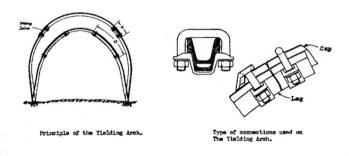

Principle of the Yielding Arch.   Type of connections used on The Yielding Arch.

TH Arches Installed at the Bakyrchik Mine in Kazakhstan

The standard of workmanship at the Bakyrchik mine was comparable to that found in Canada.

Why Mining?

Chapter 6

Grow (IMC's maxim)

IMC One October day in 1962, when taking the afternoon off from the job at Rockiron to watch the World Series game on TV, a long distance telephone call was received from Merv Upham. *(In those days, the World Series games were always played in the afternoon. The TV networks didn't dictate the time of game then, nor the number and length of the time-outs during the game. The revenue from TV and other sources has greatly changed professional sport.)* Anyway the call was from Merv Upham who was the manager at International Minerals and Chemical Corporation (IMC) Esterhazy. He said the mine had just started production the previous month and he wondered if I would be interested in the position of chief mine engineer at the Yarbo Potash Mine and if so, would I come out that way and see the job firsthand. It was already planned that the following week I would travel by automobile through Esterhazy to the Western Meeting of the CIM in Vancouver. When Merv was informed of that, he said, "Fine, I'll see you then".

It so happened that the plan was for me to drive alone in my own car, west, making stops at Steep Rock Iron Mines in Atikokan and International Minerals and Chemical Corporation in Esterhazy on the way to the convention. Tom Kierans would be met in Vancouver and we would tour B.C. while out there and drive home to Sudbury together.

As mentioned, the trip started out from Sudbury by driving to Atikokan with stops in Wawa and Manitouwadge on the way. It took the better part of two days with a night-over in Schrieber Ontario. After a day and a night in Atikokan, I drove on to Winnipeg and stayed over-night there before driving on to Esterhazy, arriving early afternoon of the following day when I called on Merv Upham. Merv Upham introduced his assistant, Bob Lindberg who, it appeared, did the hiring for the operating personnel. For some reason, I was rather leery of Lindberg. He seemed to be a bit patronising and to top it all off, he low-balled the salary offered. His big offer was $750 per month. It was mentioned that Rockiron, a small outfit, was paying $850 per month so why should one come and work at IMC (which was in the middle of the Prairies) for $100 less. Anyway, he suggested a quick tour of the underground regardless, saying he would get back to me later.

The mine superintendent was introduced next and if the job as chief mine engineer was accepted, the mine superintendent would be the immediate superior. The mine super was Andy Kyle, also a graduate from U.B.C. I hadn't met Andy before even though he was one class behind me. He must have been impressed as, finding out later, he lobbied Lindberg to agree to my salary terms. *(Andy was a very honest and decent man. This will be elaborated on later.)* The tour guide was none other than Gerry Hamilton formerly of Steep Rock who, incidently, had recommended me for this job. Gerry was the underground superintendent (general foreman). While changing into underground clothes, I met Jim Sadler who had also worked at Steep Rock. Another ex-Steep Rock man was Greg McDonald who was one of the shift captains at IMC. This was just like old home week.

Hamilton toured me through the mine on foot as the mine was not that large, having only begun production the previous month. At the time the mine was being developed by two or three 426 Goodman Borers. All these borers were developing the main entry, a seven-entry system, which ran towards the north and then swung 45° to the northeast. At the time, two other borers were being assembled in a very temporary shop just off the shaft. Anyway, the mining looked very impressive, no blasting, no track-haulage, no water, just clean white dust. It was an eye-opener to see all the mechanization in the mine. If it were not for the low-ball salary offered by Lindberg, the job would have been accepted right then and there.

When arriving on surface after the tour underground and returning to Lindberg's office, I expected that he would make a final offer. Lindberg apologized saying that Merv Upham was called away and he didn't have a chance to discuss a salary figure with him. *(One could see right there that Merv was the boss and called all the shots.)* Anyway he did ask me to give him a phone call when in Vancouver next week and by that time he might have an offer that would be acceptable.

In the meantime, the drive on to to Regina was with an unexpected passenger, Ernie Isaac, whom I had known from my single days in Malartic Quebec. Ernie was interviewing with IMC as well. At the time, he was out of work and was interviewing for any type of job that was available as he had recently left Haiti in the Caribbean where he managed a small mine there. Our drive to

Why Mining?

Regina was very pleasant and jovial. We talked about old times in Malartic and what both had done since. For the full 140 miles *(40 miles on gravel)* there was no letup in the conversation... reminiscing the whole way. After checking into the Hotel Saskatchewan *(the only decent hotel in Regina at the time)* we had a very late but enjoyable meal in the dining room. Our goodbyes were said after dinner as Ernie had to catch an early flight to Toronto and I had to get up early and drive to Cranbrook B.C. the next day. This had been a very busy day.

The next day consisted of a long drive to Cranbrook where my folks had recently taken up residence. On the way, on the straight stretch between Regina and Moose Jaw I did something that was out of character... I opened the car right up *(The car was a new Ford Galaxie 500)* and it reached nearly 120 miles per hour. There wasn't any worry about the RCMP as it was really very early in the morning and there was no one on the road nor in sight. After an overnight stay at my parents place in Cranbook, the journey was continued with my mother and grandmother to Vancouver. In Vancouver, Tom Kierans and I met at the Western Meeting of the CIM which was held for the first time at the Bayshore Inn.

While at the convention, Tom Kierans made an arrangement with Jack Greene of Nelson Machinery to act as a distributor for Rockiron on yielding arches. This arrangement didn't last long because Nelson Machinery didn't have the knowledge as to how and where the arches could be used. Also, the percentage Tom was offering wasn't worthwhile for them to really push this product.

It was planned that, after the convention, the two of us would travel across Southern B.C. visiting the various mines again as we had done the previous year. The new plan was for Tom Kierans to catch a plane in Cranbrook and fly back to Sudbury leaving me to drive home alone. Fate, however, made a slight change in plans. While in Vancouver, as promised, a phone call was made to Bob Lindberg of IMC to find out if he could increase the $750 per month offer he quoted when visiting the IMC mine the previous week. He said he could offer $850 which was the same as the Rockiron salary but there was also a three-bedroom house at very low rent in the offer as well. After mentioning that ...my wife and I have a very comfortable home in Sudbury,

I said, "My wife might not like moving to Esterhazy". Lindberg said, "When you drive back from Vancouver why not have your wife fly out to Regina and meet her there and both of you come to Esterhazy. You and she can see the house and we will introduce her to the local ladies." That was an offer that one could not refuse. He was told he could expect us the following week in Esterhazy to look over the place. When hearing of Lindberg's suggestion of a trip to Esterhazy, my wife Mary was thrilled to have an expense trip paid but she wouldn't fly on her own to Regina. She finally agreed to take the train *(The Canadian)* which was not too shabby. After her arrival in Regina, we drove the 140 miles to Esterhazy. On reaching Esterhazy, Kathryn Upham took Mary in hand *(They knew each other from Atikokan.)* and showed her around. Kathryn even threw a tea-party where she invited a number of the wives of staff members to meet her. Kathryn even showed her through two houses and said Mary could have her pick. These were very well built homes with full basements, three bedrooms, two bathrooms and brick veneer siding. The company even promised it would finish-off the basement as a recreation room. *(This they did by finishing-off another bedroom and bath in the basement, a first-class recreation room complete with wet bar and a tyndall limestone fireplace as well as a playroom for the kids.)* The tea-party and tour was a good selling job and it paid off. My wife was sold and the job was accepted.

(Kathryn Upham, like her husband Merv, is a wonderful person to know, as well as being a lady, she was as a consequence an asset to Merv. She certainly did considerable entertaining in Esterhazy which went along with Merv's position.)

Getting back to Sudbury, Rockiron was given a month's notice of my departure which worked out to the end of November 1962. Tom was disappointed but he could see that this was a better opportunity and accepted the decision gracefully. He even presented me with a camera as a going away present. Fortunately, before my leaving Rockiron, Tom was lucky to find a replacement in none other than George Carr who was at that time mine inspector in New Brunswick. It should be mentioned that a couple of years later John MacIsaac of MacIsaac Mining bought out Rockiron and changed the name to Baycarr Steel. He retained George Carr as the manager of Baycarr. Tom Kierans ended up at Memorial University in St Johns

Why Mining?

Newfoundland as professor of civil engineering teaching hydraulics. *(Tom has since retired and is living in St John's.)*

Engineering

The job at IMC was very interesting. As this was a new mine, it was up to all new department and section heads to organize their respective groups. My section consisted of mining engineers, surveyors, mine planners, technicians and clerks. All told, the group consisted of about 25 individuals and everyone was an individual. The surveyors were led by a chief surveyor who had a drinking problem. You could not trust what this guy said or did. He told or bragged about some very disturbing things. Finally after talking to him on a number of occasions, he had to be discharged. Fortunately we were able to find a good replacement for him almost immediately in Elmer Ohtichenwa *(sp)*. It was surprising how the crews' morale changed for the better after the first chief surveyor was discharged. *(The junior people like to look up to their superiors and when they cannot, as in this case, their morale suffers.)*

The ventilation was another problem. Ernie Isaac who was hired at about the same time, was placed in charge of this area. The face ventilation was very poor at the borers especially when boring on first-pass. The face setup consisted of a fresh-air fan about 200 feet back from the face with flexible tubing extending from it up to the borer and an exhaust-fan mounted on the borer with 100 feet of exhaust tubing extending from it. The problem was to prevent the mixing of exhaust air with the fresh air. This meant trying to direct the exhaust air into another entry, if available. *(Line brattice wasn't used at this time. It came later.)* Another problem was to shroud the cutters of the borer off so that the dusty air would be directed to the exhaust fan. A number of various shroud combinations were tried. Ernie Isaac did have some help from engineers from head office in Skokie Illinois but mostly he was on his own.

Another problem that Ernie was up against was the lack of fresh air entering the mine. The K-1 mine was serviced by a single shaft in 1962. It wasn't until the shaft for the K-2 mine was sunk and the headings of both mines (K-1 & K-2) were connected underground that there was adequate fresh air for both operations. For the first half-dozen years, K-1 mine had to get by with fresh air supplied via a four-foot diameter aluminum duct installed in one of the compartments in the shaft. The most that this aluminum duct could handle was 100,000 cfm. Even so there was leakage due to the aluminum being attacked by the corrosive salt environment in the shaft. Alcan sold this duct with the guarantee that it would withstand the corrosive environment found within the K-1 shaft. It just didn't work out.

Ernie's problems were not altogether in ventilation. Because Ernie had been in supervision as a shift boss in Peru and Malartic Quebec, a mine superintendent in Chapais Quebec and a mine manager in Haiti, Merv Upham and Bob Lindberg thought bringing this guy on board would be good insurance just in case Andy Kyle (mine superintendent) and Gerry Hamilton (underground superintendent) failed in their positions. This didn't go over too well with both Andy and Gerry. They scrutinised everything Ernie did to see if they could find fault...and a lot of fault they found!

In one particular instance, in the spring of the year there is a lot of humidity in the area around Esterhazy. This humid air, when it reaches underground, condenses leaving the fresh-air entries wet and slippery. The big problem is the corrosion of salt and water on steel. Ernie came up with an idea that if he could knock the water out of the air before it entered the mine, it would go a long way in solving the problem. He was right! What he did was have the mechanics drain the glycol from the heat-exchangers in the ventilation plant and circulate cold water through the heat-exchangers in order to chill the air prior to it entering the shaft. This did the job, but he did nothing to remove the water that condensed. Instead it collected on the floor of the ventilation room and eventually shorted out one of the fan motors. As a consequence, the mine had to curtail the use of diesel equipment and therefore this affected production until a spare motor could be installed. Merv Upham gave me hell on account of this.

Ernie had never worked in the technical end of mining and therefore was not up on his technical skills. He required a lot of help. In the end I had to move him out of the line of fire of ventilation and place him in a job where he would not be so close to the operation. I was able to find a young engineer by the name of Mike Berthelson to take his place.

```
Why Mining?
```

Even when Ernie was not at work, he would always put his foot in it when he was around Andy Kyle. On this one occasion, it was at a cocktail party at the Kyle residence when Ernie made another faux pas. We all had to have water-softeners and charcoal filters in Esterhazy because the water from the water supply well was heavy in iron and H_2S. If one didn't backwash their water-softener regularly, it would in fact turn a drink of alcohol a very dark colour. While at this particular party, we were all discussing the merits of our water-softeners and how often we were required to backwash, add salt, etc. Ernie piped up in a very jocular vein and said, "I know when it's time to backwash when my pre-dinner drink of whiskey turns black". Andy, who was in the group, answered very piously, "I wouldn't know about that because I don't drink that often".

After the ventilation job, Ernie then took over the job of heading up the rock mechanics section. This was a very highly technical job but at that particular time, IMC relied on consultants such as Shoisi Serata of the University of California for guidance in this area. Ernie's job was really to direct the technicians to see that the rock mechanics stations were installed correctly and the readings from these were recorded in the proper manner. This kept him out of the line of fire and made my job easier and thereby I didn't have to defend a friend to the boss. Ernie could see the writing on the wall though and left IMC for a position with U.S. Borax who were sinking two shafts at Allan Saskatchewan, a town 35 miles east of Saskatoon.

IMC was fortunate in replacing Ernie with George Zahary who had recently completed his masters in rock mechanics at the University of Toronto. I knew George from Steep Rock and therefore he was not an unknown entity. George was a comical guy. He would explain an idea by way of comparison..."on the one hand, it could be this ...but on the other hand it could be that". This would go on with every statement until he had you all confused. He did a good job though and was a good guy to have around.

Mine planning was handled by Tibor Chizmazia, a Hungarian engineer, who had escaped to Canada during the Hungarian Uprising against the Soviets in the late '50s. Technically Tibor was very good but needed assurance on everything he did. He would be halfway through a proposal and then stop and wouldn't draw another line until he showed it to me. I was interrupted at least three or four times a day to assure Tibor that what he was doing was correct. He finally was seconded by Tom Braithwaite, Manager of the K-2 shaft sinking project, to prepare the drawings for K-2 shaft project which had just commenced.

Tibor Chizmazia's place in mine planning was taken by Angus Scott who had recently arrived from Cominco in Kimberley. Angus was a dynamic little guy who caught on very quickly. After a few months in planning he transferred into operations, first as a shift boss and later as a captain. It was as a captain that things got a little rough both for Angus and for me. More about that later. The planning section was then taken over by Guy Lokhorst who was a recent U.B.C. grad. Guy was probably the best planning engineer that IMC had working in this position.

When I first started in the job as chief mine engineer, I noticed that there was no mechanical design talent within the department. This phase of the job I had to do myself or farm this phase of the work out to the central engineering department in the main office building. It was easier to do the job myself rather than try to get the central engineering department involved. It was like pulling teeth to get anything out of those guys in central engineering as they had their own priorities. At the time, central engineering was headed by Tom Braithwaite who was only concerned with designing a second shaft which could only enhance his own career and that was the problem. This meant limping along for awhile doing my own design work which added an extra load onto the job. Fortunately for all concerned, IMC hired my good friend, Stuart Johnson, from Steep Rock to come to Esterhazy and be the designer under the pretext of an electrical draftsmen to assist Art Schwandt the electrical-mechanical engineer to draw the as-built drawings for the underground electrical distribution. I allowed Stu to draw a few electrical drawings (enough to satisfy Schwandt) but then got him to get on with the backlog of mechanical drawings that were needed. This worked out fine until central engineering found out what a great *(and I mean great)* designer-detailer that Stu Johnson was. Almost immediately, they seconded him for their own department but agreed that the mine would be able to call upon Stu to do its mechanical work. This worked out fine as Stu, being my friend, would see to it that the mine engineering got priority service whenever it had design work to be completed.

Why Mining?

There were two good mine clerks at IMC. A young fellow by the name of Dave Feniuk was very good. He could do accounting, type and take shorthand. A good all-round guy. Dave was Ukrainian and followed the Caesarian calendar for Christmas and New Years. This was great as the mine worked Christmas and New Years and Dave didn't mind working these days as long as he got time off for Ukranian Christmas and their holidays. The other clerk was Joe Phipps, an ex-RCMP. Joe was very good at listening and one could bounce various personnel ideas off him and he would give an honest assessment. Dave and Joe were a good team.

Talk about working Christmas and New Years, that's what I had to do the first month after arriving at Esterhazy in December 1962. There were very few people about the mine who had previous mining experience, as the labour force was recruited locally; therefore, we had to do an extra shift each week as third-line supervisors; that is, act as underground superintendent on afternoon shift. It was all scheduled on a ten-day cycle so that we would have different days each time. My first December at Esterhazy, I drew (lottery draw) Christmas and New Years afternoon shifts. Christmas Day meant that Christmas dinner had to be eaten prior to going on shift at 4 PM and to forget about going partying on New Years Eve. We did this sort of double duty for the first six months after arriving at IMC. It seemed like one was always working as we were required to work six days a week at IMC plus this third-line supervision as well as be part of the weekend duty roster. What this amounted to was working six days each week. Then, every ten days, work an afternoon shift besides working the regular day-shift. Finally every sixth Sunday, one had to work the day shift acting as superintendent in-charge of the mine. *(Incidently, the mine operated every day of the year except the day of the company picnic in the summer.)* For this extra time, we were given a bonus of 20% on our salary which was paid at the end of each quarter.

Because of our move to Esterhazy at the end of November and first week of December, my wife and I found that we didn't have anything prepared for Christmas, not even presents for the kids. I asked Andy Kyle for a Saturday off so that I could drive my family into Regina that weekend and do some Christmas shopping. It was agreed that I would work a double on the Friday and leave for Regina the following Saturday morning. Andy agreed and so I worked the double and got the following day off. In the following February when we were paid our bonus for the previous quarter, I noticed that I had 10% deducted from my bonus for December. When enquiring about this shortfall Andy said, "Dave don't you remember that you took a day off". There was no questioning the circumstances or the fact I worked extra time...case closed.

Operations

When I was promoted to underground superintendent, the company installed a telephone in my bedroom that was directly connected to the mine telephone system. With this phone, it was possible to call the shift captain, hoistmen or underground warehousemen directly without going through the regular exchange. This also meant that these same people could call me any time in the middle of the night and they often did. At one stage, we had a young Hungarian (Jimmy Webb) loading pocket operator who always dialled the wrong number, getting my number instead. It was very aggravating to get wakened up by someone speaking broken English asking to speak with so-and-so. It seemed that, while working at IMC in Esterhazy, one was never off the job.

A Miner Operating a Potash Boring Machine on a Double Pass

Potash mining was very different compared to experiences in previous mining operations. The big boring machines were fascinating and so were the gathering arm loaders and shuttle cars. I had seen lots of conveyors at Steep Rock but they were still fascinating all the same. Finally, I asked Andy Kyle if it was possible for me to go underground after day shift and learn to operate a boring machine. He was very pleased and gave the OK. Arrangements were

Why Mining?

made with Greg McDonald, formerly from Steep Rock who was one of the shift captains, for instruction on operating the borer. He turned me over to one of his best operators, Leo Hnatyshan, who gave me my first lesson on the operation of the Goodman Boring Machine. This went on for a few weeks where I would go underground after day shift and work another shift on the borer. Finally, I would sneak off during a slack time on day shift and operate the borer. I got pretty good at it too.

It should be mentioned at this time a story about Leo Hnatyshan's brother "Joe". One spring, Andy Kyle sent an number of us on tour of various coal mines in Pennsylvania and West Virginia to observe operations there and perhaps pick up some good ideas. The group I was with was composed of Ron Busch (mine captain) and Joe Hnatyshan (shift boss). Joe was a shy quiet prairie boy who had never been further than Alberta and Manitoba before this trip to the coal mining area of the eastern United States. Busch on the other hand, had been all over and was very quick-witted. Our trip meant an overnight in Detroit where we were given hockey tickets by the IMC sales staff there. We attended the game in the old "Olympia" where Gordie Howe broke Maurice Richard's all-time goal scoring record. *(Of course that record has since been broken by Wayne Gretzky.)* The Red Wings were playing the Canadiens that night which made the game that much more interesting. Joe looked over the programme that he bought which showed that all the players were born in Canada. He remarked, "What's a matter with the Americans, aren't they athletic?" *(Times have since changed as there are a number of Americans in the NHL now and in the World Cup Series in 1997, actually beat Canada but fortunately not in the 2002 Winter Olympics.)*

After the game we went to the nearest bar which was called the "Brass Rail" to celebrate Gordie Howe's achievement. Apparently there was too much celebration as we all had terrific hangovers the next day. I remember leaving the guys at the hotel thinking that they would go to their rooms like I did; however, Joe Hnatyshan went to a bawdy house instead. A couple of days later, he told us about it while we were winding up our trip. Busch who was quite a kidder told Joe he probably picked up some venereal disease and he should go to a doctor right away before passing the disease on to his wife back home. After much prodding Joe skipped out that afternoon and went to a doctor there in West Virginia where he got a shot of something-or-other.

Both Busch and I chipped in to pay the doctor's bill so that Joe's expense account wouldn't be too far out of line.

As mentioned, we all had terrific hangovers the next day. Although we didn't drink that much. It must have been some really poor American liquor that some bars pass on to unsuspecting customers. We flew to Pittsburgh the next day, picked up a car there and stayed at a motel near the airport outside of Pittsburgh as the first mine we were to tour was located at a town called California in Pennsylvania which was not more than 50 miles from the motel. It was surprising that the mine (coal) did not have a visitors' dry so we had to change at the engineering office. I had brought some mine clothes for the trip but Hnatyshan and Busch didn't bring anything with them which they regretted. Because of this, they had to take whatever the mine had available. At one place they were given smocks and had to go underground wearing those over their street clothes. Anyway, the first mine visited was a Jones and Laughlin mine which had a very fancy cage; in fact, we were told the cage was similar to the elevator cage in the Empire State Building. The main conveyor system, about 5 miles long, was top-of-the-line also. All the transfer points were TV monitored which was very new in 1963. At the mines in the States, one sees the two extremes, the sublime and the ridiculous. Here at the California Mine of Jones and Loughlin one could see a first-class elevator cage and conveyor system but no change-house for the employees.

Getting back to the conveyor system, the one at the California Mine took the raw coal (run-of-mine) from underground to a coal washing plant on surface. This coal-washing plant was adjacent to the Ohio River where the clean coal (treated) was loaded onto barges. What impressed me in the States was and still is the use of their rivers. Any river that has any width or depth is used for transportation. *(I often wonder if Canada could not investigate the use of its many rivers for transportation.)*

The next day was a travelling day to Logan, West Virginia. The first mine visited was a very haywire place but it had the best productivity I had ever seen in my whole mining career. *(American miners work harder than their counterparts here in Canada.)* But first of all, before getting to the mining part, one should describe the surface plant. *(Surface plant is used liberally.)* The mine superintendent was a tall rangey guy *(mountain man)* with a West

Why Mining?

Virginia accent. He showed our group around the surface plant which consisted of a steel-framed buildings covered with corrugated iron sheeting. Some of the buildings didn't even have concrete floors. I remember going through the shop where the master mechanic had the walls of his office papered over with "Playboy" center-folds. The toilet was an outhouse over a dried-up stream. One didn't have to go in to see it but, from observing the three piles of human excrement below in the stream bottom, you knew it was a "three-holer". In the springtime when the stream is replenished with the snow-melt, the stream then carries this waste away. *(This is all very sanitary and environmentally friendly???)* While touring the surface, the superintendent in his West Virginia accent said, "This is the best goddamn place ah's ever worked." *(I wouldn't want to see the worst place he had ever worked as the conditions at this place were damned scary.)* As there was no changehouse or dry, it was necessary to change into our mine clothes in the superintendent's office. There wasn't a clean place to put our clothes so we had to open up a couple of magazines, place these on a table with our clothes on top. Fortunately, I had brought my own mine clothes, complete with gloves, and therefore was able to stay reasonably clean except for my face which was the only part of me exposed. The other two guys (Busch and Hnatyshan) had to take whatever attire the mine offered and that wasn't much.

The ride underground was by jitney *(a four-man trolley)* with Busch sitting up in front, Hnatyshan and I sitting in the middle while the superintendent drove, seated at the back. It was my job to hold the trolley pole from jumping off the line as we bounded down the track. And bound we did! The superintendent was out to scare us and he succeeded. The adit into the mine was narrow, low and had to be timbered most of the way. Some of the timber sets were just inches over our heads and you can be sure we all kept our heads down while riding into the mine. After travelling about a mile, we came to an intersection. The superintendent stopped the jitney and reached out for a phone there and hollered into the phone saying, "Hello Jones, is it ah clear on north a hundred?" On receiving the affirmative, we proceeded down North 100. This procedure took place a couple of times until we actually reached the dispatcher's booth where we finally met Jones. Jones and the Super began a conversation where Jones said, "I bought maself a newgun". "What kinda gun yo bought", asked the Super. "I just bought an ol' shootem-up gun" answered

Jones. *(Americans love their guns.)*

The jitney journey finally ended at the mine workings where the coal was being mined. The mining method was room and pillar where the miners were advancing a front of six 15-foot wide by 5-foot high rooms. In the first room, a two-man team was undercutting the face with an undercutter; in the second room, a two-man crew was drilling and loading the face for shooting *(blasting)*; in the third room, a one-man crew was installing ventilation fans and extending the brattice wall; in the fourth room, where the coal had being previously shot *(blasted)*, a loader operator and two shuttle car operators loaded and hauled the coal to the conveyor loading point; in the fifth room, a two-man crew, one scaling and another operating a bolting machine, secured the back; the sixth room was an idle room ready for the undercutter. After each activity was completed the crews would move from one room to the next in sequence. *(I had never seen such productivity in all my days of mining. The crews were not on bonus either nor were these men at this particular mine in a union. All in all, I do think Americans work harder than Canadians, certainly the ones that I saw working in the mines visited in Pennsylvania and West Virginia.)*

In West Virginia, the coal miners have a very different-shaped lunch-bucket from those found in Canada. Their buckets look like big double-walled soup pots about 10 inches in diameter with a 6-inch top that unscrews and forms the inner wall where the lunch is placed. The outer wall of the pot is filled with water or whatever the miner has for a beverage. Most of the miners fill this outer pot with drinking water and carry it with them to the face.

As is the rule in coal mines, there is no smoking whatsoever underground nor around the surface plant either. As a consequence all miners chew plug tobacco and it seems like they always have a wad of chew in their mouths. When these same miners have a swig from their water container *(lunch pot)* some of the tobacco juice is mixed back with the water in the container. After a few swigs, the water in the container is pretty well ripe with tobacco juice.

After visiting the first section where they were advancing a six-room entry, we next visited another section of the mine where the men were just sitting down for a morning break. The Super sat down on a coal pile beside these

Why Mining?

men and started to swap stories. One of the miners offered Ron Busch a drink of water from his container but Ron just passed the container on to Joe Hnatyshan. Joe took a big swig and then held it in his mouth instead of swallowing it as it was full of tobacco juice. Joe was too polite to spit it out and too afraid to swallow it. Poor Joe he was always having tricks played on him by Busch.

Another mine was visited in the Logan area and it was about the same as the first except that it did have a shaft and it didn't require taking a scary ride into the mine as was the case in the first mine. Anyway, our group did get back home to Esterhazy with much appreciation for our dry facilities and sanitary conditions at the IMC potash mine. *(Potash Mining in Saskatchewan is akin to coal mining except the potash is cleaner. The Saskatchewan mines don't have methane gas nor do they have the bad ground conditions either.)*

This might be a good time to discuss the boss, Andy Kyle. Andy was a very honest and faithful employee of IMC. He was very dedicated to his job and to his family. There wasn't a mean bone in Andy's body. He said what he appreciated in me was my hard work and he liked the way I followed through. When becoming underground superintendent after spending six months as chief mine engineer, Andy and I worked very closely together. Although Gerry Hamilton, who was previously underground superintendent, worked long hours (he bragged that he spent over 48 hours underground in one stretch when IMC first commenced production) he was not one to follow up or at least that was one of his failings. One could talk to Andy. For instance, one of Andy's habits was to go underground and make a note of every thing that he found wrong. Then he would have it typed up and distribute it to the Managers: Production, Operations and General. Not liking that, I told Andy I didn't mind him pointing out the discrepancies but it wasn't appreciated having these short-comings pointed out to senior management. I suggested that both of us should work together and correct these things ourselves and, if senior management wanted to know how things were going, they could come and be toured around to see for themselves if things were right or wrong. Andy agreed and from then on when he found something out of order, he and I would discuss it and find a solution for it. It was enjoyable working for Andy Kyle as he was a person who would listen and was one-hundred percent honest.

Grow

Christmas time working at Esterhazy was a very busy time for senior IMC staff members. During my second year at Esterhazy, I was promoted first to underground superintendent (general foreman) and later to mine superintendent. Along with this extra responsibility, came the command performances demanded by the company. The manager always had a Christmas Party. This was followed by the New Years Party thrown by the Operations Manager. Then each department head had to have a party for all their staff. In order to accommodate all the supervisors (as the mine worked every day), it was necessary to hold two parties. At the first party, there would be the staff (captain, foremen and engineers) from the dayshift and the staff who were on days off. This would be followed the next week with staff who were on dayshift that week and staff who were on their days off for that week. Finally, each shift captain would hold a party for his men. The captain's or crew parties were handled in the same way as the department heads' parties; that is, two captains would get together (day shift and days off) and hold a combined party for the two crews. All in all, there were six parties that a department head had to attend, the two managers' parties, the two department head's parties and the two crew parties. By the time Christmas and New Years was over, one was very much partied out.

IMC Esterhazy at the time of Merv Upham's tenure, was non-union. This meant there had to be good communications with the crews and much participation in community work with the residents in Esterhazy and vicinity. To begin with, the residents of these little prairie towns were very suspicious of outsiders and therefore, to overcome this, IMC favoured hiring staff either from the Prairies (preferably) or from Western Canada. It was the exception that anyone not from either the Prairies or Western Canada was hired. For instance, Jim Sadler came from England but his (former) wife Freida was born in Esterhazy; therefore, he was legitimate. Andy Kyle, my boss, was originally from Kyle, Saskatchewan; Gerry Hamilton, the underground superintendent, was from Winnipeg; George Zahary, rock-mechanics engineer, was from Edmonton; Tom Braithwaite, chief engineer, was from Red Deer; Angus Scott, planning engineer, was from Wilkie Saskatchewan and on and on it went. Merv Upham believed that, to make a good impression in the community, everyone should go to church. As a consequence, attendance at the United Church services in Esterhazy were very well attended while Merv was manager at IMC. *(Attendance fell quite dramatically after Merv was*

Why Mining?

transferred to Skokie, Illinois.) The staff was encouraged to join the local clubs or charitable organizations: some staff members even had their dues paid by the company.

The pay-off were the visitations at Christmas time with the work crews. Each shift boss and captain had to make a visit to the homes of their men over the holidays. The department superintendent, like the mine or mill superintendent, had to accompany either the captain or shift boss to those crew members that might prove difficult in visiting. What this meant at Christmas time was that attending all those parties and the visitations, besides working long hours, one didn't have much time for one's own family. It put a lot of stress on the superintendents and their families.

Another stressful thing that one had to endure were the seminars held at the Esterhazy Motor Inn on the subject of "staying non-union". These seminars would begin in the morning, then break for lunch (paid by the company) and carry-on until 4 o'clock in the afternoon. Sometimes the seminars would last for more than one day, sometimes even three days. The crew schedules consisted of 7-days graveyard shift followed by 1-day off, then 7-days afternoon shift followed by 2-days off and finally 7-days day shift followed by 4-days off making for a nice long weekend. It was the 4-days off that the crews especially looked forward to where they could travel into the city (Regina) or go fishing or camping with their families. Often times senior management would organize a seminar on staying non-union for those supervisors who were on their long weekend off. This didn't sit too well and there was much grumbling among those supervisors. At one such seminar, Angus Scott, who was then a shift captain, didn't show up. When Stan Williams, the personnel manager came to me *(I was the mine superintendent and Angus' boss)*, he complained that Angus should be severely disciplined. I agreed that Angus should have attended but I said,"I could sympathize with Angus and would let him go with just a warning". That was a mistake on my part as Stan had an "in" with Bob Lindberg who had recently replaced Merv Upham as manager. On another occasion, one of the shift bosses, the notorious Don Hay, *(This man, some years later, was sentenced to 25 years for kidnapping, holding captive and sexually molesting a little neighbour girl for six months in a dungeon he had dug in his garage in New Westminster.)* walked out of a seminar and went down to the beer parlour and sat there all afternoon. Again, Stan Williams came to me but this time he accused me of undermining the seminars as it was always my men who were the non-conformists. One had to agree with Stan that these men were in the wrong but I said, "Perhaps management should be a bit more sympathetic of the supervisors' needs and try to schedule the seminars during working hours and not on their days off." After that incident, yours truly was not in Stan's good books and it would go against me later on but, looking back, I would perhaps say the same thing to Stan Williams. *(If I had to do this over again, perhaps I should have been more sympathetic to Stan's point of view to his face, but that wouldn't be honest.)* At another seminar, not being all that prepared for the subject matter that I was supposed to lead in the discussion was another black mark against me. Jack Mitchell, the production manager, who was my boss at that particular time (as Andy Kyle had been promoted to project manager at K-2) came down on me very heavily. We had a shouting match afterward about his seminars. My days were numbered at IMC. Two months later, I was an employee of Cominco.

Merv Upham laid down three directives when promoting me to the job as mine superintendent. These were:
- Not to touch the tubbing unless he gave written permission.
- Not to increase the extraction percentage unless Serata (the rock mechanics consultant) and all agreed that this was good rock mechanics.
- There will be only one entry system connecting No 1 and No 2 mines *(K-1 and K-2 were not the designations at that time)* and these entries must be equipped with water-tight doors.

(It's a pity that Merv Upham's edict was not followed by subsequent mine managers. If those set of rules had been followed there would not have been the water problem at the K-2 or at least it would not have affected the K-1 Mine.)

This is probably a good time to look into the background of the mine managers that followed Merv and see what went wrong. Immediately following Merv was Bob Lindberg, a mining engineering graduate from Minnesota. Although Bob was out to make a name for himself after being so long in the shadows of Merv, it would be out of character for him to give permission to increase the extraction percentage nor have anyone touch the tubbing in the shaft. During Bob's stay, the mines were not that far along to make more than one entry system between the two mines. Bill Huston, a geological engineer with a masters degree in mineral processing, was the next mine manager. Bill, whom I knew from Steep Rock, wasn't the most honest fellow. He was fired from his president's position at Steep Rock for the alleged way he undercut the former president, "Pop" Fotheringham. Pop got back at him a year later at a shareholders' meeting, when Pop had enough proxies to turn the tables and oust Huston. At IMC, Huston wouldn't hesitate about increasing the extraction percentage nor would he hesitate about driving multiple entries between the mines as long as it showed a good bottom line **immediately**! Following Huston was John Nightingale, a mechanical engineer who was very cautious and timid. He would never change anything on his own. If anything was changed at IMC during his tenure, it was done by someone else with or without John's approval. Following Nightingale was Dave Kelland, a graduate mining engineer from McGill. Dave was always a very cautious fellow and had no part in the catastrophe that happened at K-2 mine. The damage was done before Dave took over, even though the water breech occurred and the flooding took place while he was manager. Had Huston been more responsible and Nightingale been more forceful, this occurrence would undoubtedly not have happened. Perhaps the board of directors are responsible for placing geological and mechanical engineers in charge of a potash mine as neither one had rock mechanics experience. *(Allan Potash had an accountant in charge as general manager at one stage and the mine wasn't very successful. Anyone can run a mine once all the systems are in place but you have good people subordinate to you that you can trust. However, a developing mine is another matter. The man at the top must know what should be done and how it should be executed. Good examples in my career are Merv Upham at IMC, Bart Thomson at Kidd Creek and Johnnie Waugh at MGF. These guys set the standards for those that followed. Unfortunately, some of those that followed at IMC tried to cut corners.)*

Merv Upham called me to his office one afternoon and said he was promoting me to mine superintendent, reporting now to Jack Mitchell who had been promoted from mill superintendent to production manager. Knowing a little bit about Mitchell, it was not a happy situation having him as the boss. He thought himself knowledgeable in both milling and mining and looked down on miners particularly. The man taking my place was Ken Donald, the chief industrial engineer, who had been reporting directly to Merv Upham. Ken Donald was a rather wishy-washy guy who didn't seem to achieve much within industrial engineering at IMC. Apparently Merv Upham didn't want him around and foisted him on me. Merv said to work with Ken Donald and make a manager out of him as he was being given this last chance to make it. Shortly after this, Merv was transferred to the head office of IMC in Skokie, Illinois which was too bad as the place was not the same without him.

The fact of the matter was, one couldn't believe a word Ken Donald ever said. He would delay going underground each day where finally I had to tell him that he was the underground superintendent and that his place was underground and he should be underground on the first cage after the crews are lowered at 8:15 am each morning. I found out later that Ken Donald, when he did go underground, would go directly to the warehouse and call his captain and shifters there to give him the lowdown of what was going on underground and not move a muscle to visit the faces and see for himself. In other words, he did his inspections the easy way. He was always requisitioning manpower without authorization. When asked how many men he had on each crew he would invariably give one a somewhat reduced number. One couldn't believe his reports. When the cost statements came out, they always showed a high manpower cost. Finally in order to get the truth, I had the lampman report the number of miners' lamps used on each shift and in that way one knew the correct number of men underground. Finally, in desperation I told Jack Mitchell, the production manager, that Ken should be replaced and that perhaps he would be better off at head office as an industrial engineer or in some other capacity. Fortunately at the time, head office needed someone for a couple of months to do some industrial engineering work so away went Ken at least for a month or two.

Unfortunately, Merv Upham left IMC for Newmont in New York in the spring of 1965, so my friend and mentor was no longer around. Shortly after Merv's departure, my time with IMC abruptly came to an end after Jack Mitchell and I agreed to disagree over bringing the supervisors in for seminars on their long four-days off shift change. *(I think to this day that it was not right to ask the frontline supervisors to give up their only weekend in the 28 day cycle to come and listen to how the company planned to keep the union*

Why Mining?

from becoming certified at Esterhazy.) The blow came one Monday afternoon in March 1965 when Jack Mitchell came to my office and said that I was being terminated because he and I could never work together as a team and he didn't think I was capable of managing. He said that the Ken Donald problem wasn't handled very well *(He wanted me to fire Ken but there weren't enough grounds to justify that. The irony of it all is that Mitchell placed Ken Donald as my successor.)* He told me that IMC was offering me the choice of leaving immediately and drawing six months salary or staying on for six months and leaving quietly. I had a month to make up my mind which was fair even though I hated the way Jack Mitchell treated me while under his supervision. It should be mentioned that if I stayed on for the six months my position would be that of "superintendent of new developments" *(a made-up job)* and reporting to Bob Lindberg, the general manager, with an office next to his.

Bob Lindberg treated me fairly, allowing me to travel on an expense account to any job interview that could be arranged...no questions asked. He even sent me to the CIM Annual General Meeting in Toronto. It was a nice send-off from IMC, and very much appreciated. Thinking back though, Bob could have heard my side of the story and how Mitchell treated his subordinates. It was either I go or Mitchell go and, Mitchell being the senior, carried more weight. It should be said that Mitchell was a very clever man and later on rose to the top of Kilborn Engineering which was a very prestigious mining and metallurgical design consulting firm at that time. Like many little men, he had the little man's syndrome and therefore could be very difficult in his dealings with taller people. He was a tyrant with a definite mean streak in him, like the time he went out to the mill early one morning at 2 am when the screening plant was having difficulty and took an axe to all the top screens on one row, swearing at his subordinates all the while. The plant produced a high quality standard grade potash product on that particular shift because of the excess of larger particles which increased its sieve size or SGN rating. He knew what he was doing but it was the manner in which he did it that was questionable.

IMC K-2 Mine and Refinery

Chapter 7

Cominco

House and Home

The mine superintendent at IMC had the dubious honour and duty to tour various delegations through the mine on several occasions. On one particular occasion, it was my good fortune to tour Frank Goodwin, Bruce Hurdle, and a geologist *(whose name has been forgotten)* all from Cominco. *(In those days, Cominco was known by the very long name, "The Consolidated Mining and Smelting Company of Canada Limited".)* These gentlemen were very impressed with the cleanliness and the mechanization of the potash mine (as were many others who toured the operation) in contrast to the Sullivan mine in Kimberley or any other hard rock mine that Cominco operated. Later on when on my way out after the Jack Mitchell fiasco, by chance Robin Porter, Frank Goodwin and Dave Kelland, again all from Cominco, visited IMC a second time to tour the mine and mill. Cominco had only recently announced that it was interested in developing the potash beds at Vanscoy Saskatchewan. This was my chance. I cornered Frank Goodwin and asked him for a job. He was taken by surprise and said he'd get back to me. Later in the month, Bruce Hurdle, who was the general manager of Cominco's potash division, called saying he wished to interview me. We agreed to meet in Cranbrook as I had planned to travel there to visit my parents. We met in one of the downtown cafes in Cranbrook where Bruce offered me a job at $1000 per month, the same salary as mine superintendent with IMC. The plan was to first move to Kimberley and then to Saskatoon once construction of the mine got underway.

On returning to Esterhazy immediately, arrangements were made for a mover to come and pack as soon as rental accommodation could be found in Kimberley. When arriving in Kimberley soon after *(but staying with my folks' in Cranbrook which was 20 miles from Kimberley)*, Bruce Hurdle said that the job would be there working off and on in Kimberley but for the most part in Trail where much of the planning for the new potash mine would be taking place. Robin Porter, who was manager of the Kimberley operations told me that he had an apartment available in Kimberley. This, he said, is the best they could do as housing was very scarce in Kimberley. I didn't like the idea of the Cominco apartments as they were converted bunkhouses from the old days. My parents suggested looking for a house in Cranbrook and commuting to

Why Mining?

work in Kimberley. This was a good idea, especially when there was first-class housing available in Cranbrook for rental. After finding a beautiful house for rent on the northeast side of Cranbrook, I phoned my wife and told her about this nice house and told her to tell the movers to start packing. When Robin Porter was told that accommodations were found in Cranbrook, he was upset and said that Cominco wanted its senior staff to locate in Kimberley. *(First, I didn't know I was considered senior staff; second, I didn't know Cominco had the rule about where their staff were to reside, but most importantly, neither I nor my wife were enthusiastic about living in a converted bunkhouse.)*

Anyway, the next week when working at the mine division office in Trail, Bruce Hurdle said that they thought it best if my family moved to Trail instead of Kimberley, and therefore, we should look around Trail for a place to rent. Again luck was on my side as there was a chap being transferred to Pine Point NWT who had a very fine home in Glen Merry *(an area of Trail)* which could be rented. After putting a deposit on the first month's rent, I again phoned my wife to get the movers packing as the location was changed to Trail. When reporting this good fortune to everyone in the mine division office the next day, they were all very pleased. Later that day, Frank Goodwin came all apologetic and said that they had changed their minds and that it would be best to move to Kimberley after all as I would be working in that location more often than in Trail. *(In the span of one week, I was told to move to Kimberley, then Trail and finally Kimberley again.)* Again I phoned home that night and announced that the plans had all been changed. Hoping that the house that had been lined up in Cranbrook previously was still available, I phoned the realtor who said it was no longer available as it had been rented. Fortunately, luck was still on our side as a college friend from Applied Science '50, Stan Hodgson, who had just built a very fine home *(two months old)* was being transferred from Kimberley to Pine Point. Stan gave me a year's lease and again I phoned my wife to get packing for Kimberley. You can imagine her plight. She didn't know if she was coming or going. She probably thought everyone had gone bonkers and the company officials of Cominco were out to lunch. *(It should be mentioned that the employee relations of Cominco at that time were pretty poor to say the least. It seems that Cominco could transfer its employees, especially engineers and geologists, anywhere on a moments notice without much regard to family, housing or financial situation. Every time these engineers and geologists were transferred it meant selling their homes and then buying new ones at their new location at higher mortgage rates than they had previously, as not all operations had housing for rental.)*

As it turned out, the work was in Kimberley all that summer. This worked out fine as it allowed me to get my family and belongings settled in the Town of Kimberley. It also gave me the chance to see a lot of my parents as I had never had the opportunity to live close to them during my working days in the past. My father and I spent a great deal of time together on the weekends in the autumn, hunting. *(This was to be my father's last year as he died the following June of cancer.)*

The work in Kimberley that summer and fall was mostly writing reports for Robin Porter regarding mining activities about the Sullivan mine. These reports consisted of comparing rockbolts and methods of installation, mining with yielding and rigid arches, methods of placing fill in the relatively flat stopes, looking at the conveying bottlenecks at the Sullivan, and my impressions of the Sullivan mine in general. There were other reports written as well. While in Kimberley, it meant reporting to Ken Davies, senior engineer, whom I had met before when working in Kimberley as a student. Anyway, Ken thought I worked for Cominco Potash and therefore left me alone. Frank Goodwin thought, when I was in Kimberley, I was working for Ken Davies and as a consequence, he left me alone also. It was sort of like being left in limbo. While in Kimberley, I more or less had the pick of jobs which amounted to writing reports for Robin Porter, the manager. He treated me like a visiting expert. *(An expert is someone who is five miles from home base or maybe off base!)*

Commuting

Towards late fall, Frank Goodwin sent work over from Trail which I could do in Kimberley. This worked out well as it allowed me to stay at home rather than commute to Trail. But later in the winter, more and more of my time was spent in Trail. Winter is a bad time to be commuting back and forth between Kimberley and Trail because of the snow slides on the Skyway Highway between Creston and Salmo. When the snow slides close the Skyway, it means that one has to drive the extra mileage between Nelson and

Why Mining?

Creston along the shoreline highway of Kootenay Lake. Going by plane between Cranbrook and Castlegar was no better as nine times out of ten in the wintertime, the Castlegar Airport was closed because of fog or low clouds. *(The Castlegar Airport is nestled in the narrow valley at the confluence of the Kootenay and Columbia Rivers.)* It was necessary to drive around Kootenay Lake more than once because of snow slides, or ride the bus between Castlegar and Cranbrook a number of times because of low clouds at Castlegar on these commutes that winter.

Early in the spring, Cominco awarded the design of its potash plant to Kilborn Engineering in Toronto. This meant that besides commuting back and forth between Kimberley and Trail, it was necessary to attend design meetings in Toronto every four to six weeks with Frank Goodwin the project superintendent of Cominco Potash. The sinking of both potash shafts by the Cementation Company had begun in late summer of 1965. Therefore, on the way back from trips to the design meetings in Toronto (which lasted for a couple of days), Frank and I would return home via Saskatoon where we would observe the progress of the shaft sinking with Dave Kelland, who was Assistant Project Superintendent. It should be mentioned here, it was on one of these trips in flight to Toronto that Frank Goodwin and I drew the rough layout of the surface plant for Cominco Potash. We both had our folding tables down in the aircraft and drew small boxes that represented the various buildings (concentrator, loadout, shops, warehouse, powerhouse, headframes and crude storage). When we got to Toronto, we passed these sketches to Kilborn for refinement. Although these sketches were crude, they did give Kilborn Engineering an idea of what Cominco had in mind and how the plant could be expanded if required.

Cominco Potash Plant

Vanscoy Saskatchewan

Spring of 1970

Photo by G Hunter

Cominco

Kilborn, at the time, was headed by Ken Dewar, with whom I worked during my early days at Steep Rock. Ken was a very gracious host to everyone at these meetings where he would take the group to the Boulevard Club on the Lakeshore for lunch nearly every time we came to Toronto. Ken would always reminisce about someone or something which seemed to get Frank Goodwin's goat as Frank, who spent all his career in Kimberley, didn't know any of the people that Ken would be discussing. I enjoyed Ken's stories as I had worked in the Val d'Or and Steep Rock camps as well as having visited many of the mining areas in Canada as a salesman with Rockiron. I was familiar with the areas that Ken would mention and at the same time I knew the people. Other personnel in the Kilborn organization at that time were John Dew, Carl Frietag, Dick Roach, Hans Schmitt, Bill Scott and Bunny Crocker.

John Dew, a civil engineer and second to Ken Dewar, was a gentleman. No one ever had a bad word to say about him. Carl Frietag and Dick Roach were the metallurgists who, along with Cominco's Tony Banks, developed the process and specified the metallurgical equipment. Hans Schmitt, the Kilborn architect, was very temperamental. On one occasion, Frank Goodwin asked about movable partitions between the offices instead of having fixed plastered walls which were recommended for the proposed office building. Hans jumped up and said he would wash his hands of the whole project if movable partitions were to be used instead of the "plaster and lathe" partitions that he, Hans proposed. *(I would have told Hans to fly a kite if I had my way but Frank Goodwin, the perfect gentlemen, gave way to Hans and his tantrum and went for the plaster and lathe partitions.)* Bill Scott was the Kilborn project engineer who was assigned to the Cominco Project, while Bunny Crocker was the metallurgical sage that everyone looked up to when a metallurgical question was in doubt. Kilborn had a very good team with much practical experience at their Toronto office and all of them were first-class individuals.

On returning from a design meeting in Toronto via Saskatoon, there was a very serious meeting with Jim Black the project manager for Cementation Company and Frank Goodwin. *(I say serious because what was decided at this meeting was responsible for the flooding of the No 2 Cominco Potash Shaft in 1970. This flooding will be discussed a little further on.)* The shaft contract called for the removal of all the metal casings in the ground that

Why Mining?

housed the freeze pipes once freezing was concluded. It further called for flushing out the holes with high pressure water once the casings were removed and finally grouting the freeze holes with neat cement grout. What Cementation proposed instead was to flush out the pipe casings with high pressure water, then perforate the freeze hole casings and finally grout the casings in place rather that pulling these out from the holes altogether. Unfortunately Cementation did not flush out the casings as well as they should have nor were these casings properly perforated or thoroughly grouted either.

The trips to Toronto were not that easy. When flying from either the Castlegar or Cranbrook airports, it was necessary to fly by Canadian Pacific Airlines to Calgary then change planes there and fly the rest of the way by Trans Canada Airlines to Toronto. Canadian Pacific flew Convair aircraft for a while but changed to DC-6 planes later. The connections at Calgary between the two airlines were not very convenient as there was a long wait just to make this connection. The flights from Calgary to Toronto were all in DC-6 planes but flying between Saskatoon to Calgary on the way back was by Viscount aircraft. After being transferred to Saskatoon, the trips to Toronto were a bit easier distance-wise. The Trail-Saskatoon trips were still bad with the poor connections in Calgary. On the trips from Saskatoon to Toronto, I would leave for Toronto in the evening after a full day at work and fly Viscount to Winnipeg where there was a six-hour layover before catching the "red-eye special" which was a Vanguard type aircraft to Toronto. Generally ,when waiting for the "red-eye special", I would rent a roomette at the airport. The roomettes at Winnipeg were fairly roomy compared to the ones at the Toronto airport. The "red-eye special" would get into Toronto around seven in the morning in just enough time to deplane and get checked into the motel in Mimico which was close to the Kilborn office on Park Lane Road and then attend the early morning meeting at nine. The journey back was not as bad, as one could leave for Winnipeg at about 4 pm Toronto time, wait only an hour between the Toronto-Winnipeg and Winnipeg-Saskatoon flights and arrive home about 9 pm Saskatoon time.

The job called for a lot of travelling after moving to Saskatoon. For the first while, there were a few trips to Trail as my boss Frank Goodwin was still located there and hadn't moved to Saskatoon yet. But mostly, the trips were to Toronto for design meetings at the Kilborn office. Also while in Ontario,

Frank Goodwin and I would travel to Dundas, Ontario to visit the John Bertram Plant where the shaft tubbing was being manufactured. Cominco hired the Foraky Company of Nottingham and Bruxelles to act as inspectors for the tubbing. The inspector's job was a very difficult one as he had to reject the iron-castings for no other reason than that they did not sound true when he struck them with his sounding hammer. At the same time he, being all alone, would have the plant foreman and superintendent breathing down his neck demanding that he rescind his verdict when he rejected a particular casting. Also, the fact that he was a foreigner, a long way from his home in Belgium, made it doubly difficult. He really welcomed our visits to the Dundas plant which gave him a chance to unburden his problems which we would talk-over with the plant manager. Bertrams didn't make much of a profit from the Cominco order because the inspector was extremely fussy and rejected a goodly number of the castings, perhaps in the order of 25%.

Travel

Other trips were with Ed Yates who was hired as the mechanical engineer and later became the maintenance superintendent once the operation commenced production. Ed and I were sent on a number of trips into the States to view various types of mining equipment. On one particular trip to Carlsbad New Mexico via El Paso Texas, we happened to run into Bill Schultz who was at the time working for Dave Stark and Associates. (*Dave was formerly vice-president operations for IMC and Bill had worked for me as a mine captain at IMC Esterhazy.*) Anyway, Ed and I, along with Bill Schultz, after landing in El Paso Texas and staying over-night there, headed out in a rented car the next morning for Carlsbad which was about 175 miles away. The purpose of this trip was to view the new 10-foot-high Goodman borer. The type of borers used at Esterhazy were the $7^1/_2$-foot-high machines. Our guide on this trip was Ed Roth, an employee of the Goodman Company, who arranged our visits to the various mines as he had worked in the Carlsbad Area for the Potash Company of America and knew the various mine operators in the district. He arranged trips to the US Borax mine to see the new Goodman borer; the IMC mine to see the newly installed Ramcars; the PCA mine to see the ore handling system at the bottom of the shaft and the extendable conveyor system; and finally to Duval mine to generally see the best operation in the Carlsbad Region. While I visited the IMC mine, Ed Yates visited the Kerr-Magee mine to see their four-rotor borer.

```
Why Mining?
```

All four of us, Ed Roth, Bill Shultz, Ed Yates and I visited the US Borax mine to see the Goodman borer that was on trial there. The idea was that this machine, which was a prototype, would be tested for about a year in Carlsbad. If successful, it would be sent to Allan Saskatchewan where US Borax was developing a potash mine which later became Allan Potash Mines, a consortium of US Borax, Swift Company of Canada and Homestake Mining. Anyway, it was David Porteous who was sent down from Allan Saskatchewan to oversee the testing of this machine. This was the first time I had met Dave; however, I had heard about him before. *(David would later become my best man at my marriage to Irene six years later.)* While underground viewing this new borer, I asked Dave if I could try operating the machine and he said, "Go ahead". I climbed onto the borer and started it up on a double-pass face and proceeded to give it all it could take. Pretty soon the machine started to slew about and it wouldn't go straight. It so happened that when pushing the machine hard into the second-pass, the cob-cutter mounted to cut the lower cusp between rotors had snapped off. It meant stopping the machine and getting the mechanics to reinstall another cutter. Anyway I did get a feel for the machine and for what it could do. Dave, although he didn't say anything, was not amused.

While on the trip to Carlsbad, Nestor Fillip, an old UBC classmate, came over from Los Angeles with a designer from Hughes Bit Company to interest us in a revolutionary design for a potash cutting machine. Nestor, who was a salesman for Hughes, (but prior to joining Hughes, he had worked on boring shafts and tunnels for the US Atomic Energy Commission in Nevada) was familiar with mechanized excavation methods. What his designer friend (*his name escapes me*) was proposing was a bucket wheel excavator capable of being operated automatically and fully adjustable for any height of room within reason. It looked very interesting but we were not able to make a commitment and Hughes was not prepared to develop this machine alone without some financial commitment from Cominco. As a consequence, the good idea was dropped.

The National Mine Service Company (NMS) organized a trip into the coal mining areas in West Virginia where one could see borers as well as their NMS Torcars operating. Again Ed Yates and I made this trip together and visited three coal mining operations. What puzzled us about these visits to the coal mines in West Virginia and Pennsylvania was the lack of change-room facilities at these places. Generally all that was provided for the miners was a hangup area for their hand tools and coveralls, and often times no shower facilities. Many of the men just removed their coveralls and would go home in whatever they wore under them.

A very interesting trip that Ed and I took was one to Northern Michigan visiting the White Pine Copper mine. This mine was a fully trackless mine producing over 25,000 tons of ore per day. The mine employed the room and pillar method of mining. The faces were drill-off with rubber-tired jumbos. This mine used 25-ton teletram trucks (loaded by gathering-arm loaders) to haul to the various conveyor dump-points in the mine. What was fascinating were the conveyor transfer-points. The conveyor system was designed to take run-of-mine ore of about 24-inch chunks. The transfers were not equipped with impact idlers but just a piece of conveyor belting supporting the live conveyor belt. It seemed to work so one cannot argue with success.

On another trip, Ed Yates and I visited the National Mine Service Company's (NMS) Nashville Illinios plant where the company manufactured the Marietta boring machines. After flying to St Louis, we drove from there to Nashville. *(That's Nashville Illinios not Nashville Tennessee.)* I did the driving while Ed did the navigating. We got lost and drove through the middle of East St Louis just one week after a "race riot" there. It was very scary to be the only white faces in the area. One could feel the ill-will and hate that the black people had for the likes of us. On the way back, East St Louis was by-passed by a wide margin by driving to the north. On another trip we visited the same company's Ashland Kentucky plant to see the manufacture of the "NMS Torcar". It was there where I enjoyed my first slice of the famous "Kentucky Pecan Pie".

While visiting the National Mine Service plant in Ashland, Bill Dawes , their vice-president of sales, told us a story about his experience in selling NMS's first torcar that the company had manufactured some years back. Apparently, all that Bill knew about this new product was the load it could carry (10 tons) and price ($50,000). The customer (the president of a coal company) was a tough old bird and wasn't too friendly; but all the same, he did show some interest in the Torcar but questioned Bill as to why it should cost so much. He

Why Mining?

said to Dawes, "...you mean to tell me that if I put five Cadillacs in a row *(Cadillac cars cost only $10,000 at the time.)* and then put your Torcar beside them, that it would be equal?" Dawes said he scratched his head and then gave the only answer he could think of at the time, which was, "Well that all depends if you are hauling cock or hauling coal." He told us that remark broke the ice and he got the order. *(Cominco didn't buy Torcars.)*

On another occasion we toured the Joy Manufacturing plant in Galt Ontario to review Joy's plans for their boring machine. As a sales pitch, the Joy sales people flew a number of us down to Pittsburgh to talk to their design engineers and at the same time tour their manufacturing plant. We, Ed Yates, Bill Schultz and about ten others, met at the Toronto airport and were flown by charter to Erie, Pennsylvania where we cleared customs. Bill Schultz, having been born in the States, was the only American national. When the immigration officer asked Schultz to spell his name, Bill pronounced the "Z" in his name as a "ZED" like Canadians do and not as a "ZEE" sound as an American would. The immigration officer said, "...you've been in Canada too long". It was on this same flight to a small airport outside of Pittsburgh that our plane was buzzed by a small single-engine plane just as we were coming in for a landing. *(That was a very close call.)*

Ed Yates and I, along with Frank Goodman, visited the Goodman Manufacturing plant in Chicago to discuss the Goodman borer. We brought along Darle Dudley (a gear specialist), who had written the text book on gearing. Frank wasn't sure that we should have hired Dudley as he didn't appreciate the problems one could get into regarding gearing. *(I knew from firsthand experience at IMC what one could get into in that regard. At IMC we budgeted one gear box failure at $50,000 a month until the problem of gearing was finally solved.)* Anyway, the Goodman engineers began to give Dudley a bad time until Dudley, in order to prove a point, asked if they had a copy of the gear manufacturers' handbook. They said they had and produced the book. They noticed Dudley's picture was on the fly-cover as he was the author of that handbook. After that the whole atmosphere changed as they were very respectful, where Dudley acted as the schoolmaster and the Goodman engineers the students. He went into theory and answered a number of questions and gave them a lot of free engineering. Frank Goodwin recanted his objections in choosing Dudley as the consultant.

One of the very important trips was the one to Cleveland Ohio to attend the Coal Mining Show. At this show many new ideas and new equipment for the coal industry were on display. As potash is mined in a similar manner as coal, it was natural that we (Ed Yates and I) should travel there to see what the manufacturers had to offer. It was a worthwhile trip as there was everything one could imagine at this show. The outcome of all these trips was that Cominco purchased three National Mine Service Marietta borers, three Goodman loaders, four Jeffery ramcars, two Jarvis-Clark teletram trucks and six Meco conveyors. Besides these, there was a used Joy undercutter and a number of Joy fans were also purchased. This and a few Ford tractors set Cominco up for underground production.

At this point in time, the Cominco travel policy should be mentioned. Everything one wanted to know about travel was written in the Cominco General Practices. Entertainment was a "no-no" unless you had prior permission. It was a must that you stay at CPR Hotels if there was one in town. Frank Goodwin had to get special permission to stay at the motel in Mimico when travelling to Toronto instead of the Royal York. Each city had a specified amount that one could spend on meals. For instance, one could spend $8.50 for meals while in Toronto, $9.20 when in Chicago, $7.50 when in Trail and Saskatoon. Generally one spent over the designated amount and would have to absorb the extra themselves. On one of my trips to a CIM Convention in Quebec City, I had purchased a number of tickets to the various dinners and luncheons before leaving for the convention. On my return, I claimed the full specified amount $7.50 per day for each day away. A junior accountant came to see me about that expense account and reminded me that I was claiming the full amount as well as having previously purchased luncheon and dinner tickets for this affair. I told him to get lost. I never heard any more about it, so probably one can say for this one occurrence in all the time being employed with Cominco, the system was finally beaten.

Construction

At the time, Cominco was sinking its two potash (16-foot and 18.5-foot diameter) shafts, the Duval Company (now Cory) Cominco's nearest neighbour was sinking two 16-foot diameter shafts. The Potash Company of America (PCA) at Patience Lake had its shaft repaired and was back in production. Later on, PCA began sinking a second shaft. Two 16-foot

Why Mining?

diameter shafts were well underway at the US Borax mine at Allan. Noranda (now Central Canada) started two 16-foot diameter shafts shortly after Cominco started its shaft sinking. Also, one 18-foot diameter shaft was underway at the Alwinsal in Lanigan. About a year or so after Cominco started sinking its shafts, Sylvite now PCS Rocanville, began sinking two 16-foot diameter shafts. The only two shafts successfully completed up to this time were the IMC Yarbo 18-foot diameter shaft and the 14(?)-foot diameter PCA shaft. IMC did start a second shaft ($18^1/_2$ foot diameter), called the Cutarm shaft for the K2 mine, but it was only half completed at the time that Cominco was starting to sink its shafts. During the planning of the Cominco shafts, management couldn't make up its mind what diameter to choose so they followed the industry with the 16-foot diameter for No 2 shaft and IMC with the $18^1/_2$ foot diameter for No 1 shaft. In the 1960's there was much activity in Saskatchewan with 14 new shafts being sunk at a cost of approximately $12 to $15 million each. Besides these shaft-sinking costs, there were eight new potash processing plants (Cominco, Duval, US Borax, Noranda, Alwinsal, Sylvite, IMC K2, and Kalium) and a major expansion (IMC-K1), all costing on the average of $50 million or more each. There was a lot of money spent in the 1960's. One could see this everywhere in the province, especially the roads and highways as these were all being paved, and the Gardner Dam was completed forming the Diefenbaker Lake on the South Saskatchewan River System during the 1960's; in fact during this period, the province became a "**HAVE**" province along with B.C., Alberta and Ontario. These were indeed exhilarating times for Saskatchewan.

In constructing the Cominco plant, it was decided to attempt to build the plant using non-union labour. The Webber Construction Company from Yorkton Saskatchewan, which had experience in operating non-union, was finishing the last remnants of the K-2 plant when Cominco was ready to begin construction. This was a natural fit and, as a consequence, Webber got the contract. This was a labour contract where Webber supplied the labour and Cominco purchased the supplies and equipment required for the job. One of the very unique aspects of this job was the fact that Cominco gave a generous transportation allowance of $5 per day which saved the company constructing and operating a camp for the construction personnel. *(Not to brag but I will do it anyway. I was the one who talked management into offering this allowance in lieu of a camp.)* The construction went very well with no work

Cominco

stoppages whatsoever. There were a couple of union attempts to picket and to shut the job site down, but Webber received prior knowledge of this and notified the RCMP who came out in force and prevented any labour strife that the union organizers had in mind.

Another very unique feature of this non-union operation was the personnel man (or rather recruiter) in the name of Greg McDonald. Having worked with Greg at Steep Rock and at IMC, I found him very quick witted and utterly dishonest. He was the ideal type to sort out who should be hired and who should not for a non-union job *(I should say a union-free job)*. He seemed to be able to smell a union sympathizer within minutes of meeting an individual. It was Greg's job to recruit the tradesmen required and find accommodation for them if they were from out of the Saskatoon area. He did an excellent job in doing this as the Cominco construction site was never without the necessary skilled tradesmen. In fact where the other job sites were always short of journeymen electricians and instrumentation men, the Cominco site always had the proper ratio of journeymen to apprentices. *(It goes to show that when labour has the chance to work non-union, they will always choose the non-union site, if given the opportunity.)* Other Webber Construction people who were on site were Gerry Hamilton and Kurt Kunkel. Gerry was Webber's construction engineer while Curt Kunkel was their structural superintendent. Gerry Hamilton had a gift for reading a plan very quickly. I have never seen a man who could just glance at a plan and get so much out of it. It was uncanny what he could see. Gerry's main failing was that he would not follow up, perhaps it was because he had a drinking problem at the time. Curt Kunkel on the other hand was slow and methodical and did a thorough job in supervising the steel erection. He went on later to start up his own very successful construction business.

Webber Construction continued to stay non-union all through the four years of construction. The management at Cominco Potash, seeing the success that Webber Construction had in operating non-union, thought it a good idea to remain non-union also. However, there was one big difference and that was Cominco (Trail and Kimberley) were union organizations and many of the foremen at these operations had come up through the union. Unfortunately, those supervisors that had been seconded from Trail and Kimberley to work at Cominco Potash were strongly in favour of a union and couldn't see why

Why Mining?

an operation should operate without one. Frank Goodwin, manager operations and Bruce Hurdle, general manager, wanted to operate non-union but the mine, mill and maintenance superintendents did not have the same commitment. As a consequence, Cominco Potash was unionized without a vote. Those of us, like myself and Ron Emes the personnel superintendent at that time, who had previous experience in potash at IMC Esterhazy, were considered outsiders in the Cominco family and were held in somewhat contempt. I remember when one of the Montreal head office personnel types came to Cominco Potash in Saskatoon, he questioned every item in the personnel practices handbook that Ron Emes had put together, saying, "Is this another IMC idea". Ron Emes had written a handbook that was very creative and made use of "Theory Y". This man from head office finally scrapped the handbook and instructed the potash operation to use his handbook which was very "Theory X" oriented that was put together by the old guard of Cominco and that was that. Ron quit shortly afterward. However, before Ron left, he and I were able to introduce some very innovative operating practices at Cominco Potash. We borrowed the "Non-Punitive Punishment Procedure" from Cominco Fertilizer of Regina and made that work. Also from the same operation, we introduced the "Positive Evaluation Procedure" and had modest success with that. But the most radical operating procedure at Cominco Potash was the "Skills and Progression System (SPS)" which was much like Ontario's "Modular Training" introduced some ten years later. The SPS equated the mine and mill operators with the tradesmen by giving a number of points for each skill mastered. There was a time function in this system where a workman would have to work the equivalent of four years on these skills, similar to what a tradesmen does. One other plus in this was that when Cominco was unionized, the Steelworkers bought into the system and therefore did not demand a seniority clause in the contract except in the case of layoff.

Startup

Generally when one starts up a new mine or plant, one brings out the big guns and over-staffs the crew with people with considerable experience. Then, when the operation is producing successfully, the excess personnel are withdrawn. Cominco Potash did the very opposite. Cominco started with three superintendents who had never had the responsibility of fulfilling the position as superintendents let alone as general foremen. In other words,

Cominco Potash was going into production with three green superintendents and many inexperienced foremen. Cominco Potash was asking three untried engineers to hire a crew and train that crew to operate a new mine and plant. At the same time, these new superintendents: Dave Kelland, mine superintendent; Tony Banks, mill superintendent and Ed Yates, maintenance superintendent, were asked to operate in a non-union situation and, at the same time, start up a new mine and plant. As stated, Cominco Potash failed miserably and became unionized very quickly, also they failed to meet production schedules. The mine couldn't sustain production, partly because of machine availability and partly because of the lack of good supervision and organization; therefore, Dave Kelland, the mine superintendent, was replaced by Fritz Prugger who was hired because he had previous potash experience. Ed Yates, the maintenance superintendent, quit shortly after Fritz came on board because Fritz demanded all underground maintenance to be placed directly under him as mine superintendent. *(This was what I had been advocating but I was a voice of one and when Fritz Prugger came on board he confirmed what I was recommending.)* In the mill, the housekeeping was shoddy and the recoveries were down, especially in the crystallizer section. There was also a fatality in the product storage building which hurt morale. Finally at this point, Cominco sent its two top mill personnel to get the plant going. The maintenance department was strengthened by sending Clint Albright (ex-New York Ranger), a top maintenance superintendent from Trail, to take charge of surface maintenance, and Mick Henningson, another maintenance man from Pine Point to act as assistant mine superintendent and take charge of underground maintenance. To head up the operations, Ted Fletcher was brought in to act as general superintendent over both the mine and mill. The unfortunate part about this, it was a year too late. These men should have been on site at the commencement of production, not one year afterward when the place was falling down.

During the construction period in the mid-sixties, we would get together with personnel from other mine projects for curling in the winter and golf in the summer. On the site next door at Duval, most of their top brass were from the States, particularly from southern California and New Mexico. In other words, they were from a very warm climate as compared to the cold climate around Saskatoon. I must hand it to those guys, they were game to try anything, even curling with us in the winter. Harry Shively, Duval's mine

Why Mining?

superintendent, when curling, always complained about the cold in the Saskatoon area. One time when we were having coffee after curling, Harry said, "You know what I'm going to do, I'm going to put this curling broom over my shoulder and head "**SOUTH**" and when someone asks what's that funny-looking broom I've got over my shoulder used for, I'll know that I've gone far enough".

As well as these sporting activities, we held information meetings in which we would all compare notes as to the progress of our development and about our personnel policies. Regarding development progress, much of the information presented was of the general nature as no one wished to divulge their company's secrets. *(This was being very naive as there was considerable personnel movement between the mines and therefore it was hard to keep anything exclusive to any one operation.)* Regarding personnel policies, these were quite open. The personnel reps would compare wages and benefit schedules as well as union activities. All management of the new potash operations had hopes of operating non-union but this was not to be. PCA was already unionized with the United Stone Workers *(a very weak union)*, and therefore, were not too active in the discussions; in fact, they were very secretive about everything. *(It was their company policy.)* US Borax (later Allan Potash) was determined to operate non-union but they were the first to be certified by the United Steel Workers. Duval were more realistic; they said they would like to operate non-union but they said they would let the chips fall where they may. As a result they were unionized before they went into operation. The two operations, IMC and Kalium, were non-union because their top management made it a policy that it would place everyone on salary as well as give the employees the same benefits as those in senior management and in this way strive to stay non-union. IMC later on when hiring Bill Huston as its manager, became unionized when Huston invited the Oil and Atomic Workers into the plant. *(Huston was not considered a very honest person. He had to leave Steep Rock Iron Mines for what was considered underhanded dealings and had to leave IMC for much the same reasons.)* Alwinsal was able to stay non-union because top management wanted it so, but perhaps it was through the efforts of one "Greg McDonald", their mine general foreman, who had a way of double-crossing the union and beating them at their own game. McDonald would talk an employee into revoking his union card as fast as the union signed up the employees and then lie about it saying that he had

Cominco

in no way coerced any of the employees. Alwinsal had to let McDonald go eventually as he was an embarrassment to them. Alwinsal Potash was not unionized for quite some time until the Potash Corporation of Saskatchewan bought that operation from the German-French consortium. Cominco tried to stay non-union but the superintendents were inexperienced both in operations and in personnel and, as a consequence, the operation was unionized without a vote. Noranda was also unionized without a vote.

It wasn't very long after operations started at Cominco that the mine experienced heavy ground conditions. The potash beds, at a depth of 3500 feet on the Cominco property, were deeper than any in the Saskatoon area. The beds at Duval, the neighbouring mine to the north, were about 200 feet shallower and therefore that mine did not experience as severe ground conditions as those found at Cominco. The beds at PCA and Allan Potash, being on the same horizon and also much shallower at 3200 feet depth, were less prone to collapse in the short term. All mines in the Saskatoon Region did have some ground conditions eventually when the back would collapse. All five mines, *(Cominco, Duval, PCA, Allan and Noranda)* mined the top ten feet of the "A" potash bed including the three clay bands at the very top of that bed. Above the third clay band were four feet of salt and then a thin parting of clay which caused a weakness in what was originally thought to be a strong beam of salt. The horizontal component of pressure tended to buckle this 4-foot beam of salt above the openings at the thin parting of clay causing the whole roof of the entry to collapse. Once the back *(roof)* of the entry started to show signs of failure there was no way to support the back. Grouted rockbolts were tried but these were only a temporary solution.

Cominco tried single pass entries (which were only 16 feet wide) to no avail. Also, the single pass entries were difficult to excavate the breakthroughs which were necessary for ventilation and travel between entries. Even support props were tried. What had to be done in the short term was to excavate the 4-feet of salt above the entry in order to maintain a strong back. This four feet of loose salt rubble had to be handled in one of three ways, namely: 1) hauled and stowed into an abandoned entry, 2) hoisted to surface, or 3) left there on the original entry floor and a new roadway smoothed and graded over the loose salt. Most of the mines elected for the latter. In any case, mining the entries and then taking down the backs to obtain a stable entry roof was expensive. A better way of mining had to be found.

I suggested that Cominco call Dr. Shoisi Serata of the University of California (Berkley) for assistance. Again, Cominco showed its reluctance but after some convincing on my part and with the support of Fritz Prugger who had at that time replaced Dave Kelland as mine superintendent, Frank Goodman capitulated. Not known to us at the time was the fact that US Borax (Allan Potash) also requested the services of Serata for the same problem. Serata was happy to come to Cominco and gave us hope the first day on site. He said the problem was the horizontal component of the stress which had to be eliminated or minimized. His solution was to drive two parallel entries and allow these to collapse and then drive a permanent entry between the two collapsed entries. He called this the "stress relief" method, and thin pillars on either side of the permanent entry, he called his "magic pillars". Experiments were run on Serata's theory where various rock mechanic instruments were in place recording the results of these trials. Everyone was pleased at how well the trials went and with the results of this method of protecting the main entries. The stress relief method was a success and it made mining at Cominco Potash possible. However, what disturbed me and others at Cominco was a comment that Serata made, saying that because the stress relief method was successful, he thought he should patent the idea. The comment at the time, made by Bruce Hurdle, the general manager, was that a mining method could not be patented. Hurdle was wrong. Cominco finally had to pay a fee to use Serata's stress relief method to mine although they and some of the other mines balked at the idea at the time.

Things were looking up in the summer of 1970, the mine had reached and sustained 5000 tonnes daily production and seemed to be getting closer to full production of 8000 tonnes daily. The mill was improving on its recovery but was somewhat handicapped by having a limited number of compactors (only two). The maintenance underground improved with the addition of Mick Henningson, who answered directly to Fritz Prugger, the mine superintendent. Also hired at this time was Don Barrila who was an expert with boring machines, especially with the hydraulic controls and drives. I might add, the maintenance planning system that was put in place by the former regime was scrapped and a less cumbersome system was put in place. As mentioned, things were looking up.

Cominco

Flooding

Production was curtailed during the first two weeks of July 1970 when all operating crews were sent on an annual summer vacation. Everyone, staff and hourly-rated employees (except for some maintenance personnel) were off for this two-week period. Unknown to me and many others, Fritz Prugger, the mine superintendent called the Cementation Company to come to the site and commence grouting in the No 1 production shaft which was part of their original shaft contract with Cominco. His idea was to grout the No 1 shaft while the crews were away and then move over to the No 2 ventilation shaft once the vacation period was over. This was a good idea except that it was all done by means of a phone call with **"no paper work"** whatsoever to back up what was to be done or how. Fritz did assign a young engineer (Keith Durston) to oversee the operation. Unfortunately, Durston had no shaft experience. *(I guess that's what Fritz wanted to give him with this assignment.)* Anyway, Durston was on the stage with Steve Winters of Cementation when all hell broke loose. The grouting job required drilling short holes, about four or five feet long, through the concrete lining of the shaft and into the wall rock behind where it would be possible to pump neat cement grout into these holes in order to stem the flow of any water seepage into the shaft. This was called "wall drying". The job went very well in No 1 shaft where all precautions were taken; that is, short over-sized diameter holes two feet long were first drilled into the concrete shaft wall and sleeves of pipe were grouted into these holes, then valves were screwed onto the sleeves before any further drilling was done. In this way, if a sudden flow of water was encountered in the wall rock, the drill rod could be removed and the valve closed which would prevent any further inflow of water into the shaft. Because no sudden inflow of water was encountered in No 1 shaft nor was there any in the initial holes at No 2 shaft, Winters decided to forego the placement of sleeves and valves in No 2 shaft as he could see the shaft walls were fairly dry and proceeded to drill into the walls without any protection against an inflow of water.

The decision to forego this safety practice was indeed a bad one. What did happen was that Winters drilled into one of the abandoned freeze pipes (freeze hole #18) that was thought to be perforated and grouted. Apparently this particular pipe was neither perforated nor was it grouted. When Winters drilled into No18 freeze pipe, all kinds of drill mud and fine sediment came

Why Mining?

shooting into the shaft followed by gallons of water under high pressure. The decision made by Frank Goodwin of Cominco (*who was very trusting*) and Jim Black of Cementation (*who wasn't altogether forthright*) not to remove the freeze pipes after the Blairmore Formation was buttoned-up was perhaps an even worse decision than that made by Steve Winters to not install high pressure shut-off valves on all grout holes. It was No 18 freeze hole that channelled water under pressure at 600 psi from the Blairmore Formation down to the Nisku Formation (about 2200 feet below surface) and then through the newly drilled grout hole into the shaft, causing the mine to flood. (*What a pity!*)

It was about noon on this July 1970, Steve Winters and Keith Durston came to my office door with their slickers covered with black drill mud and excitedly said that they had hit a grout hole that was out of control and they feared that it would cause the mine to flood. I was not aware at the time that Cementation was on site as I, like many others, had taken my annual vacation the two weeks prior and had just returned to work that very Monday. The reason Winters and Durston came running to my office was because my office was next to the front door. Apparently they tried to phone Fritz Prugger and then Frank Goodwin but neither of these men could be reached. When hearing what had happened, I went running to find Prugger and Goodwin to notify them of the bad state of affairs and then I went over to No 2 shaft headframe. There was very little that could be done. We tried to manufacture and install a drop shield with an open valve which could be dropped into place and then when the shield was secure, close off the valve and stop the water from flowing into the shaft. For the rest of the day, shaft crews continued to attempt the installation of the drop shield while the underground crews moved the underground equipment into the higher ground underground. The idea was that perhaps some of the equipment might be saved by leaving it in the high areas where an air-bubble would be created in case of flooding thus the equipment would be relatively dry. The force of the water was too much for the shield and we had to accept the fact that the mine would be flooded. About 10 o'clock that evening, the shaft crew came to surface with the news that it was impossible to attempt the installation to the drop shield any longer. I went looking for Prugger and Goodwin again, to tell them the bad news as orders must be given to abandon the mine. I found both of them in the office men's room holding each other and sobbing. Upon seeing this, I retreated to

Cominco

the No 1 shaft headframe and called the mine captain on the mine phone and told him to bring all the underground crew to surface as we are abandoning the mine. *(Yes, it was I who gave the order to abandon the mine.)* I was both surprised and ashamed at the same time. I will say in Frank Goodwin's defence that he was a very honest and mild mannered man who did his best to consider other peoples' views and feelings but, in this situation, he was shattered by this development. On the other hand, Prugger was very forceful but not altogether considerate of others, but in this particular case he seemed to have gone to pieces.

Frank Goodwin 1979

Frank Goodwin, the first manager of Cominco Potash, was a very honest and a trusting person. He managed by consensus as he liked to have everyone involved and committed to a decision. Frank was later promoted to manager of the Kimberley Operations where a few years later he retired and continues living in Kimberley B.C.

With the flooding of the mine which took six days, all kinds of officialdom came to the mine. Our superintendent of administration, Bill Ellis, tried to get Cementation officials to sign a work contract for the grouting in the shaft. Of course they would not sign and were very closed-mouthed about who was responsible for the accident. It was not Ellis's fault that no work contract had been written or signed as he too did not know that Prugger had contacted Cementation to do this work while the mine was down on its annual summer vacation. Many Cominco officials, president and vice-presidents, came to the site as well to see first-hand the catastrophe.

During the time that the mine was flooding, we determined which crews would be needed in the next few months while the mine was being recovered. All the mill crew was laid off except for a few persons who worked in the loadout as we did have quite a bit of potash product in storage which could be shipped. All the mine crew except the shaft and hoist crew were let go. The staff in most cases was also laid off. Some of the senior staff were transferred

Why Mining?

to other Cominco operations. I was asked if I would consider a transfer to Kimberley but turned it down as I thought I could be of use here to rehabilitate the mine. On looking back, I am glad that transfer was rejected.

It was during this time that the remaining staff were assigned various tasks to recover the mine. Prugger and Cementation were assigned to stop the water which, at one time, had flows up to 14000 igpm and to repair the breach in the shaft which had widened from a small $1^{3}/_{4}$ inch diameter drill hole to an enlarged opening measuring 11-inches by 16-inches. In the meantime, I was assigned to plan and procure the necessary equipment and material to pump out the shaft and mine. Having been in a mine-flooding at Steep Rock Iron Mines some years before, I was familiar with what a Byron-Jackson (BJ) submersible pump could accomplish and therefore made enquiries where one could find a number of these pumps. It was a lucky break to find six of these pumps which had a head and capacity of 1000 feet and 1000 usgpm in Cementation company's yard in Brampton, Ontario. These pumps had not been used in years and had been stored on their sides thus damaging their mercury seals. Fortunately the manufacturer had a serviceman available who came on site to refurbish the pumps and restore the seals so that they could be used in the dewatering of the shaft. It was again a lucky break to find a couple of transformers at Fording Coal in Sparwood B.C. that could handle the power requirements. Pipe and couplings were obtained from the oil fields in Alberta. The local fabricators manufactured the tanks used as sumps for the pumps. The closest we came to finding suitable mineral oil with the correct specification for the submersible pumps' motors was that used in the bakery business which was tried and was successful. With everything almost immediately available, the method put in place was ready to start dewatering the shaft.

The plan was to lower a submersible pump (called the lead pump) 100 feet below water surface in the shaft and pump the shaft dry in stages of 100 feet. When this lead pump reached its dynamic head, which in this case was 750 feet because it was pumping saturated brine at 1.25 specific gravity, a vertical tank would be installed along the side of the shaft wall with a BJ pump inside. The lead pump would pump to the tank and the tank pump would pump to the next tank pump and finally to surface. Ultimately there were five tanks with BJ pumps and one BJ lead pump all connected in series. The key to all this was whether a water-proof electrical plug could be made to work in conditions such as these with the shaft dripping with saturated brine. Fortunately, Cominco had on site a chief electrician by the name of Bill Wolfe who came up with an insulating method that could be used. Bill was very helpful in this regard. Another person, by the name of George Rochon, a pipefitter foreman was seconded to the job from Cominco Trail. That man came up with a number of good ideas as to how the pipes and pumps should be connected to achieve the best results. Between the three of us, Bill doing the electrical, George the pipefitting and me doing the engineering and procurement, we came up with a system that worked.

The next chore was to see if the main friction hoist in No 1 shaft could be used to service the activities of lowering the pipe and pump installations in this shaft. No 2 shaft, serviced by a single drum hoist which raised and lowered a stage, would be occupied by the shaft crew whose activities consisted of repairing the breach and monitoring the water flow between the freeze pipe and shaft. What was of concern were the tail ropes for the hoist in No 1 shaft. If the tail ropes had become somewhat tangled as a result of water and debris pouring from the mine into this shaft, it would mean using the rope-changing winch to service the pump and pipe installation activities. Divers were brought to the mine from Vancouver and were lowered down to the shaft bottom 3600 feet below surface to observe the tail ropes and untangle them if necessary. Except for a bit of wood caught in the steel members at the loading pocket, the ropes were relatively free. We were fortunate in this instance as we could safely use the main hoist.

With all equipment in place and a method to be followed for pumping the shaft and mine, it was necessary to wait for the breach to be repaired. What this entailed was back-grouting the breach from surface. First of all the water in the No 2 shaft was allowed to rise to a point where the column of water in the shaft was equal to the pressure of the water source which, in this case, was the Blairmore Formation at 600 psi. When this point was reached, the flow into this shaft was zero. When it was determined that there was no water flowing into the shaft, back-grouting from surface to a location in the wall rock ahead of the breach was started. When it was determined that the grouting had sealed the breach, orders were given to pump 100 feet of the shaft water column. Flow tests were made to determine if the shaft should be

Why Mining?

pumped further. It was touch-and-go until the breach area in the shaft was exposed. At this point, all pumping stopped until the breach from the shaft side could be permanently sealed. Back-grouting continued all during the pumping cycle in order to reinforce the wall rock surrounding the shaft at the location of the breach. Altogether 640 tons of cement were used to grout the area of the breach from both surface and within the shaft in order to fill the void left by the flow of sand and water into the mine. Pumping started initially in mid-August 1970 and was completed in January 1971.

Decisions about water in shafts should not be taken lightly as was seen from the bad experience with No 2 shaft at Cominco Potash at Vanscoy Sashatchewan. The faulty practices which resulted in the Cominco Potash mine flooding are as follows: ●Attempting to seal the freeze hole pipes instead of removing these altogether from the holes. ●Drilling grout holes without installing high pressure shut-off valves. ●Deepening the grout holes to a depth where they intersected the freeze holes. ●Not having any formal contract or paper work to outline the grouting procedure.

The end result to all this was that in January 1971 at a low point in my life both professionally and personally, (*I was going through a divorce*.) Cominco gave me a three month's work guarantee and after that time, employment would be on a month-to-month basis. In other words, a lay-off could result after three months. I was the chairman of the local branch of the CIM and, because of that, Cominco allowed me to attend the AGM of CIM held that year in Quebec City. At the convention I made a few contacts. Prior to returning to Saskatoon, stops were made in Toronto where interviews with a number of firms were made. All in all, a total of 116 letters of application were written in order to find suitable employment. Trips were made to Unity Saskatchewan, Sparwood B.C., Vancouver, Calgary, Winnipeg, Timmins and Toronto interviewing for various jobs.

The Cominco Potash Logo Adopted in 1965

Chapter 8

Ontario Again

Ecstall

At the time when joining Ecstall Mining Ltd, a subsidiary of Texas Gulf Sulphur Company, I was separated and going through a divorce. My intention was to strive for a position close to Saskatoon where it would be easy to visit my children (Karen and Robbie) and have them visit me frequently. This did not pan out as the best job available at the time was the position of chief mine engineer with Ecstall Mining Ltd. It was indeed a comedown at the time, but as it turned out later, it was all for the best. *(Ecstall Mining afterward had a name change to Texasgulf Canada Limited and then when Texasgulf was taken over by the Canadian Development Corporation, the name was again changed to Kidd Creek Mines Limited.)*

The interview at Timmins was first with Elmer Borneman, who was the personnel assistant at Ecstall. Elmer seemed like a decent and likeable chap and in time proved my first impressions to be correct. His boss, Don Grenville, the employee relations superintendent, was anything but friendly and made me ill at ease. The third person on the interview routine on this visit to Timmins was Alan Perry, the superintendent of engineering and technical services. *(Alan had a very pretty secretary which I could not help noticing but more about that later.)* I would describe Alan Perry as, "An Aristocratic English Gentleman". He gave me the impression that he didn't know much about underground mining and wanted someone at the minesite who was versed in the technology of underground mining engineering. The last person on the interview list was Merle Marshall, the assistant superintendent of engineering and technical services, who was Alan Perry's assistant and would be my immediate supervisor. Having met Marshall before at Manitouwadge on my travels for Rockiron when he was with Geco Mines, I wasn't impressed with him when I met him at Geco and I was a bit apprehensive during the time of this interview. Marshall, as part of the interview, toured me around the metallurgical site and drove me out to the minesite where I would be working. *(The mine and metallurgical sites were separate sites some 35 miles by road and 17 miles by rail. Ecstall hauled ore from the mine to the concentrator over its own 17-mile railway.)* At the time of this tour, the pit was down to the fifth bench and, at the south end of the pit, they had just uncovered a very high grade section of the orebody that would impress

Why Mining?

anyone and probably had as much to do as anything for me making the decision to join Ecstall Mining.

One thing that gave me a clue regarding Marshall's penny-pinching was when he drove me back to the Empire Hotel where I was staying and he suggested a drink in the bar and had me sign for it as it would be coming off my expense account for the interview. *(Marshall must have got a number of free drinks that way as Ecstall was hiring a goodly number of engineers.)* Anyway, we'll talk more about Marshall later too. Marshall did offer me the job of chief mine engineer on the spot but I said that it couldn't be considered at this time until I could see the offer in writing. Anyhow, he said he would get onto it first thing the next day. After flying to Toronto the following day and interviewing with a number of other firms, I then flew on to Saskatoon. Sometime later, perhaps three weeks, Elmer Borneman phoned saying that a letter would be forthcoming from Ecstall with an offer. He said they had met my price which was $25 less than what they were paying Marshall who was to be my boss. To say the least, the offer was accepted and I was on my way to Timmins one month later and joined the company on May 24, 1971.

Elmer Borneman the personnel assistant, signed me up for all the various benefits and gave me the usual company pamphlets. Ray Clarke, the mine manager, came into Elmer's office during this time. He was very friendly and very low key. One had to like him from the beginning. Elmer then took me down to the engineering section but, before going in to see Merle Marshall, he introduced me to Alan Perry's attractive secretary, Irene Clark, whom I had noticed when interviewing some two months earlier. Irene, as well as being secretary to Alan Perry, was also in charge of other stenographer/clerks in the engineering section. *(Irene and I were married some sixteen months later and Dave Porteous, whom I met in Carlesbad New Mexico six years before, was my best man.)*

The mine engineering office *(digs is a better name),* composed of a temporary building and two trailers, was where the mine engineering group worked for a year until the new and permanent office and dry complex was built. In my trailer, there were four engineers as well as myself. One of these engineers was John Matousek (a Czech) who was the ventilation engineer; another was Alex Sarnavka (also a Czech) who was designing the pumping system; a young Roy Young (a Haileybury and South Dakota Tech grad) was designing the crushing facility and Rod Doran (also a Haileybury and South Dakota Tech grad) was investigating stoping layouts. The adjacent trailer housed the geology section which at first was composed of one PhD chief geologist (whose name escapes me) whom I terminated a year later, Angelo Matulich who later became the chief geologist; Paul Simunovic, a technician, and Ron Brouliard a probationary technician whom I had to terminate one month after I arrived on site. Later that year two other geologists, Hans Wagenknecht for core-logging and Allen Amos to act as the right-hand man for Angelo Matulich; and a technician, Ron Cook (to replace the technician that was terminated) were hired. *(It was strange how everyone buckled down and worked hard after the probationary technician was terminated. I guess they thought I was a bit hard-nosed.)*

The temporary building housed the mine planning, surveying, and grade control sections. The mine planning was divided into two groups; namely, the open pit production and underground development planning. Tom Dowe headed up the open pit production planning as well as the grade control sections. He did a very good job at both, even though he was not a graduate engineer. He did a much better job in mine planning than a lot of engineers that I had met in my career. Tom went on to head up the underground production planning group when the underground mine went into production. The open pit planning entailed laying out the size of the blasts, both for ore and waste, and then seeing that the right balance between "A ore" (copper-zinc) and "C ore" (silver-zinc-lead) were maintained each day. Dowe's grade control section, whom he headed up along with his right hand man Ed Vukovich, was very accurate at predicting the grades of both ores on a daily basis. Ed, while being a supervisor, was very frustrated because Don Grenville, the employee relations superintendent, would not approve his designation as an exempt employee and therefore Ed was not eligible to attend the monthly supervisor meetings at the Empire Hotel (and later at the Porcupine Inn) where the supervisors were served free beer and a wonderful buffet dinner. It took two years (with the help of Bart Thomson, the manager of mining) to convince Don Grenville to have Ed's classification changed.

As mentioned Tom Dowe supervised the open pit and underground production group. A few of the names of employees who worked there were

Why Mining?

Red Sage, Doug MacPhee, Doug Manning, Wayne Barabas and Gerry Charon all graduates of that very good college, Haileybury School of Mines.

Steve Bizyk, a graduate from Haileybury and Michigan Tech, headed up the underground development planning. Steve was a strange individual. He was very clever but he lacked inspiration. He could not visualize how or what was needed. Steve loved statistics. If you gave him a number he could do all kinds of things with it but first, **YOU** had to give him the number. I remember giving Steve the task as project engineer in planning the second shaft and mine at Texasgulf-Kidd Creek where he was completely lost as to where to start. I suggested to him that he start with the yard on surface and determine the drainage pattern. After a full morning of fiddling around, Steve came to me and asked what elevation should he make the yard. I suggested that it should be a foot or two below the shaft collar elevation which he thought was a good idea. His next question was, "...what elevation should I make the shaft collar"? I asked him, "...what elevation was No 1 shaft collar"? He said, "...it was 1320 feet". I said,"...why not make it the same." You had to lead Steve by the hand on everything. He just couldn't come up with anything original, but turn him loose on a set of plans where he had unit costs for the various items, he would come up with a very accurate estimate as to what it would cost. The last time I saw Steve in 1997, he was working for Kvaerner Metals in Toronto as a cost estimator and statistics keeper where he was doing excellent work.

Harry Coott, the chief surveyor, headed the survey section of four surveyors and four helpers. Harry was as bald as a cue-ball and was very conscious about his baldness to the extent that he wore an outlandish wig. Harry was in charge of the South Porcupine kids' hockey in the winter and the men's fastball in the summer to a point where his job at the mine was almost a sideline. For the first part of my tenure at Ecstall, I more or less had to organize the survey department as Harry was too busy with his extra-curricular activities to do anything new at work. He did take an interest in ventilation dust surveys and wrote good reports on these surveys. I wanted him to write a survey manual similar to the Steep Rock survey manual. Finally, I gave him a copy of a similar manual from Kerr Addison mine which he could follow as a guide. After much prodding, Harry's survey manual was written and copied. Harry was very good with the English language and he

Ontario Again

did quite an acceptable job on his manual. The next project for Harry was a map of all the pipelines on the property. That job was never quite completed under Harry's tenure. I finally moved Harry out of the survey section and into the ventilation section as he seemed to be more interested in ventilation and dust surveys than mine surveying.

There were four surveyors and survey assistants in mine engineering. The surveyors were: Wayne Totten, Art Mick, Bob Taylor and Corkie Hall. The assistants were Chip Merritt, Don Grenville Jr, Bob Monaghan and Otto Vaclavek. Andy Gallagher, Keith Gage and Mark West came on board sometime later as survey assistants. *(The reason I remember Otto's name so well is because he and I would go fishing together each summer.)* Wayne Totten took over from Harry Coott, when Harry was transferred to the ventilation section. Wayne did an exceptional job in organizing the survey section and he completed the map of the pipelines and other installations about the property with no prompting. Unfortunately, Wayne died suddenly of a heart attack brought on through stress from over-extending himself financially in purchasing a good amount of Hemlo gold stock. His widow, it was rumoured, was anything but faithful to Wayne while he was living, benefited from his speculation. *(Sometimes there is no justice.)*

Art Mick was one of the surveyors working in mine engineering. He was finally sent to the metsite and promoted to a construction supervisor. However, while at the minesite, he worked as a surveyor. One time when I was giving out the annual raises to everyone, Mick's raise was $25 per month which I thought was pretty good at the time; however, Totten's and Taylor's were $35 and $30 respectively. I called Art Mick in first as his raise was somewhat smaller than Totten's and Taylor's. He left quite happy and then later I called each of the other two surveyors in separately and gave them theirs. Soon afterward there was a knock on the door and it was Art Mick quite upset that his raise was not as much as the others yet only minutes before he left my office thrilled at receiving a $25 per month raise. *(Everything in this world is relative.)*

Bob Taylor was probably the smartest man *(and very sarcastic)* who worked in mine engineering at the minesite. It was probably because he was frustrated in his job as he had the brains to accomplish much more if he had had more

Why Mining?

education. He definitely should have been an engineer. Bob ended up in stope planning where, later, he finally headed up that section. As for Corkie Hall, I didn't get to know him too well as he was the construction surveyor at the metallurgical site and I didn't get down there too often.

An interesting group was the drafting section who were also housed in the temporary buildings. At this early stage, there were only three draftsmen but later, when Texasgulf was designing the infrastructure and facilities for the No 2 mine, there were over 30 draftsmen, mostly contracted from Cambrian and Kilborn Engineering. At this early stage, there was Ed Spehar (a Croatian refugee) who was then the chief draftsman and later was designated as the model-maker and photographer. *(Ed was the photographer at our wedding.)* Ed was an artist and a temperamental one too, but an all-around good guy. He was very good at his work to a point where people took advantage of him as he never said "no" to any request. Even head office in New York would request things from Ed, where I'm sure there were good people in New York who could easily have done what was requested. Working with Ed were Tony Defelice and Seppo Sorsa. Tony left Timmins about a year after I arrived as he wanted to get away from the very strict parental control that his Italian parents imposed upon him. The Seppo Sorsa saga is a sad story. Seppo was very talented and was the key man in designing the new shaft steel, headframe floor layouts and other key No 2 mine facilities. Unfortunately, when he was promoted to chief draftsman he went all-to-pieces as the responsibility was more than he could stand. He had a nervous breakdown and committed suicide by walking in front of a transport truck on the highway that runs past the psychiatric hospital at North Bay, Ontario.

The construction at the minesite was originally handled from the metallurgical site some 35 miles away via road by my boss, Merle Marshall. Merle had as his own construction engineer, Leon Laforest, who was a first-class civil engineer. Under Leon was a mine technician by the name of Ray Pigeon and a student, Phil Donaldson who, years later, became manager of mine services. Also, supervising the construction, (particularly the powerhouse and hoisting installation), was a consulting electrical engineer by the name of Ralph Adams. I knew Ralph from Esterhazy and Saskatoon as Ralph worked for me temporarily as a mine captain at the K-1 mine. After much lobbying on my part, Merle finally put all the minesite construction under me. Leon Laforest didn't like this too much but times do change. Regarding Ralph Adams, he was a strong silent type and didn't believe in putting anything down on paper for fear of losing his job as a consultant. He wanted to be indispensable and, as a consequence, no field changes were recorded on paper as they were all in Ralph's head. I finally had to let Ralph go and obtain from the metsite, Vaclav Vicslo, a Czech electrical engineer, to revise the electrical drawings showing the field changes made by Ralph Adams. Vicslo did an exceptional job but was never given the credit he deserved for the work he did at the minesite for the simple reason ...he was an immigrant.

Texasgulf

At about this time the move was made into the new office and dry complex, the company had a name change from Ecstall Mining Limited to Texasgulf Canada Limited. Also at the same time, it was possible to get all the mine engineering group under one roof. The industrial engineering group, (under Bill Yanisiew), which was originally at the metsite came under mine engineering when moved to the minesite. Yanisiew had worked for me before at Steep Rock Iron Mines as a surveyor. He was an excellent surveyor but a very unproductive industrial engineer as he never achieved anything worthwhile. On an old Wang desk-top computer, he would turn out all kinds of efficiency and regression graphs that no one used nor gave a second look. Bill was confused... he thought he was an operations research engineer but was designated as an industrial engineer. As a result, he was neither. The three people who worked for him, Rick Pyror, Gil Reashore and Jerome McManus more or less did their own thing. I finally moved these people under Roger Harris, the operations research engineer, and left Bill Yanisiew on his own. Roger Harris was a man ten years ahead of his time. He designed and inaugurated computer stope design and drafting at Texasgulf-Kidd Creek ten years before anything like it was on the market. He had a lot of opposition to his vision. Fortunately he had people like Ray Clarke, manager, and Dick Mollison, senior vice-president, who were open-minded and could see the possibilities of Roger's vision. Roger Harris later ended up as a vice-president and operations manger of Texasgulf's trona mine in Wyoming. He now resides in London, England.

After moving into the new offices, it meant consolidating a number of activities and hiring some more people. The mining layouts of the new

Why Mining?

underground mine had to be firmed up and a system of designing the stopes and the blasting that goes along with the design had to be started. Fred Wrona, who was also a Haileybury grad, was hired. I would say he was probably the best employee I ever hired. He and a young technician by the name of Norm Fournier did a tremendous job in setting standards and procedures on the stope blasting layouts. Fred was then given the task of determining the method of stope backfilling. He determined the design mix of the consolidated fill as well as the flow-sheet starting from where the coarse rock was excavated at the mine dump to the actual filling of the stope. It was he who determined the rough specifications of the crushing and screening plant for the preparation of the backfill. Fred then went on to operations, where he was the mine captain during the early development of No 2 mine. Later he headed up the training section for both No 1 and 2 mines. In this position, he really excelled. He revised **all** of the existing training manuals and wrote new ones where required. After I left the Kidd Creek Timmins operation for a head office position in Toronto, all those who had worked for me or showed me any loyalty at No 2 mine were squeezed out by Eric Belford, manager of mining. Unfortunately, Fred Wrona, for all he did for Kidd Creek, was one of those terminated.

It should be mentioned at this time, Dave Porteous was hired as my assistant. It was the same Dave Porteous of Carlsbad, New Mexico in 1966 where I damaged the bottom cusp-cutter of the Goodman prototype boring machine that he was testing. Anyway, Dave must have put that all behind him as he and I had a good working relationship. This was proven as Dave acted as my best man when Irene *(Alan Perry's attractive secretary)* and I were married in 1972.

Maybe this would be a good time to discuss my relationship with my bosses. Bart Thomson was the mine superintendent when I was hired at Ecstall Mining but was later promoted, first, to manager of mining and then to general manager when Ray Clarke was moved to Toronto as president. Bart seemed to take an immediate dislike towards me perhaps because I fired his best friend's son, Ray Brouliard. Anyway I greatly respected Thomson, especially for his knowledge of mining. He set the standards for underground mining at Kidd Creek where one would have to go a long way to see their equal anywhere, except perhaps the Kiruna LKAB mine in Sweden. Thomson

Ontario Again

was very stubborn to a fault. He believed in rotary drilling even though Inco proved that I-H-D hammers could out-perform the Robbins mini-borers used at Kidd Creek. He spent $1 million (plus a lot of capital on revisions) on four mini-borers requiring two operators, where the same amount of money could have purchased ten drills requiring only one miner to operate each which would out perform the mini-borers by a factor of two. Thomson also played favourites. In his eyes, Clarence Amyott, who was hired from Cementation Company as assistant superintendent of No 1 mine and later promoted to No 2 mine superintendent could do no wrong. The fact was that everyone, except Thomson, thought Amyott was rather irrational.

Now, Eric Belford was something else. I first met Belford and his wife Margaret in Saskatoon at a party at Ernie Isaac's home. He was, at the time, chief mine engineer supervising two surveyors and a draftsman at Potash Company of America. This was hardly a very significant position. Anyway somehow Eric impressed Bart Thomson when he came to Timmins and that's all that mattered. Eric said to me, one time when he was manager of mining, and I, being superintendent of No 2 mine at the time and reporting to him, "...Dave if the truth be known I should be reporting to you." From that time on, our relationship was not quite the same. One time when MacIsaac Mining, through neglect, damaged an electrical cable they were lowering into No 2 shaft, Belford asked me what I was going to do about it. I mentioned that Clause 17 of the contract would be exercised and a letter would be written requesting compensation. Three days later, Joe MacIsaac was in to see Belford asking that they not be held responsible for the damage. Belford called me into his office and, in front of Joe MacIsaac, had me explain my action. Then again in front of Joe MacIsaac, he said, "...well Joe I can see how it could happen and therefore we'll take responsibility for the damaged cable". *(Guess how much control I had after that with MacIsaac Mining?)* After that incident, I always thought of Belford as the **"little prxxx"**!

Merle Marshall was the complete opposite. Merle was honest with me and accommodated me in my requests. He was a slob but a nice slob. Elmer Borneman would joke about Merle wearing this stock-car racing tie as it was all souped up. One time in the geology office, Merle decided he would lunch with the geologists. His lunch consisted of a slice of bread and a can of sardines. He placed the bread down on one of the sandpaper blocks used for

Why Mining?

sharpening the points of pencils. Then he proceeded to open the can of sardines and sop up the sardine oil with the bread which was black on the underside from the pencil graphite. The sardine oil, which was now black, drooled from both sides of his mouth where he looked like he had a "Foo Man Chou" moustache. The geologists had a hard time to keep from laughing. The guy was stingy with his money but very generous with the company's money as he liked the expense account living. Merle had a large ego and would pick peoples' brains and use the information for himself. He had a talent of surrounding himself with very competent people. He also knew how to get the best out of people where others would rather give up on a slower-performing individual. A case in point was when the Zinc Plant gave up on Tony Hill as a production engineer, Merle took him under his wing and made him the project engineer of the Concentrator Expansion. Tony did a superlative job there and went on to become operations manager of St Lawrence Cement Company. Bob Skrobica was another one. The maintenance department gave up on Bob but Merle put Skrobica into supervising construction where he did an excellent job. Merle was a compassionate person where he made a job for Ken Stubbs who had MS and could hardly stand up. You have to admire him for that. One of the funniest things I saw Merle do was when all the managers were sitting up front on a dais at one of our awards dinners with all the brass from New York in attendance, Merle who was at the very end of the head table moved his chair too far near the edge, where it fell off the dais and Merle took a complete somersault onto the floor.

Ray Clarke was the manager when I first arrived in Timmins. He later became president when CDC took over the Canadian assets of Texasgulf and renamed the company Kidd Creek Mines Limited. Ray was a wonderful person. Everyone liked him, I guess because the man was so friendly and very approachable. He would come out to the minesite every Wednesday for the full day and for the morning on Fridays. On these visits, he would make his rounds to everyone's office and generally socialize. When he left one's office you felt really important and proud to work at the Kidd Creek mine. Had it not been for Ray Clarke, I would not have been made superintendent of No 2 mine nor vice-president and director of Kidd Creek Potash Company.

Another great guy was Dick Mollison. I first met Dick while I was living at

Ontario Again

the Senator Hotel waiting for my furniture to arrive from Saskatoon. Mollison approached me one morning and said he had seen me the previous morning and wondered if I was just passing through Timmins or was I moving there. I told him I had just accepted a job at Ecstall Mining. He told me he was senior vice-president and then proceeded to welcome me into the company. My first impression was, "...is this guy for real?" He seemed so genuinely sincere. It was true, the man was very sincere and very fair with everyone. When he discussed safety, he meant every word. Once when Ed Gonce, who was the first superintendent of the Zinc plant, offhandedly answered a concern of one of the foremen at the monthly supervisors meeting, said that one didn't have to worry about drowning in the sump in the basement of the zinc plant as the specific gravity of the acid there was greater than that of a human body and therefore one would float. Dick Mollison took offence at this lack of safety and compassion on the part of a senior and fired Gonce the next day.

Over the years with Texasgulf-Kidd Creek, I enjoyed a number of trips as I did with Cominco. One of my first trips was to Las Vegas to attend the AMC Equipment Show which was a tremendous display of hardrock mining equipment similar to that of the Coal Show that was attended in Cleveland some years before, except that this was for hardrock while the coal show featured coal (*or soft rock*) equipment. Bart Thomson was there at the same time and he and I did some socializing together. We were invited to breakfast at Caesar's Palace with Jack Clark of Jarvis-Clark Equipment where, at the last minute, Clark weaselled out of paying for the breakfast. That hurt the Jarvis-Clark Equipment Company as Thomson wouldn't have any of its equipment on the property. He even went so far as to demand some years later, that I return a 5-yard Jarco-Scoop that had been delivered for No 2 mine. *(Thomson had a long memory.)*

On two other occasions, with Thomson, were trips to New York, where we flew to Toronto and over-nighted there, and proceeded to New York the next day and back to Toronto the same day. These trips were to offer suggestions on the new trona mine Texasgulf was planning in Wyoming. The reason we stayed over-night in Toronto was because Thomson liked to party. I, of course was a willing partner. I seriously believe Thomson wanted to get to know me and perhaps wanted to get to like me as well. Anyhow, it didn't

Why Mining?

work. I must have failed somewhere along the way as Thomson never did have much faith in me.

One trip to Sudbury was made with Thomson, Clarence Amyott and Tom Dowe where we visited two Inco mines. The idea was to observe the Inco method of cut and fill and underhand cut and fill mining methods. We spent one night and two days in Sudbury. The night was one big party at the Sheraton Hotel. The next day, Inco showed us their drilling results using a one-man "In-the-Hole" drill which Thomson wouldn't believe nor would admit that the In-the-Hole drills could outperform his two-man rotary mini-borer drills. He was stubborn and wouldn't admit defeat. Anyway on the way back from Sudbury to Timmins, he insisted on stopping at every little town that had a bar and having a drink; in fact, he wanted us to stay overnight in Gogama. I was the only one to voice my objections, whereas, Amyott was all for it and Dowe was not saying one word for or against it. I guess that is why Thomson didn't particularly like me as I was not afraid to state my opinion. *(When I look back, I think I could have gone much farther in my career if I had been less vocal regarding my opinions. There is a time to be silent.)* Because I wanted to get home that night, Thomson handed me the wheel for the rest of the way home so that he could drink with the other two guys.

The very first trip made after joining the company was with Merle Marshall to New York and to Princeton NJ to observe a computer graphics demonstration. When Merle and I arrived in New York City late at night, Merle had forgotten the itinerary sheet made out for us as well as forgetting at which hotel we were to stay that evening. After taxiing around Manhattan for an hour we finally found a hotel next to Central Park that would put us up. Our rooms that Irene Clark *(the engineering secretary and my future wife)* had arranged for us were guaranteed payment and therefore our lodging that night cost us double. As I said previously, Merle was very generous with the company's money.

Somehow, I think, by phoning head office in New York City, we caught up with Roger Harris who had put together the computer graphics demonstration that we were to witness at Princeton University in Princeton, New Jersey. Roger wheeled around to our hotel *(he had rented a car and knew the city well enough to chauffeur us about)* and drove us out to Princeton University.

Ontario Again

Later that morning, Dick Mollison and Ray Clarke flew in by company plane and joined us to view the demonstration put on by Howard Olpen and Bill Stiel from Salt Lake City, Utah on the Princeton University computer. *(This was in 1972 and the computer at Princeton University was one of the only computers at that time to have the capability of showing 3-D in real-time.)* The demonstration was a success as it showed six diamond drill holes that could be rotated about and viewed from various angles. I was somewhat sceptical but could see some possibility in using this tool especially for plotting diamond drill holes which at that time were plotted by hand and often inaccurately. The outcome of all this was an AFE for $750,000 of which $500,000 went towards a DEC mainframe computer and the balance to Olpen and Stiel for software programming. Marshall and I flew back to Toronto via the company plane with Mollison and Clarke, leaving Roger Harris with Olpen and Stiel to further update their computer programme. At Toronto we disembarked, and Clarke, Marshall and I returned to Timmins via Air Canada leaving Dick Mollison in Toronto. It was an interesting trip all around.

Texasgulf Awards Dinner

The Author at an Awards Dinner is shown receiving his 10-Year Award Shares shaking hands with Dick Mollison (Vice-President) with Ray Clarke (General Manager) looking on at the left and Dr Chuck Fogarty (President) on the right.

Texasgulf awarded their employees very generously in Safety and for Long Service. The company's officers would always be in attendance at the various company functions such as annual company picnic and Christmas parties.

Why Mining?

No 2 Mine

In the autumn of 1973, Texasgulf Mine Engineering *(my group)* was given the job of doing a preliminary feasibility study in designing the infrastructure and materials handling system for a second underground mine at Kidd Creek. The operating group at No 1 mine was given the task of outlining the mining method and the equipment needed for this project. The two estimates were melded, where the coalition came up with a reasonable figure to present to corporate management in New York. These estimates were accepted and the go-ahead was given to begin the design in earnest on this new mine complex. Texasgulf Mine Engineering would design this new mine, which was given the very original name "No 2 mine", with its own resources and act as its own prime contractor. This meant that my department would design the shaft and station development, headframe and hoisting plant, powerhouse addition, electrical distribution, and office and change-house facilities. The total project to bring this new mine into production was estimated at $125 million and I am proud to say we came in on budget.

Clarence Amyott was immediately promoted to Superintendent of No 2 mine. It was my duty as Assistant Superintendent of Engineering and Technical Services to supply Amyott with the necessary drawings and specifications to complete the project. What this amounted to was that I would be responsible for the design and the surface construction and Amyott would be responsible for all the underground work and would have to approve the design and the surface construction. It was not to my liking as I felt that having done all the conceptual study and project design, I should have been made Superintendent of No 2 mine. Of course, Thomson didn't see it that way.

The shaft site was chosen after the drilling showed that a good bedrock ledge existed 65 feet below surface at a distance of 350 feet west of No 1 shaft. The 65 feet of overburden consisted of muskeg, till, brown clay and 40 feet of grey varved clay which had little or no strength; that is, it could not be excavated without endangering the adjacent buildings such as the existing powerhouse and office and change-room complex. After studying a number of methods of excavation and support, it was decided that freezing was the way to approach this problem. At the same time that all this was going on, the group was determining shaft and hoist design. Gerry Tiley, who had just started a mine hoisting consulting firm, was hired as a consultant to critique the shaft layout and hoist design. Gerry talked us into increasing the shaft size from 24 to 25 feet. We also used Gerry to help in choosing the four hoists that were purchased. Texasgulf was Gerry Tiley's first client.

Drilling Freeze Holes for Excavating the Unstable Till at the Collar of the Texasgulf No 2 Mine Shaft April 1974

Slip-Forming No 2 Headframe

No 2 Mine September 1974

No 1 Mine Headframe is the Large Structure on the Right

Photos by E Spehar

Why Mining?

The headframe, hoists, hoisting conveyances and ropes were all chosen in a very unique way. We chose manufacturers whom we believed were capable of supplying quality products at a reasonable price. Normally, one goes out for quotations on things like this but it was thought it would be best to negotiate with those suppliers who, in our mind, offered superior equipment as Texasgulf had explored the waters, so to speak, regarding quality and price prior to selecting the final suppliers. The firm of V B Cook Company was chosen for the headframe design; Canadian General Electric for the hoists; Dorr Oliver Long for hoisting conveyances; Wire Rope Industries for hoisting rope and attachments and finally Gerald Tiley & Associates to act as the hoisting consultant. All participants were invited to the minesite office where all would meet together in the boardroom to iron out the final details for the hoisting plant. At these meetings, which carried on over two days, the participants went through a number of hoisting combinations, where a small change in the size of any one of the four hoists would cause changes to be made in the shaft configuration which would affect the floor plan of the headframe as well as the size of one of the conveyances or diameter of rope. By having all parties there in the same room, there was instant reaction to any change. By having Gerry Tiley at the same meeting, it gave Texasgulf the assurance that it was getting the best combination of all components involved. The Texasgulf Purchasing Department didn't like the way this was handled as they said we didn't go out for quotations on these components. They were wrong as Texasgulf Engineering did negotiate with all suppliers before settling with the ones chosen as we were looking for a quality product, not only price. As far as price goes, there was no doubt Texasgulf did get a firm commitment which was probably as good, if not better, than if Purchasing had gone out for formal quotations.

Engineering was pretty well left alone on the collar excavation and headframe construction as Amyott was busy supervising the drifting from No 1 mine levels over to the underground location of No 2 shaft on each level horizon (200, 800, 1200, 1600, 2000, 2400 and 2800 feet) below surface. It was planned that the first stage of shaft excavation (surface to 2800 level horizon) all muck handling would be done through No 1 shaft. The plan was to drill 6-foot bore holes from the underground location of No 2 shaft at each of those levels listed and slash these bore holes to full size of the shaft which had an excavated rock diameter of 27 feet and finished inside diameter of 25 feet

Ontario Again

when the concrete was poured. This work took from mid-1974 to mid-1975 to complete.

In the meantime, under Engineering's direction, the overburden at the collar was frozen and excavated and the 210-foot concrete headframe was slip-formed. Regarding the headframe, it was the engineering crew who drew the layouts of the floor plans and turned these over to V.B.Cook Company Ltd for final design and detailing. Shortly after getting the contract to design and detail the headframe, Jim Cook came into my office with his preliminary design drawings. *(This was before our meeting of all hoisting plant suppliers.)* At a glance one could see that his group had not followed the floor layouts given them and had drawn the headframe well over 250 feet high. I told Jim that Texasgulf wouldn't even look at his design or any design that was over 200 feet as I knew Engineering couldn't sell a 250-foot high headframe to management. Of course he was upset and left. Finally there was a compromise on a height of 210 feet which I knew could be sold. Jim said a couple of years later that "..you sure know how to hurt a guy, Dave". I know that, had I not been firm with V B Cook Company, Texasgulf would have ended up with a very costly headframe with much wasted space just like the recently built headframe located at Brunswick Copper. What Jim Cook was trying to sell Texasgulf was the identical design as the Brunswick Copper headframe in New Brunswick saving some engineering time and expense for Texasgulf-Kidd Creek. In my mind the Brunswick headframe was and still is a "dog" and would be completely unacceptable. After the Kidd Creek headframe was built and had been equipped with all the hoists and electrical gear in place, Jim Cook said to me that this headframe was not one bit congested and in fact it was an excellent design. Of course he was in a way congratulating himself, but all the same he used Texasgulf's floor designs, and because we were adamant about the height that Texasgulf-Kidd Creek received a commendable headframe.

The collar freezing didn't go as well as expected. Texasgulf hired Foraky Nottingham of England as its freezing consultants but the persons they put on site were not that much help once the ground was frozen and excavation was underway. Leo Allarie & Sons of Timmins were given the contract for the excavation of the overburden which went well for the first 20 feet and then the walls began to cave-in. The problem was, the correct insulation was not

Why Mining?

used for the walls to maintain their frozen condition. To make matters worse, it rained a lot that June as well. Styrafoam sheets and hay were tried as an insulation medium to no avail. At one point it was thought that the collar excavation should be postponed until winter as we couldn't come up with anything that would preserve the frozen walls. Luck was with us when Alfie Norkum of Gorf Contracting, happened by and said he had the perfect material to save the walls. Alfie was about to build a new shop in South Porcupine and had these insulating blankets which he was planning to use to insulate the walls of his shop. He was given a purchase order on the spot to take charge of the insulating of the walls while Allarie would continue on with the collar excavation. Alfie Norkum of Gorf Contracting saved Texasgulf's skin by directing his crew to build small sheds all around the collar, covering the freeze manifolds and draping dark plastic sheets over the insulating blankets to keep the sun and rain from damaging the insulating blankets. *(Alfie Norkum was one of the most ingenious construction men I have ever met in my career.)* After getting Gorf Contracting on site, the project went very well.

After the collar was excavated preparations were made to slip-form the headframe. In order to save time and square up the bottom ledge of rock, a mass-pour of concrete was made. By doing this, it got the project back on schedule as it had lost three weeks in the collar excavation freezing problem. The rest of the headframe construction went like clockwork. The slip-forming took a month to complete and then the fixed steel was installed and finally the concrete floors were then poured. Once the first floor about the collar was poured, the headframe construction was isolated from the shaft sinking and therefore shaft sinking and headframe construction could proceed simultaneously.

At the time the headframe floors were being poured, MacIsaac Mining was busy setting up their sinking and galloway stage hoists. It should be mentioned at this time that MacIsaac Mining should not have received the contract in tackling the largest shaft (at that time) in North America as they did not have the expertise, except for Ernie Yuskiw their consultant, who had a limited-time contract with them and acted as their initial manager to commence the project. First of all, MacIsaac crews had never worked from a galloway stage prior to this job; second, they did not know how to design a proper sliding form for the pouring of concrete; and lastly, they had no idea how they would handle the galloway stage hoist other than to re-rope it after it reached 3000 feet which was the limit of the galloway hoist drums. Texasgulf Mine Engineering redesigned their sliding forms because the ones supplied by MacIsaac went egg-shaped after pouring 50 feet of shaft concrete. Gerry Tiley was brought in to design four magazine drums to hold the excess rope in order for the galloway stage to access the shaft below 3000 feet. MacIsaac at that time did not have the engineering talent to manage a large project such as this. The blame for choosing MacIsaac would have to go to Amyott. *(Remember Amyott could do no wrong)*

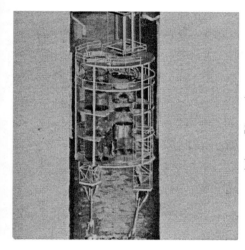

A Five-Deck Galloway Stage

Showing Two Cyderman Shaft

Muckers Below the Stage

A Section Through the

Texasgulf No 2 Mine Shaft

Showing the Galvanized Steel

Guides and Buntons

Drawings by E Spehar

Why Mining?

Besides the poor engineering on the part of MacIsaac, their performance in the actual work of sinking the shaft wasn't that good either and consequently, they came under a lot of heat. Everyone blamed their manager Rod Doran a former employee of Ecstall Mining *(Before the name was changed to Texasgulf)* who replaced Yuskiw once Yuskiw's contract with MacIsaac was completed. *(Rod was a good kid but was certainly out of his element in this situation.)* Thomson and Amyotte put pressure on MacIsaac to replace Doran (which they did) with Gerry Pigeon, a former shaftman with Paddy Harrison and Cementation, who had impressed them when he worked as shaft captain sinking No 1 shaft. Gerry was perhaps a good shaft supervisor and not a shaft manager. He did not have the technical know-how and neither was he very forthright. Things still did not go any better. Finally, corporate management demanded a change where Amyott would be replaced. As the story goes according to what Merle Marshall said to me, Thomson resisted but Ray Clarke said to Thomson, "...do you think this time next month when we review the sinking performance that things will be any better"? Thomson's answer was, "No". After that, I was put in charge of No 2 mine, taking over all the remaining construction of both the surface and the underground. Amyott was moved sideways to the open pit department, where he directed the mining of its remaining reserves.

In the first month after taking over, the shaft advanced 200 feet down to the 2800 level in a two-week period. This was a coincidence as I had done nothing to contribute to making that achievement but all the same it was a tremendous start. At the 2800 level there was a lull in the sinking as there was much slashing to do at this level where a mid-shaft loading pocket was to be installed. Once sinking began again, the progress was not much better; in fact, there was a fatality when one of MacIsaac's men, by the name of Larry Whissel was hit by a descending bucket. After that Pigeon's days were numbered. Joe MacIsaac, the owner's brother, was placed in charge to act as interim shaft manager until they could hire someone suitable to assume the responsibility for completing the shaft. Joe lasted a month and then a man by the name of Jim Doherty or "Gentleman Jim", as he was known, came on site. Jim was a mild-mannered man who didn't seem to be able to inspire anyone, especially those working for him. Things didn't change much with Doherty. How he got the job was a mystery to all. Somehow Gentleman Jim had impressed folks in the Paddy Harrison organization who recommended

Ontario Again

him to Joe MacIsaac. *(The truth was that someone in the Harrison organization wanted to get even with Joe MacIsaac and by unloading Gentleman Jim, it was probably a good way to do it.)*

Things still didn't get any better. Finally, Eric Belford and I went to Sudbury to meet John MacIsaac face-to-face and request a replacement for Gentleman Jim. We took a lot of abuse from MacIsaac that day but ended up with the "Grand Tour" of Sudbury Downs that was owned and operated by John MacIsaac and his family. Within a month there was another shaft manager, which happened to be Ernie Yuskiw again who seemed to be able to put some life into the project. His first job was to fire a couple of the shift supervisors and replace these with people he knew and trusted. As a consequence, the project moved ahead. The problem was that Yuskiw, who was a consultant at the time, was under a limited-time contract with MacIsaac's and would stay on the job only temporarily as he was not about to move his family to Timmins. When his time-contract was up, Yuskiw was replaced by Karl Pieterse who was the seventh manager, if one counted Yuskiw's two tours separately. The project didn't have much in the way of continuity. *(When granting a contract, one should insist on a continuity clause.)*

Pieterse stayed on to the end of the project which was fine although, under his management, the job experienced two fatalities. One miner was squeezed between the bucket and galloway stage, while another man, who was acting as top lander, fell out of the headframe onto the steel shaft doors below. Karl was an engineer and had a fair knowledge of what was required to keep the project moving. Being a South African, he was a bit stubborn and wanted to have things his way even though I represented the customer and therefore was his ultimate boss. Even though he was an engineer, he had some weird ideas as to how factors of safety should be applied when calculating the choice of cappels and wire rope. It was a difficult time getting him to see that the method of roping-up the friction hoists at the end of the job was in no way putting extra strain on the sinking hoists. It was finally necessary to call in Gerry Tiley to convince Karl that what was proposed, was correct and safe.

The actual sinking phase was finally completed at the end of the third quarter of 1977; however at that time the mining industry was in a recession with the price of commodities at a new low. Texasgulf was feeling the pinch and were

Why Mining?

contemplating layoffs. Layoffs were something that had never happened before at the Kidd Creek mine. It was proposed instead that Texasgulf would complete the shaft with its own forces and lay off the MacIsaac crews. This was done.

The shaft project was entering the equipping phase where the shaft cables (13 miles of cable) and steel shaft sets and loading pockets (5100 tons of steel) would be installed before the four friction hoists could be roped up and commissioned for production. The personnel records at the minesite were searched for miners who had previous shaft or steel erection experience. It was surprising that a full crew was found without too much trouble. Three of MacIsaac's hoistmen and three of their shift supervisors as well as the MacIsaac manager and captain were kept to round out the crew. As part of the equipping crew, Texasgulf appointed a technician to each shift to monitor the work, making sure that the installation met specification. The plan was to complete two full sets of steel in a period each 24 hours, which amounted to 30 feet per day not counting the two loading pockets (on 2800 and 5000 levels) as well as the shaft bottom steel. After a month (which was a learning period), where persons like Doug Duke (production engineer) and Karl Pieterse (shaft manager) worked 12 hours cross-shifting each other in training the crews, the work of installing the shaft steel and equipping the shaft reached the projected 30 feet per day. One could say the equipping phase was a success as there were no lost-time accidents in the 14-month time to transform the stage from one designed for sinking, to one dedicated to equipping, installing 13 miles of electric cable and positioning 5100 tons of shaft buttons and guides. As a result of all this, I wrote a paper on the sinking project and delivered it first at the CIM Annual General Meeting in Toronto in 1981 and then at the American Mining Congress in San Francisco in the same year.

All the while the shaft sinking and equipping was underway, the two top levels at No 2 mine were being developed gaining access through No 1 mine. By developing the first block of stopes from No 1 mine, the No 2 mine could go into full production once the crusher on 4900 level and ore passes between 2800 and 4800 levels were completed. It was interesting and coincidental that the design of the pumping system at Steep Rock A-2 mine and Kidd Creek No 2 mine resulted in identical volume and head. Also at the time, Steep Rock had pumps and many pipe fittings up for sale. Jack Ramsay, Assistant Superintendent No 2 mine, went to Atikokan and found that these pumps and fittings were in excellent shape and purchased them on the spot. *(These were the same pumps I had bought some 25 years before when I worked as a young engineer at Steep Rock.)* The first year of production (1981), the No 2 mine produced its 100,000 tonnes per month as budgeted. The following year, it not only made its budget but the mine won the coveted Ryan Award for being the safest mine in Canada for 1982. Its mine rescue team also won the Campbell Trophy as the top mine rescue team in Ontario. These were indeed great accomplishments but still Belford and Thomson seemed to favour Keith Youngblut, Superintendent of No 1 mine, over me for some reason(?). It was during 1982 that Texasgulf conducted an opinion survey by independent consultants from Chicago to determine the morale of the various departments in the company. The Kidd Creek Operation of Texasgulf came out on top and No 2 mine department topped the list for Kidd Creek as having the best morale and attitude. Ray Clarke came to my office and congratulated me personally and thought I had done a wonderful job. It was strange though, only a month or so earlier, Don Grenville Sr of Employee Relations said to me that he had word that No 2 mine was on the verge of becoming unionized. At the time, I told Grenville that was absurd and inconsistent with what was known of the department. The Texasgulf opinion survey proved me right. Many years afterward when visiting Ray Clarke at his home in Oakville after he retired, Ray told me that Gerry Larocque, Safety Supervisor, would make it a point, after visiting the minesite, to report to Thomson on all the critical things he could find about No 2 mine as a way of ingratiating himself with the top man. Thomson would take it all in as he had formed a low opinion of me in the first place so anything that Larocque cared to report was taken as proof of my incompetence. Ray Clarke went on to say, "...guess who was the most surprised person and taken aback when the news came out that No 2 mine was at the top of the list in the opinion survey and No 1 mine came out somewhere near the bottom?" *(I didn't need three guesses.)*

At this point one should mention some of the fine people who worked at No 2 mine making that operation the success that it was. My right-hand man was Jack Ramsay, Assistant Superintendent, who was a non-graduate professional engineer with considerable mining experience in hardrock mining. Jack was

Why Mining?

a gentleman and one can't recall him having any enemies. He assisted me a lot and was good at picking out details that should be observed and corrected. He was nobody's fool. Jack and I divided the mine into three natural sections called blocks. No 1 block consisted of those stopes between 2600 and 2800 levels, No 2 block spanned between 2800 and 3400 levels and No 3 block included everything below 3400 to the shaft sump at 5100. Three production engineers, Doug Duke, Mike Lalonde and Bob Curry were assigned to No 1, No 2 and No 3 blocks respectively. Duke was a rambunctious sort of a fellow who was very hardworking which was very apparent during the equipping of the shaft. Doug had one failing in that he was not altogether forthright when he was in trouble. He tended to cover up if any of his section or crew was in difficulty. As a consequence, Jack Ramsay had to keep an eye on him when things didn't add up. *(I was saddened to learn that, Doug, while he was managing a mine construction project in Mexico in 1997, developed cancer and died very suddenly.)*

The 1982 Annual Smith Tool Bonspiel

**McIntyre Community Centre
Schumacher ON**

Representing:

No 2 Kidd Creek Mine

From Left to Right:

Jack Ramsay,

The Author,

Vic Prokopchuk

& Doug Duke

Ontario Again

Mike Lalonde was altogether different. Mike...you could read him like a book. He was completely honest and very loyal. He was hard working but at the same time a bit naive. Mike went on to become a successful mine manager at a number of operations owned and operated by the Barrick Corporation. The third production engineer was Bob Curry from Northern Ireland. Bob was a mechanical engineer who had worked in mines in South Africa before emigrating to Canada. Bob was short on practical mining experience but he was quite open when he did have a problem. As No 3 block was made up of the lower mine which comprised the ore pass system; crusher and loadout pocket and conveyors; hoisting facilities; and surface yard and concrete batching plant, it was a natural fit for a mechanical engineer like Bob Curry.

Each engineer had, under his authority, a mine captain and a number of shift supervisors. In the case of Doug Duke in No 1 block, he had Ron Price, a very quiet individual. I used to come down on Ron quite hard as he seemed to be a bit lax when it came to housekeeping in his block. In any case, Ron did seem to get the production and safety that we asked of him. I can't recall anyone I would have had other than Ron so he couldn't have been all that bad. John Dasovitch was the captain for Mike Lalonde in No 2 block. Dasovitch was a hard worker when he was a miner and later as a shift boss, and he carried on in the same manner when he was appointed mine captain. He used to like to see if he could out-walk me going up the ramp but I happened to be running three miles each day at that time so I was in pretty good shape and could keep up with John. Jacques Verreault was the captain in No 3 block after Greg McDonald was fired for attempting to cover up a lost-time accident. *(More about the "McDonald Incident" later.)* With regard to Jacques Verreault, he was a gruff sort of individual whom everybody loved. Yes, loved. He knew what was required in No 3 block and he and Bob Curry seemed to compliment each other perfectly.

My left-hand man was Chuk Buchar, the administrative assistant. Chuk was with me in engineering and naturally I wanted him at No 2 mine. Dave Porteous who took over Texasgulf Engineering was good enough to let Chuk Buchar go to No 2 mine. Chuk did everything to lighten the load of paper that seemed to gather on my desk. I felt my job was where the action was and that was underground, so I would go underground as often as I could which was a minimum of three times a week. Chuk would sort out the mail and take

Why Mining?

what he or Jack Ramsay could do leaving me only that which could only be handled by myself. He wrote AFE's, drafts of letters, created budgets, collected costs and analyzed them, as well as performing a multitude of other duties. He was a good listener and would offer good advice. *(I have been very fortunate in my career to have good assistants like Dave Feniuk at K-1 mine at Esterhazy and Chuk Buchar at No 2 mine in Timmins.)*

Mine Model of Kidd Creek/Texasgulf

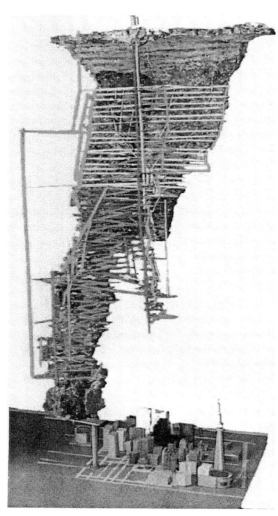

Open Pit

No 1 Mine

No 2 Mine

No 3 Mine

CN Tower & Toronto City Centre Area

Note contrast in height between the CN Tower to the depths of the Kidd Creek Mines

Photo & Model by E Spehar

Ontario Again

As for the "Greg McDonald Incident", (yes it is the same Greg McDonald that was at Steep Rock and worked at Esterhazy and Cominco), the guy seemed to follow me around. Greg was good during the construction phase when he was supervising contractors as he seemed to be able to determine when a contractor was being dishonest. *(It takes a rat to catch a rat, so the saying goes.)* However, when it came to him or his section to perform the work when the mine entered the production phase, Greg would slough-off a poor job to a point where he had to be watched carefully as he was not at all honest. One graveyard shift, two of his men were blasted when they were slow in vacating a sublevel when the shift foremen in charge of ore haulage warned them of an impending blast in a chute below. The two men were thrown against the wall and were slightly shaken up. The accident was reported when these men came off shift and a letter was filed with the mine inspector regarding this incident. Both men reported for work the following shift so it was not reported as a lost-time accident; however, a few days later, these men were absent. Jack Ramsay asked Greg McDonald why these men were missing and Greg answered that they had the flu. Nothing was said about the matter until the two men returned back to work and then all hell broke loose. The men brought doctor's slips stating that these men had suffered a concussion as a result of a blasting accident. McDonald took these slips but not before Mike Lalonde (who was at the time the production engineer in charge of that section) had seen the slips. McDonald asked the two men to go back to the doctor and have him change the reason for their absence to "flu". On hearing this from Lalonde, I immediately held an inquiry with the Don Grenville Jr, safety supervisor; Jack Ramsay, my assistant; Mike Lalonde, production engineer; the shift boss involved as well as McDonald, who was captain of the section where the accident occurred. McDonald had the shift boss so afraid for his job that he lied on behalf of McDonald. I didn't believe the shift boss nor did the others for that matter. So I suspended McDonald immediately and recommended to Employee Relations that he be terminated. This they did but not before hearing from Greg McDonald's lawyer (his son-in-law Chris Knutsen) who threatened Kidd Creek with a $1,000,000 lawsuit and incriminated me with a $250,000 suit as well. Kidd Creek offered, over my objections and those of Don Grenville Sr, a year's salary to McDonald to drop the matter and go away which he did. Looking back, it is just as well that the company took the easy way out even though the company would have won; however, the publicity would not have done

Why Mining?

the company any good as these things get blown away out of proportion. *(As the saying goes,"when you get into a pissing match with a skunk, you're going to come out smelling".)*

While sinking the shaft and developing the ore handling facilities for No 2 mine, we were also developing a mining method that could be used successfully at the mining depths that were encountered in No 2 mine. We knew that a stope pillar configuration would experience stress build-up in pillars which would result in failure in mining these and therefore a loss in ore recovery. We also did not know at what height the backfill would stand. The mining trials began at what was called No 1 block which included those stopes between the 2600 and 2800 levels. The first stope was excavated where all four walls were bound by solid ore/waste. This stope was then filled and, after a short period to allow the fill to cure, a second stope was mined up against the fill of this first stope. Observations were made to see how the fill in the first stope reacted while mining the second stope. After the second stope was completed, a third stope was mined and filled adjacent to the first stope in the next panel of stopes. A fourth stope was mined next to the third stope in the same panel and adjacent to the second stope. This fourth stope had two walls butting up against fill, ie., the walls common to the second and third stopes. In all cases, the fill stood up well with only a minimum of sloughing. From there on it was known we had a mining method where no pillars were required and therefore no ore would be lost in pillar failure. What was created was a stress arch that extended from foot to hangingwall and from one end of the orebody along strike to the other end. As a result of these experiments Douglas Duke, production engineer, and myself co-authored a paper, "Mining with Backfill". This paper was delivered by me at the "Proceedings of the International Symposium on Mining with Backfill" in Luleå Sweden in June 1983.

Scandinavia

One of my most memorable company trips was the one to Scandinavia where Irene and I enjoyed a combination business and pleasure trip. The symposium, which drew 180 delegates from 37 countries, conducted all communications, both written and verbal, in English. *(There were people from all over the world communicating in only one language. Even the French spoke English!)* As part of the symposium, after delivering the paper at Luleå

Ontario Again

University of Technology, the delegates were split into two groups for tours to Kiruna and Boliden. I chose to visit the mines in Kiruna because they were famous for their size and for their up-to-date materials handling facilities. At Kiruna, the group first visited the Forskningsstiftelsen Svensk Gruvteknik, which roughly translates into Swedish Technical Research Mine. At this mine, the personnel were conducting experiments of accuracy of drilling, stope configuration and cable-bolting equipment. The following day, the group visited the famous LKAB Iron mine which is probably the largest underground mine in the world. This mine reminded one of an auto assembly line as everything was of a grand scale and, at the same time, very highly mechanized and automated. After a week in Northern Sweden, where I saw for the first and only time the midnight sun from the top of the Fermann Hotel in Kiruna, I caught up with Irene in Stockholm where we spent the weekend. While in Stockholm, a very remarkable thing happened. Irene and I were walking near the Stockholm Opera House when we bumped into two Australian couples that we had met in Timmins the year before when they were touring mines in Canada. *(It just goes to show you that this is indeed a small world especially when you're in the mining industry.)*

Map of Scandinavia

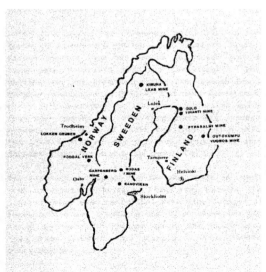

The sketch shows the location of the various mines and plants the Author visited while attending the International Symposium on Mining With Backfill" in Luleå Sweden in June 1983.

Why Mining?

From Stockholm, we took an overnight train to Trondheim, Norway where two underground mines, Foddal Verk and Lokken Gruben were visited. Both these mines produced copper and zinc ores and both were very efficient operations employing sublevel stoping methods. At the Foddal Verk, they had a rather unique method of lowering equipment into the mine as they, with the assistance of the army camped nearby, lowered and raised the equipment in the ventilation shaft using a couple of Centurian tanks as motive power. *(I enjoyed the tour of Norway, especially the wonderful salmon dinners we had at the Britannia Hotel on both evenings spent in Trondheim.)* From Norway, we flew back to Stockholm and were met by personnel from the Sandvik Company of Sandviken and were their guests for the balance of our second week in Sweden. The next mine visited was the Garpensberg mine owned and operated by the Boliden Group. Here was seen the famous Boliden undercut-and-fill mining method which this company promotes even though productivity at this particular mine was only 4 tonnes per manshift. At the Garpensberg mine all the miners come to surface for lunch in the company's cafeteria.

The next operation visited in this area was the manufacturing plant and experimental mine operated by the Sandvik Company to manufacture and test drilling bits and rods. Both operations were of interest, especially the Bodas mine as the operators were drilling with higher air pressures than is generally recommended in order to impose greater stress on these components than normally found in an operating mine. As a diversion, the old Fallun mine and museum were toured. Here one could see how mining was done in the 1600's.

Again Irene and I returned to Stockholm for the weekend before sailing to Turku, Finland on the Sunday. We were met at the dock in Turko by Tonelli Hallanero, a sales engineer of the Tamrock Company, who then drove us to Tampere. After spending the day in Tampere touring the Tamrock manufacturing plant and experimental test mine, Irene boarded the train to Helsinki that evening as I would be touring mines in Northern Finland with Tonelli Hallanero for the next couple of days. Irene was quite happy to spend the next few days by herself as it gave her the opportunity to do some shopping without having me bug her about how much money she was spending. Anyway she did have Tonelli's wife meet her on one of those days and was toured around Helsinki.

Ontario Again

From Tampere, Tonelli Hallanero, my guide, and I flew to Oulo and then drove south from there to Laminsaari where the Vihantti mine was located. The Vihantti mine at 900,000 tonnes per year was somewhat smaller than the 1,200,000 tonne Kidd Creek No 2 mine, but all the same, it was interesting to see the similarities, especially how they mined and filled their stopes. This mine used blasthole mining and cemented backfill to mine 85% of its ore, while cut-and-fill mining accounted for the remaining 15%. The binding agent for the cemented backfill was blast-furnace slag ground at the headframe with the addition of lime that acted as an activator. The mine was toured in the morning after the usual slide presentation and coffee-and-donut refreshment. After the underground tour, we came to surface for the excellent lunch that seems to be the custom in Scandinavia. I would say that the Vihantti was the best operation visited in Scandinavia and had the best slide presentation as well.

From the Vihantti mine we drove on to Pyhasalmi to visit the Pyhasalmi mine the next day. After arriving at the town of Pyhasalmi, our hosts took us out to one of the rivers nearby to experience white water rafting, except that the raft was a 4-meter row boat. As the date was approaching mid-summer, (the Nordic summer solstice), a number of Finnish families were working along the shoreline preparing wood for bonfires that would be lit on June 21st and 22nd. After the ride down the river, I enjoyed a beer and a pancake cooked by the Finnish women over oil drums that were brought there for the upcoming celebration. *(I read in an English newspaper, printed in Finland, where it berated the behaviour of the Finnish people, comparing their conduct to that of the Swedes. Apparently, the Swedes celebrate mid-summer by going to their home towns or in the country where they sing and dance around a decorated pole similar to that of a May pole. The Finns, on the other hand go to a lake or river, light a bonfire and drink as much as they can and then try various water sports with dire results as many drowning accidents occur at that time.)*

The Pyhasalmi mine, which was slightly larger than the Vihantti mine, was also shallow (500 meters deep) as compared to our mines in Canada. The orebody was first mined by open pit and later, when the surface ore was exhausted, the orebody was mined by sublevel caving but because of the high

Why Mining?

dilution (33%) the operators were in the process of switching to a modified vertical crater retreat and blasthole mining. Shotcrete and rebar bolts were used extensively.

After touring the mine in the morning, we drove for quite a long distance to the town of Joensuu near the then Soviet border in the Karelia district. At Joensuu considerable manufacturing is done. While at the hotel, before dinner that evening, I met a young engineer who had spent one summer at Cominco Potash in Saskatoon. After making his acquaintance, we discussed mutual friends. You would have thought we were old buddies as he would not let me pay for anything, buying me drinks and paying for my dinner that evening. He insisted that the next day we visit his manufacturing plant where they made Normet equipment. I found this plant tour very interesting and was given the V.I.P. treatment again.

Later the same day, we drove to Outokumpu (which was only a short distance away from Joensuu) and visited the Vuonos mine. This mine was on its last legs, or one should say it was a salvage operation. The mine being quite shallow was accessed by a ramp, and mining was accomplished by room and pillar, sublevel caving, benching (modified room and pillar) with fill, and inclined wall stoping with backfill. This latter method reminded me of the Kidd Creek No 2 mine stoping method. Like the other Outokumpu-owned mines (Vihantti and Pyrasalmi), this mine brought hot meals underground to the miners at a cost of 85 cents a meal as compared to Sweden where the miners come to surface for their lunch which cost them 12 Kronas or $2.75.

After touring the Vuonos mine, my guide Tonelli Hallanero and I flew on to Helsinki where I caught up with Irene at the Marski Hotel which was right across from the famous Stockman Department Store. While I was away up north, Irene had a great time shopping which I later began to realize when the shipments started arriving at the post office in Timmins some four weeks later.

The evening I caught up with Irene in Helsinki we went out for a late night walk as it was still light at 11 o'clock at night at that latitude. There were drunkards on every corner celebrating early as mid-summer holiday was only a day or so away. There were so many of these celebrants that the police paddy wagon had no trouble finding a load. We saw one paddy wagon travel down the street and return five minutes later with a wagon full of wild celebrants. This cinched the idea that we would get out of Helsinki before the city became too wild.

Tonelli and his wife had made arrangements for us to spend the mid-summer holidays with them and others on a lake steamer cruise. From what we heard and saw in Helsinki, this would be one big drinking session so we thought the better of that and declined the invitation saying that we had made other arrangements in Stockholm which we could not cancel at that late date. Immediately we arranged flights to Stockholm where we knew we could spend a quiet weekend and rest as the city centre there closes down for that holiday. We arrived back in Stockholm and got a good deal on a full hotel suite at the Wellington Hotel (Best Western) for the mid-summer holiday at half price. The only inconvenience was that the diningroom of our hotel closed for the holiday so we had to make other arrangements.

Right after the mid-summer holiday, I visited the Atlas Copco manufacturing plant and experimental mine. The mine was right underneath the plant located on the outskirts of Stockholm. We entered the mine by elevator from one of the offices in the plant. In the mine, I was shown Copco's latest mining equipment that it was testing. Also shown, were the trials that a piece of equipment would go through in its development before final manufacturing, which I found rather interesting. That evening we were guests of Atlas Copco where we were taken to Kallar Aurora which translates into Aurora's Kitchen. *(We enjoyed the best desert we had every tasted...Swedish cranberries in caramel sauce.)*

When leaving Stockholm, I absentmindedly forgot my camera in the taxi on our way out to the Stockholm airport. This was noticed while waiting in the airport just prior to take-off. I immediately phoned the hotel where we had been staying to request the name of the taxi company that the hotel had called for me as I would then call the taxi company and perhaps retrieve my camera. The lady at the hotel said not to worry as she would call the taxi company for me and I should call her from Copenhagen when I arrived there and she would leave information for me regarding the camera. When arriving at the hotel in Copenhagen, I called the Wellington Hotel in Stockholm and true to

Why Mining?

her word the lady at the desk said that the taxi driver had called around to the hotel with the camera requesting the flight we were on so that he could return it before we flew off. *(The Stockholm airport is some 50 kilometres from the city. How many taxi drivers in Canada would go out of their way and drive 50 klicks to return property to a passenger from out-of-the country or within the country for that matter?)* She said not to worry, as the hotel will send the camera by mail and I should have it in about a month. About a month after arriving home, my camera showed up no worse for wear. *(This experience just showed us how honest people were in Sweden. Another thing which showed us the honesty in Sweden was that bicycles were not locked when they were parked outside of a bar or restaurant as in contrast to those in Canada. We need more people like the Swedes in Canada today.)*

Kidd Creek

A sad thing happened in the late summer of 1982 in New Jersey which affected the company. The executive plane with eight senior Texasgulf people, including the chairman, Dr Charles Fogarty, crashed in a storm on landing, killing all eight as well as the two pilots. Dick Mollison, the president was not on that plane. The vice-president exploration, George Mannard, had disembarked from the plane in Montreal and therefore escaped this crash. This was a tremendous blow to the company, one from which it never did recover. Three months later the Canada Development Company (CDC) along with Elf Aquataine (Elf) of France, between them, purchased all the shares of Texasgulf (Tg) and spilt the company in two. CDC took control of the Canadian assets of Tg while Elf took charge of the US and overseas portions. Dick Mollison tried to prevent this take-over by phoning President George Bush, who was formerly a director of Texasgulf and a good friend of Chuck Fogarty to see what could be done to prevent two foreign companies from taking over an American company. Bush would not take his calls. We all know that, had it been Dr Fogarty who called, President Bush would certainly have taken the call and would perhaps have done something to prevent that unfriendly take-over.

We all had mixed feelings about this take-over. First of all, Tg was such an excellent company, we were reluctant to be cut adrift from it, but on the other hand, because a number of us held shares and options which quadrupled, we were somewhat elated. My share paid the outstanding mortgage on our home plus the cost of some renovations and new furniture.

Things in the company changed shortly afterward. The new company was now called Kidd Creek Mines Limited with Ray Clarke as chairman and George Mannard, who was formerly vice-president exploration, as president. This certainly did not please Bart Thomson as he thought he should have been made the president and therefore he resigned as a result of his disappointment. The company was losing a good man in Bart. It was too bad. My only concern at the time was Eric Belford, who was still my immediate boss. Anyway, shortly after this take-over, I had a call from George Mannard asking me to go out to Saskatoon to attend the monthly participants' meetings at Allan Potash. At the time, Kidd Creek held 40% of the shares of Allan Potash through its subsidiary Kidd Creek Potash Company while, at the same time, the Potash Company of Saskatchewan held the remaining 60%. Unbeknown to me at the time, I was made vice-president and a director of Kidd Creek Potash Company. This didn't give me more money but it did give me some prestige in that I travelled to Saskatoon each month for four days attending the monthly participants' meetings. At these meetings, I held the voting rights of 40% of the Allan Potash shares.

The meetings were held on the third Wednesday of each month in the conference room at the Allan Potash. Prior to the meeting which was held on Wednesday morning, I would usually arrive in Saskatoon on Monday night and go out to the mine on Tuesday morning with Doug Mahood, Kidd Creek's resident traffic manager, to tour the underground in the morning, and the mill in the afternoon. The meeting part was held Wednesday morning while the afternoon was dedicated to review some specific item that came up in the morning discussion. Sometimes we would tour one of the neighbouring mines to see some new method or piece of equipment that could be used at Allan. When I first went to these meetings, Jim Padden, who was the former president of Texasgulf Potash Company, and Rudy Schauffer, manager of Texasgulf Potash Operations in Utah, would attend these meetings assisting me in the change-over from Texasgulf Potash Company to Kidd Creek Potash Company. When Texasgulf was handling things, Jim would always invite the Allan seniors (manager and superintendents) out to an evening dinner following the Wednesday meeting. At these dinners Jim would tell the waiter to put a bottle of wine in front of each attendee and when it was empty to

Why Mining?

bring on another. I asked Jim how he could do this and he said, "...when Chuck Fogarty (chairman of Texasgulf) asked me go to these meetings, he made it clear that I should not be niggardly about expenses when I was entertaining". *(Boy, Jim sure followed Fogarty's advice.)*

On one occasion shortly after taking over from Jim Padden, I flew to Toronto where I was to change planes for Saskatoon. I was met by a number of Kidd Creek head office employees, namely Dave Baskin, Eric Dison, Ron Garva, Bert Scragg and others who were also attending the participants' meeting along with myself. When we boarded the plane for Saskatoon, I noticed that those head office guys were all flying first-class while I was seated in the tourist-class section of the aircraft, yet I was an officer of Kidd Creek Potash and they weren't. After that, I always flew first-class or business-class if first-class was not available.

Originally, Brian Kaukinen, vice-president PCS Mining, would chair the participants' meetings but was replaced later by Jim Sadler, an old buddy of mine from Steep Rock and IMC. Kaukinen who was not an engineer, yet occupied a very high position with PCS Mining, left their employ to attend the University of Saskatchewan where he successfully obtained a degree in law. I was somewhat pleased when Jim Sadler replaced Brian Kaukinen as I found Kaukinen was a strong sympathizer of the then Premier Allan Blakeney and his NDP party which formed the government of Saskatchewan at the time. Everything that came up at the participants' meetings (when Kaukinen was in the chair), had to be in line with socialist thinking or it would be over-ruled by him. Kaukinen, representing PCS, held 60% of Allan Potash shares.

Shortly after Kaukinen had left, an old buddy of mine (again from Steep Rock) by the name of Steve Harapiak, was made president of PCS Mining. Steve and I always got along fine. With Sadler chairing the participants' meetings, and Harapiak attending these meetings as well, I had no trouble in representing Kidd Creek Potash. However, a provincial election brought in a change in the government of Saskatchewan when the PCs and Premier Grant Devine took power in that province. As is the case when governments change, there were a few changes in the Potash Company of Saskatchewan, a crown company. The president of that company was terminated and Steve

Ontario Again

Harapiak was moved up from president of PCS Mining to president of the parent company, "Potash Company of Saskatchewan". Steve's replacement as president of PCS Mining was no other than Don Matheson, whom I knew from IMC. Don was a big fellow, whom we nick-named the "Friendly Giant" and a jolly guy he was too. He helped me a lot especially when Kidd Creek was over-loaded with standard potash product and we had expended our allotted space in the product warehouse at Allan.

Sometimes when corporate positions change there are undoubtedly some changes in the operation of the company. In this case, with the arrival of Don Matheson as president of PCS Mining, Jim Sadler, vice-president of PCS Mining, was terminated. I was sorry to see this happen but I did not know and probably will never know the full circumstances for that action.

My dealings with PCS at Allan Potash confirmed my belief that governments should never get involved with things that can be operated best by free enterprise. As an example, PCS owned a 25% interest in the potash leases where the IMC K-2 mine was mining. IMC offered PCS $95 million for its 25% share of these lands. This was certainly a good deal for both companies as IMC would have full title to the lands it was mining and PCS could use the cash for other developments. PCS approached Kidd Creek and offered $90 million to purchase the 40% of Allan Potash that was controlled by Kidd Creek. This was a good deal for both Kidd Creek and PCS as Kidd Creek was not making much, if anything, at the time from its 40% of Allan Potash and also potash mining was not one of its core activities. It was a good deal for PCS as Kidd Creek was more of a hindrance to them as they could not operate freely unless Kidd Creek was first approached and agreed to any change. The boards of directors for all three companies (IMC, PCS and Kidd Creek) agreed to this three-way deal where IMC would obtain the PCS leases presently mined by the K-2 mine for $95 million and PCS would obtain from Kidd Creek the 40% of Allan Potash controlled by Kidd Creek for $90 million giving PCS 100% ownership of Allan Potash.

All went well until the whole deal was vetoed by the Grant Devine Cabinet which stated it would not sell any of its crown assets. Two months later on December 25th 1985, a leak was noticed in the IMC K-2 mine which turned into a torrent, causing great flooding of one section of the mine and with the possibility of losing the whole operation. The total damage cost over $300

Why Mining?

million to save the mine. Because PCS still owned 25% of the IMC mine, it and the people of Saskatchewan had to fork over $75 million to cover its share to salvage the mine. Had Devine and his cabinet let things be, PCS would have had its $75 million spent saving the K-2 mine and full ownership of Allan Potash. The people who came out happy that this deal didn't go through were the people of IMC as the company had to pay only 75% of the cost to save the K-2 mine which was covered by its insurance.

In another case where socialism didn't work was the attempt by Kidd Creek to put pressure on PCS to streamline the Allan operation and reduce the levels of supervision and extra manpower in that operation. Allan Potash had six levels of supervision: manager 1st level; operations manager 2nd level; superintendents 3rd level; general foremen 4th level; captains and senior shift foremen 5th level; and shift foremen 6th level. The mill had an extra level of supervision as it had an assistant maintenance superintendent answering to the maintenance superintendent. Mike Amsden, general manager at Kidd Creek and president of Kidd Creek Potash and I met with Steve Harapiak and Don Matheson of PCS to try and convince them that some reduction in the supervision and manpower at Allan was necessary as the mine was not performing as planned. We proposed the elimination of 2nd and 4th levels, the assistant maintenance superintendent position as well as a number of operating and maintenance personnel. Harapiak and Matheson agreed with our assessment and, at the end of the day, admitted that their hands were tied until the forthcoming provincial election campaign was concluded and the PCs were re-elected. Again, politics interfered with good operation of an enterprise.

As an aside, Steve Harapiak tells the story of chastising Martin Quick, manager of PCS Lanigan, about the number of levels of supervision he had in his operation. Steve said to Martin, "...don't you know that there are only four levels of hierarchy in the whole catholic church?" Martin said, "No, I didn't know that but how could I? I'm not a catholic."

Shortly after the Devine government was re-elected in 1987, the suggestions made to streamline the Allan operation were executed but what really surprised us all was that Steve Harapiak was terminated as president. The man who took his place was Chuck Childers, formerly from IMC Skokie,

Ontario Again

Illinios. Under Chuck's direction, PCS has since been transformed from a crown corporation to a public company, trading on the TSE and NYSE and becoming the largest supplier of nitrogen, phosphate and potash in the world, something it could never have achieved had it remained a crown company with all the interference that goes along with that type of ownership.

Allan Potash 1982 Allan, Saskatchewan

Allan Potash in the Spring of 1982 as construction of the Raw Ore Storage Building was underway at the extreme left of the picture. The dark emission coming from the stacks on the Refinery Building on the right was before the hydro scrubbers were installed as part of the general plant upgrading taking place in 1982-83.

Why Mining?

Head office

In the summer of 1985, I was approached by Bruce Gilbert, senior vice-president marketing, to join his team at head office in Toronto as manager, marketing services. I was a bit reluctant at the time as I knew nothing about marketing but Bruce said it was not my knowledge of marketing he was interested in but my organizing ability and because I was a disciplined man. *(Whatever a disciplined man means.)* At the same time, Eric Belford was given the job of streamlining the Timmins operation for management and it was clear to me that he wanted me out. So on that note, I decided not to wait for Belford to push me aside (or whatever he planned) and, therefore, I decided I would give marketing a try. Bruce and I met at the Toronto Airport while I was travelling on my way to one of the participants' meetings in Saskatoon.. At the airport between planes, Bruce outlined the job as manager, marketing services. It was agreed that I would retain my duties as vice-president of Kidd Creek Potash and would continue to attend all participants' meetings, rule on potash operations as well as direct traffic for potash. Besides this, he wanted me to supervise the sales control group at head office, direct corporate traffic, administer all sales and traffic contracts as well as act as office manager for marketing. All this extra work went with no extra pay at this time as Bruce said I was being paid more than some of the other vice-presidents of Kidd Creek. However, I would occupy one of four corner offices on the 50th floor of Commerce Court on the corner of Bay and King Streets. *(Anyway, I took the job because I wanted to get away from under Eric Belford. I also wanted to see how the other people lived, especially those who worked in a big city, something that I had never experienced before.)*

The first part, which was to retain my duties in potash, was relatively easy as I had been doing this as well as managing No 2 mine for over a year. To act as office manager was also no large chore as I had supervised office personnel in engineering in the past and office management was largely administration. Except for my little disagreement with my personal secretary, whom I had to terminate, that part of the job was easy. Directing corporate traffic was a little more difficult but I was blessed with two good people at head office namely Allen Day and Maryellen Olsen who required little in the way of supervision. These two people seemed to be able to direct Kidd Creek's metal and concentrate traffic all over Canada and the States with no problem and at the same time coordinate the traffic department in Timmins. Occasionally

Ontario Again

there would be a problem with the concentrate stockpiles in Quebec City but with my survey knowledge I was able to step in and solve that sort of thing. My main problem in this section was finding warehouse space for potash product when sales fell behind production. At one time, Doug Mahood, my traffic man at Allan Potash and I had to lease 1000 railcars and a number of railroad sidings in order to store excess product. We even had several barges on the Mississippi River (during one spring) full of potash product waiting to be sold. That was a very nail-biting time. But come spring, the barges moved on the Mississippi and they were able to be emptied when sales personnel were able to sell the fertilizer for the spring planting season.

The sales control was something new. The system used at that time by Kidd Creek was inherited from Texasgulf and was largely a manual system where orders were filled and recorded by hand. The clerks on copper had started converting to mini-computers but those on zinc and cadmium were still somewhat behind. My job was to work, along with the computer programmers, to create a new up-to-date sales control system which would be installed on the mainframe computer and could be accessed at various stations in the sales control section. The new computer sales control system had fallen behind schedule prior to me being given this task. The head of the computer section was more than happy to unload the task of setting up the sales control system onto me. Then, I had the financial people breathing down my neck, saying that the implementing of the new sale control programme was approaching its budget appropriation and I should be more forceful and motivate the programmers into moving faster so that the system would not be substantially over-budget. On the other hand, I had a group of contract programmers, who were so much in demand in industry, that any intemperate word would cause them to pack it up and quit right there on the spot. It meant walking a fine line between the bean counters where everything was measured in dollars and cents and those prima donna programmers who held the power over success or failure. To make the situation more difficult, I was on a learning curve and was trying to assimilate in my mind the old system of sales control and at the same time establish a new one for the group. My own sales control group were not too fussy about learning a new system either, so there were really three groups to juggle; namely, the accountants, the programmers and the sales control clerks. *(How did a mining engineer get into such a predicament?)*

Why Mining?

The last part of my job was to gather all the various contracts and list them into one file. The contracts varied between small sales contracts to very large ones. The sulphuric acid contract (could be classified as a very large one) between CIL and Kidd Creek called for shipment of anywhere from 350,000 to 550,000 tons of sulphuric acid annually. Some of the other larger contracts included potash sales agreements between Kidd Creek and Texasgulf; New Zealand Zinc Contract for the sale of zinc slabs; Noranda Copper Concentrate Contract for the shipment of copper concentrate; Noranda Copper-Silver Concentrate Tolling Contract where concentrate was tolled through the Noranda smelter and then sent to the refinery in Montreal, then turned over to Kidd Creek for sale. Minor contracts included shipment of concentrates and refinery slimes to various processing plants in North America and Europe; copper metal sales contracts; zinc metal sales contracts; cadmium metal sales contracts; traffic and shipping contracts. Altogether there were 86 contracts that had to be monitored and updated on a regular basis. With these I was able to make a synopsis of each and monitor these with the use of a desk-top Macintosh computer.

All was not work at head office. As Irene and I lived only three blocks from Commerce Court, I was able to go home for lunch and have a short nap each noon after lunch. Many times though, I would be invited out for lunch by one of the various shipping agents representing trucking firms, ocean-going vessels and railways. Those traffic people sure knew how to live. I was continuously invited out to baseball, hockey and football games as well as lunches and dinners. There was no end to the various invitations. Being sheltered all my career back in the boondocks at such places as Timmins, Malartic, Atikokan and Esterhazy, I had no idea that these traffic and sales people in the city could spend so much. The sales people had tickets to all the baseball games which they used for their clients during the week; however, because I lived downtown, I was given the weekend tickets as the salesmen wouldn't be around on weekends as that was their time off. *(I often wondered when did they work?)*

The year I was at head office, I did attend the Traffic Convention held annually in Montreal. This was a great experience as there were railway people from all over North America. There was no limit as to what these delegates would spend. It was an eye-opener. I also attended the Canadian Golf Open at Glenn Abbey where Kidd Creek Marketing entertained its

Ontario Again

customers. This was quite an event: we started our event at breakfast in early morning and then out on the course afterward to watch the proceedings with a packed lunch and then to a big banquet that night. It was a fun day.

When moving to Toronto, Irene and I decided first to lease an apartment or townhouse in order to get the feel of the city before we committed ourselves to investing in a place of our own. As a consequence, we leased a condominium at Market Square on the corner of Front and Church Streets. This was a very well-run condo and was indeed well situated for anything we wanted in the way of entertainment. We enjoyed our life in Toronto, especially living downtown right near the theatres and the fine restaurants. We went out to dinner quite often. After awhile, we decided that perhaps we should buy a place in Toronto as I was only 58 years old at the time and retirement was a couple of years away. I did have in the back of my mind to perhaps retire at 60 and move to Kingston where we were in the process of having a townhouse condominium built for us. We decided in the autumn of '85 that we would buy a condo in the same Market Square complex where we lived, and planned to use our Kingston townhouse as our weekend getaway (as the French say, "pied à terre"). This was a great plan, except that, after buying at Market Square in October, a week later, Falconbridge bought all the shares of Kidd Creek Mines (except those of Kidd Potash Company Limited) from the Canada Development Corporation. There was some doubt as to where I stood with the new company; in fact, no one knew what would happen next.

Christmas passed without incident. Work went on in January and February as usual with no change in routine at the office. As an added insurance in case I was thrown out on my ear, I had to decide how best to protect against future adversity. The worst case scenario would be termination with a minimum severance of perhaps a week's pay for every year of service. At the age of 58, it was not our plan for Irene and I to move back into the bush, so to speak, where I would operate a mine for a living. It was decided that consulting should be tried. How does one get started? Fortunately for me at this time, there was a course being offered at Seneca College addressing the issue of starting one's own consulting business. The company paid for that course even though it had nothing to do with any part of marketing. *(I decided that to hell with principle. Why not take this course and get the company to pay for it as I owed Falconbridge nothing.)* As mentioned, work

Why Mining?
as usual went on at the office. I attended the participants' meetings in January and February. In March we began to hear rumours of layoffs but because I wore so many hats (office manger, sale control coordinator, contract administrator, traffic supervisor and potash manager) I thought my job was relatively safe.

Irene and I invited Mike Hughes, the new senior vice-president marketing (he was my new boss as Bruce Gilbert had retired prior to Christmas '85), for dinner on the Friday of March 7th. Mike phoned Irene on the Thursday morning saying that a friend from England was arriving unexpectedly that next day and would be staying at Mike's place and therefore he had to decline the invitation to dinner. When I arrived home from attending the consulting course at Seneca College that afternoon Irene, who is most perceptive, said, "I think something's up because Mike's excuse seemed to be a little weak to me". Well I didn't think much about it and enjoyed the weekend all the same. On Monday March 10th (1986) a few days before the "Ides of March", we all went into work as usual and found only "Chiefs and no Indians". Bill Fearn, the senior vice-president finance, came into my office and said that all junior staff were told not to come into work until ten o'clock in order to sort out who would or would not be laid-off. It so happened that all the vice-presidents and managers with the exception of vice-president marketing (Mike Hughes) and vice-president project development (John Heslop) were laid off. I was one of those chosen to be terminated. **What a shock!** Believing, because I was handling so much and because the sales control system (on which we had spent over half a million dollars to perfect and was close to being ready for implementation) I would have been kept on at least to see the fruits of these efforts. This was not to be. Falconbridge even scrapped the sales control system and put in their own. *(C'est la vie!)*

I was given sixty weeks pay plus all my benefits which included my car as well as counselling which included a private office and secretarial services across the street at First Canadian Place. Deciding that instead, of feeling sorry for myself, I would get up and go to work as usual as the office at First Canadian Place was only across the street from Commerce Court and it was available more or less for an indefinite period. This was done and it paid off. After making a few phone calls to let people know where to reach me, I started to make plans for the future. Three plans were made: 1) to start a consulting business, 2) set up a potash agency and brokerage company and 3) investigate

Ontario Again

a career in teaching at an engineering school. Regarding the latter, I phoned my friend Jack Frey who was Dean of Haileybury School of Mines and asked a few questions. There was some hope that if I was serious it could be tried. For the second scenario, an agency was investigated; that is, setting up a company to manage the assets of Kidd Creek Potash Company for the Canada Development Corporation (CDC)...the idea being that the agency would represent CDC at all participants' meetings as well as the Canpotex meetings *(a company formed of Saskatchewan potash producers for the sale of potash overseas)* and market its potash in domestic or North American market. Joining me in the agency would be Eric Dison (Kidd Creek's manager of potash sales) to handle sales and Doug Mahood (the traffic man in Saskatoon) to direct transportation and marketing. As for consulting, this was pursued aggressively, beginning first with an outline of services that I would be offering, then writing a current resume and having printed business cards.

Fortunately, the day after receiving Falconbridge's boot-out-the-door, I received a phone call from Mike Stewart, vice-president Canterra Oil Company, saying that the assets of Kidd Creek Potash Company (which remained with the CDC) would be managed by Canterra Oil. He asked me to consider helping him to get started within the Potash Industry and to attend the next participants' meeting in Saskatoon. To make a long story short, I delivered my proposal for a potash agency which he looked at seriously but, in the end, he decided that Canterra would manage the potash asset itself. Failing the agency bid, I requested a six month's contract to assist Canterra with potash at $3000 a month for a minimum of five days each month but he said they didn't believe in anything over three months. Brazenly I said, "Well if it is only to be for a short term, I'll have to ask you for $4000 instead". This, Mike Stewart accepted but the best news was that this association lasted for ten years until the spring of 1996. *(I'll talk more about this association later.)*

Besides these potash contacts, contacts were made with others or, as they say in the business, "networking". Colin Chapman, who occupied an office down the hall from me at First Canadian Place, suggested I give Malcolm Slack, president, Belmoral Mines a call as perhaps he might need someone with my experience in engineering. This I did and bless Colin Chapman *(Colin was killed in a helicopter crash in Siberia in 1998)* put in a good word for me when Slack called him regarding hiring me as a consultant.

```
Why Mining?
```
Anyway, before getting carried away with details of those last days in Toronto, it should be mentioned that the plans for our Toronto condominium and our townhouse in Kingston changed with the take-over of Kidd Creek Mines by Falconbridge. Irene and I decided that we would sell our Toronto condo and move to Kingston as soon as the Kingston townhouse would be completed. We made a trip to Kingston that Thursday, March 13th, to inspect the townhouse project and to notify as well as motivate the contractor to complete our unit as soon as possible as we wanted to move our belongs to Kingston that spring. While in Kingston, we visited the office of Don Good, a lawyer, to incorporate the consulting mining engineering company that was being started. I thought at first, a good name would be "David L McKay Inc" but the provincial authorities said the company name must be more definitive than that so the name chosen was "McKay Mining Engineering Inc".

Texasgulf/Kidd Creek Minesite 1985

Photo by E Spehar

PART 2

Chapter 9

On My Own

Consulting

Consulting employment can be conducted utilizing two different approaches. In the first case, an individual can work as an independent, where one works alone and takes full responsibility for collecting all the data, writing the report and signing and stamping the report. In other words, he/she takes all the flak if things don't turn out right. *(He/she also has to see that they get paid because some clients fail to pay.)* In the second case, the individual works under an umbrella of a large consulting firm where more than one engineer works on the project and the large engineering firm accepts full responsibility for the report and the results that ensue. In this second case, this is called subcontracting as, when one accepts assignments from a large consulting firm, it is necessary for the engineer to sign a contract and a confidentiality agreement with that firm. In my career, I did a bit of both but found that subcontracting took me to many parts of North America and Asia; whereas, I would not have had that opportunity had it not been for subcontracting to these large consulting firms, such as Fluor Daniel Wright, Dynatec Engineering and Kvaerner Davy.

Consulting requires one to do a bit of "networking" which is knocking on doors and talking to people who might be in the position of directing one to an individual who may be in the position to award an assignment. The first few times I was slightly apprehensive regarding a "cold call". That's a call where you have never met the person before and you go in and introduce yourself and try to impress him with your skill and knowledge. But mostly, one has to depend on friends and acquaintances to direct one to those who are in the position of assigning project contracts. It was at one of these networking sessions with Malcolm Slack (where we were having lunch at the Engineers Club in Toronto) that Leon LaPrairie asked me to write a report on mining peat from the Meinzinger bog in Northwestern Ontario. I knew Leon from working on the Vedron property, but it was just a coincidence that he saw me in the Club that day which triggered the thought that I was the one he needed to write a report on mining peat. Knowing little about mining peat, except that it is generally harvested by dredging or by similar methods, I took on the assignment anyhow. The time at Steep Rock gave me some inkling into

Why Mining?

dredging so that would be the start. I did, however, do some research at the Queens University Library in Kingston in order to obtain some added information on peat harvesting. Personal contact was made with Bill Gilmore of Wright Engineers in Vancouver, who used to work at Steep Rock and was quite knowledgeable on the subject of dredging. Bill gave me a few hints. He also suggested that Ken McRorrie be contacted as he was also well experienced on this subject. Ken was formerly chief engineer and general superintendent of Steep Rock when the two large dredges were being moved ("Operation Up and Over") from the Hogarth to the Roberts dredge pond. When I phoned Ken, he told me that one of the former 27-inch dredges was presently located at Thompson Manitoba and was immediately available. As Leon Laprairie only wanted a ball-park estimate, the report was completed fairly quickly. The information that Ken McRorrie gave concerning the availability of the 27-inch dredge was also passed on, along with some general information.

Another example of how networking paid off was when Irene and I were travelling to Arizona in November 1990. While travelling on Interstate-40 through Texas I decided to phone home to Kingston and pick-up the messages from our answering machine and there, lo and behold, was a request for potash consulting services from Northstar Energy Corporation in Calgary. These people had obtained my name through Mike Stewart of Canterra. As a result of that call and after getting settled in Scottsdale Arizona, I flew to Calgary to meet the people of Northstar Energy and received instructions regarding the task. What this job involved was a due diligence review of Central Canada Potash, as Northstar planned to install a cogeneration plant and wished to determine whether or not the project would be viable. After meeting the people involved in Calgary and then flying to Saskatoon, it was then necessary to tour both the underground and surface plants at Central Canada Potash at Colonsay, Saskatchewan. This was a quick trip, sort of in-and-out of Saskatoon and back to Scottsdale where the report was written and then typed by a secretarial service. The report along with the billing for expenses and fee were sent by Federal Express to the client. The fee, earned from that assignment, more than covered the rental on the house in Scottsdale for the next three months.

A contract was awarded with Bond Gold to act as their mining consultant through Nat Scott of Stafford & Associates, whom I met when we were both

On My Own

members of the Atikokan Lion's Club. Nat, at the time, was making a professional search for a mining engineer for Bond Gold. Answering his ad in the Northern Miner in the autumn of 1988 really paid off quite handsomely. *(Nat and I go back a long way and we still to this day exchange Christmas cards.)* Desmond Kearns, vice-president operations, interviewed me at the Bond Gold office on Adelaide Street in Toronto and afterward we went to the Engineers Club for lunch. At lunch he was saying the person they had in mind should be bilingual which of course automatically let me out. But being undaunted enough and answering with my pet phrase when asked if I spoke French, I said, "Au oui, je parle francais un peu, mais mal je crois". Desmond didn't speak French and when saying this little mouthful fast as I had done many times before, he thought this guy was perfectly bilingual. I never travelled with Desmond in Quebec so he would never find out that my French language skills were very limited. While travelling with a Dave Molloy and Peter Huxhold, geologists for Bond, they both thought their mining consultant (being me) was bilingual. At a fast-food counter in Noranda, they asked me to order for them. One doesn't have to know very much French to order a meal at a fast-food counter. This was done without any trouble.

While working for Kidd Creek, I attended a CIM Convention in Toronto where I delivered a paper on sinking the Kidd Creek No 2 shaft. At the CIM Annual Dinner that evening, Irene and I were seated with Ray and Pauline Clarke along with Dennis Bridger and his wife. Dennis, at that time prior to his retirement, headed up the Toronto office for BP Canada Inc. When starting consulting sometime later, I went in to see Dennis as well as John Haney (who was a junior there and an acquaintance from Steep Rock) and left a resume. A couple of years later, John Auston, who had replaced Dennis Bridger when Dennis retired, phoned and wanted an investigation of the BP gold mining operation at Hope Brook Newfoundland. Apparently, he had seen my resume in the files and talked to John Haney (who was then working for a brokerage firm) about me. *(It's nice to have friends.)* The work in Newfoundland covered a two-year period.

While living in Kingston, I attended a seminar sponsored by Revenue Canada on small business. At the seminar I spoke with a Michael Wilson, business auditor for Revenue Canada, who said that his department often used mining engineers to audit mining research projects and asked if this was of any interest. After giving him my card and forgetting all about it until much later,

Why Mining?
a call was received from authorities in Ottawa asking me to obtain a security clearance from the RCMP as they had a mining research audit to perform. *(All this was out of the blue.)* After filling out the papers they had sent and then being interviewed with the RCMP in Kingston, I was given a high security clearance. *(The security clearance was passed with flying colours.)* The mining research audit turned out to be seven different audits covering everything from drill rigs and bits to agglomeration of smelter dust.

Surface Plant of the LKAB Iron Mine Kiruna Sweden

This picture was taken at midnight from the roof of the Fermann Hotel at Kiruna Sweden in June 1983. The large building at right center is the famous headframe tower holding the eight ASEA friction hoists.

Chapter 10

Potash & Sulphur

Canterra

The timing for starting a consulting business couldn't have been better. First of all, the assets of Kidd Creek Potash were retained by CDC and would be managed by Canterra, an oil company, which did not have any potash expertise nor did it, at that time, have any mining engineers on its payroll. Here was the opportunity to continue what I had been doing for Kidd Creek. Secondly, the federal government had started a programme of "Flow-Through Shares" where any investment spent in mineral exploration would flow through to the investor as a tax deduction. As a consequence, there was considerable exploration and mining development in the mineral industry at that time.

Falconbridge made it clear that it would not be active in potash as that particular asset would remain with CDC. Mike Stewart of Canterra came to the Kidd Creek Commerce Court office in early February 1985 when I still worked for Kidd Creek and asked to be brought up-to-date on potash and requested access to all outstanding potash agreements. He said that, as the future representative for potash, he would be attending the March participants' meeting on the 20th of that month and asked both Eric Dison and I to attend that meeting with him. Both Eric Dison and I obliged by attending the March meeting with Mike who asked many very pertinent questions at that particular session. One could see from the start that Mike Stewart was a very thorough man and would leave nothing unanswered. Mike Stewart impressed everyone at that meeting right from the start. Afterward, I presented Mike with my agency proposal for managing the potash asset for Canterra. This proposal included potash operations, transportation and traffic as well as sales and marketing. He didn't comment on any specific item except to say that it was well written and covered all the items he had in mind. This agency agreement that Eric Dison and I prepared was for us (Eric and me) to manage the potash assets for Canterra under contract.

The first assignment as a consulting engineer, besides attending the March participants' meeting, was to fly out to Vancouver and attend the Canpotex meeting with Bob Murray vice-president sales for Canterra. *(I was formerly on the board of directors of Canpotex Bulk Terminals.)* While at the coast, a bit of networking was done but nothing ever became of this, perhaps because

Why Mining?

people thought... here was someone from the east who really had no right to be in the Vancouver area. Somehow, a call to Wright Engineers did pay off much later which was surprising. *(I did, however, visit an old high school buddy, Doug Williston, who was suffering from a malignant brain tumour. Doug was in very bad shape and finally succumbed to this terrible condition a few months later. This trip afforded me the opportunity to visit this very dear friend for the last time).*

The first week in April 1986, Mike Stewart phoned requesting I meet with Parvis Mottahed, the rock mechanics expert for PCS the following week in Saskatoon. He did not mention anything about the agency agreement that was proposed. In any case, Mike, at this time committed Canterra for consulting services for a three month period at $4000 per month plus expenses. It was hoped that a longer term would be forthcoming but as it turned out, it worked out better than one could have anticipated.

That next week (April 9, 1986), Bob Murray, Eric Dison and I met at the CDC office in Toronto to review the potash budget for the "1986-87 Fertilizer Year". At the same time we had an hour conference call with Mike Stewart who was calling from Calgary. Still no word was forthcoming from Stewart regarding the agency proposal. In the agency proposal, one could not be too impressed with what Dison offered in the way of sales marketing plan. He didn't seem to know what were the potential sales for each prospective customer. Some time later when face-to-face with Stewart, Dison was quizzed on this and was very vague in this respect. As a consequence, Mike said that Canterra would, for the time being, manage the potash themselves but they would use McKay Mining Engineering as its consultant to advise them on the technical aspects (mining) of the business. *(Had there been someone else other than Dison, that is a person who was more organized, there may have been a chance to have obtained a commitment from Canterra. The problem with Dison at the time was his drinking problem as the troubled guy was an alcoholic.)*

The last week of April 1986 was dedicated to moving to Kingston where Irene and I took possession of our townhouse in Frontenac Village on the inner harbour along Anglin Bay. After moving to Kingston, I finalized the incorporation of McKay Mining Engineering Inc (MMEI) and set up an office at home. During the first year in Kingston, a typist was used to type the reports as I didn't consider myself capable of typing these reports myself. This meant a lot of running back and forth making corrections to the various drafts. That first year, the only electronic device in the office was a combination telephone and answering machine that was purchased for MMEI. *(Talk about high tech!)* Finally in the spring of 1987, MMEI purchased an XT compatible with all of 30 mega bites of memory as its first computer. But the biggest item that was purchased or, should I say, leased for MMEI was a 1986 Buick Regal. Because of the travelling by automobile either to Toronto or Montreal for meetings or to catch air flights, it was necessary that Irene and I have two automobiles, otherwise it would be leaving her without transportation. On leaving Kidd Creek after the take-over, Falconbridge was not aware that I had a company vehicle (Oldsmobile Firenza). When being terminated after the take-over, Falconbridge offered 60 weeks salary and all my benefits which of course, in my mind, included the automobile. As a consequence, Irene used the Falconbridge car the first year and I used the MMEI car. It was a good setup as Falconbridge paid all the gas bills too. After some time Dave Hambley, the personnel man from Falconbridge called, enquiring as to what was going on. Answering him that this was one of my benefits Falconbridge gave, he said, "...it was an expense not a benefit". Anyhow, in order for Falconbridge to get out of having this automobile benefit, they suggested that I purchase the car for much less than cost which was accepted and that same car is still being driven today by Irene in 2002.

When leaving Kidd Creek, I made three plans as to what I would do for the rest of my working life, these being: 1) start a consulting business, 2) set up a potash agency, and 3) begin a career in teaching. Teaching didn't have that much appeal so it was abandoned quickly. The agency fell through when Canterra elected to market and manage its potash asset themselves and perhaps partly, it was flawed for choosing the wrong people to come into the organization. The consulting business more or less fell into place as a result of Canterra deciding to manage its 40% of Allan potash. As added insurance, it was necessary to have something to fall back on just in case consulting in the mining industry became slack. The opportunity came when a friend of mine, Duncan Prange of Midland Walwyn suggested enrolling in the Canadian Securities Course in order to prepare to become a stock broker. He mentioned that his firm was opening an office in Belleville soon and would be looking for local people who knew something about the resource industry to work for them. I immediately registered in the Canadian Securities Course

Why Mining?

that summer and passed the exams the following spring. Even though passing the exams and qualifying to work for a brokerage firm, it was never necessary to pursue that option as consulting more than kept me busy for the next fifteen years.

With regard to consulting for CDC Canterra, the three-month contract was up at the end of June and to my surprise, Mike Stewart extended it to the end of the year at the same rate ($4000 for five days each month). That works out to $100 per hour but in most cases, it worked out to be longer than eight hours per day. *(When one works for one's self, one doesn't mind working the extra hours.)* It was on those trips to Saskatoon that Mike Haug, Canterra's manager of marketing and transportation was introduced. The relationship with Mike Haug lasted for over fifteen years, first with Canterra and its subsidiary Saskterra Potash Inc, then with Husky Oil and its subsidiary Sulchem and finally to NuFarm and its subsidiary SulFer Works. *(This association will be discussed in greater detail later.)* Once Mike Stewart had more or less gotten his feet wet in the potash industry and had Saskterra organized, he turned Saskterra Potash over to Mike Haug to look after the details but not necessarily to have the final word. Everything still had to go through Stewart. Mike Haug and I hit it off, right from the start. When making the monthly visits to Saskatoon, Mike Haug would accompany me underground and through the surface plant as well. For the three years working with Mike Stewart on behalf of Saskterra, Mike Stewart in contrast to Mike Haug never went underground. There was ample opportunity for Stewart to go underground but he always declined. He did accompany me through the surface plant on a number of occasions which led me to believe he suffered from claustrophobia when it was suggested that he tour the underground. *(The fact that Mike Stewart didn't like the thought of going underground was probably the reason why he kept me around.)*

In the autumn of 1988, the assets of Canterra and Saskterra Potash were sold to Husky Oil. Because Mike Stewart was no friend of the CEO of Husky, he took his golden handshake which was later disclosed... a golden parachute and up and left. Canterra had a poison pill clause for its top management which was very lucrative. Mike Haug then became the immediate contact with Saskterra. Again, because Husky Oil had no one on board who was knowledgeable about potash or mining, the contract with Saskterra was honoured and extended. *(Mike Haug had a lot to do with this.)* As mentioned

Mike Haug

before, Mike Haug and I hit it off right from the start. Mike was one of those fellows who was never cheap with the company's money nor with his own. When Allan Potash won a safety award, Mike Haug went out and bought safety awards for all the employees at Allan. He was always buying jackets to give visitors to the Saskterra office in Calgary. As far as expense accounts go, he was always suggesting that "this or that" be included in my expense account. On one occasion, (while attending a participants' meeting which necessitated a stay in Saskatoon for over a weekend), I received word of a death of a friend in Victoria. Mike suggested, rather than staying idle in Saskatoon that weekend, I go to Victoria on my expense account. It was not my intention to do that but Mike insisted. Mike was one to make some very outrageous statements. One time when he was attending a traffic convention in Montreal, he was asked if he had ever met any French people before. Mike said, "Oh yes, but out west we call them Metis". Mike could say these shocking things with a smile and get away with them.

Husky did quite well managing its subsidiary company, Saskterra Potash. The management and staff of Saskterra numbered less than four persons: Mike Haug (who acted as manager) and Eugene Nagai (who managed the traffic) were the only full-time employees, while George Ayers (who did the accounting) and myself (who offered consulting services) were contract employees working part-time. In the years 1988 and 1989, Saskterra made a profit of $2 million and $10 million respectively when, at the same time, the profit for the parent company Husky Oil during those years was identical. In other words, had it not been for Saskterra, Husky Oil would not have made a profit in the years of 1988 and 1989. In the spring of 1990, because Husky was in need of capital, and potash was not its core business, Saskterra was sold to the Potash Company of Saskatchewan for $57 million.

Looking back at Allan Potash, when first starting the travelling to the participants' meetings, the mine could never fulfil its budgeted tonnage. At the time Floyd Pellan (who was formerly the chief accountant at Allan) was appointed manager when Jim Sadler was moved to head office and promoted to vice-president. PCS thought at the time, they needed someone who could negotiate with the union and keep it under control, thus Floyd Pellan became the general manager. At that time Jim Brandle, the mine superintendent, who

Why Mining?

was a practical man and not an engineer, was able to pull the wool over Pellan's eyes. To compensate for Pellan's lack of mining knowledge, PCS appointed Hedley Duffield, a mining engineer, as operations manager under Pellan. This was not a good move. First of all, neither Duffield nor Brandle got along. Both men resented young engineers who tried to make suggestions for improvements and as a result, many young engineers gave up and quit. As a consequence, Allan Potash lost a number of good engineers. Finally, PCS could see the error and terminated Duffield and then moved Pellan to the financial department at the PCS head office where he belonged in the first place. They then hired Jim Johnstone, a Scot, as general manager. The first thing Jim did was make a new organization chart where he could see that, in its present form, there were too many levels of supervision and too many people. He had to wait until after the provincial election in the spring of '87 before making his move. *(Politics plays an important part in all crown companies.)* When he did make his move, he terminated the mine superintendent and mine general foreman and replaced only the mine superintendent with an aggressive young engineer by the name of Barry Schmitke. Johnstone also replaced the maintenance superintendent with Ken Podwin and mill superintendent with Jim Mitchell, both good men. Finally, to make a long story short, Johnstone cut the total payroll from 450 employees to 325. The very first month (after making these various changes) the mine made its budget tonnage for the first time in years and lowered the product cost from $45 per tonne to just $25. *(A lesson can be learned from all this. Here was a mine and plant with the same equipment and the same orebody but with different people. It just goes to show that it's people that make or break an operation.)*

Unfortunately, about a year after Steve Harapiak, president of PCS, was terminated, Johnstone left PCS and joined Belmoral where Harapiak had taken over as president of that company. PCS lost a good man in Johnstone but he could not see eye to eye with the new vice-president of PCS Mining, Jim Bubnick. Bubnick, who was an accountant by training had progressed up the ladder through employee relations and didn't seem to have an appreciation of the technical end of mine management. Bubnick said to me on more than one occasion that he thought the present head of accounting at Allan Potash would make a fine general manager as anyone with a bit of management ability could manage a mine. *(I agree that anyone with some management ability can manage a mine as long as it is organized and all the technical*

Sulphur & Potash

details are in place. In the case of Allan Potash, it had operated for 15 years with little success because they had the wrong people and were using the wrong techniques. Between Johnstone and Schmitke, they got the mine turned around. Also with Podwin and Mitchell appointed to the refinery plant, they between them had that unit up and running consistently. It was people with technical ability that got Allan Potash producing. Once a system has been started, as in the case at Allan Potash, then anyone with managerial talent, even an accountant, could then take over and successfully manage the mine. A case in point is, "What ruined IMC K-2?" It was having people who were not concerned with technical details but only production that aggravated the roof condition where flooding was the result.)

Sulchem

During the last year (1990) of MMEI's potash contract with Husky, when the assets of Saskterra (Allan Potash) were sold to Potash Company of Saskatchewan, there were still a number of days remaining on the Saskterra Potash Contract. Rather than pay it off, Husky decided that I should travel to Crossfield, Alberta and review the construction of the Sulchem Fertilizer Plant which was a joint venture between Husky Oil Company, Palladium Resource Corporation and two individuals (Raymond Lee and Duncan McRae). The Sulchem Fertilizer Plant was being built to produce granular sulphur and sulphur powder for the citrus industry and for those agricultural regions that were deficient in sulphur. *(Besides the three main requirements in plant food: nitrogen, phosphate and potash, there is a fourth and that is sulphur.)* The Fertilizer Plant was in the last stages of construction when it was visited in June 1990. Startup was scheduled within the next two or three months and the consortium was very excited as this would be the prototype of other plants to be built.

The individuals in the consortium were rather interesting. Mike Haug, the outrageous one, represented Husky on the project. Hans Schmidt and his wife were the soul owners of Palladium. Hans was a deadbeat as he was reluctant to pay for contracted services. *(More about him later.)* Ray Lee was an agriculturist who was in charge of marketing the sulphur fertilizer product which was something new in the agricultural industry. Ray was raised in Crossfield where his parents, immigrants from China, formerly ran a restaurant in the Town of Crossfield. Duncan McRae was a civil engineer who had worked in a sour gas plant with Ray Lee where he and Ray Lee

Why Mining?

gained their knowledge of sulphur. McRae was a hands-on engineer who couldn't leave well enough alone sometimes. George Ayres, the accountant, called McRae "Tinker Bell" because he was always tinkering with the circuit and not always for the good either. Along with these guys, McRae and Lee brought with them a number of cronies who weren't that great. When Mike Haug finally took charge of the plant, he fired both McRae and Lee along with their cronies.

The assignment given on this initial visit to Crossfield was to determine a date when the plant could commence commercial production on a continuous basis as well as to point out what could be done to generally improve the overall operation. From what could be seen, I couldn't add much to what had been done, except to say that those in charge tried to commence production that spring (1990) by adding a 60 ACM micronizer (which was not part of the original circuit) and thus complicated the circuitry and at the same time they missed the Spring Fertilizer Market.

As it turned out, the plant started up in the autumn of 1990 and it immediately had trouble in the drying section. An additional dryer had to be added. Then, the pan granulator would not give consistent size. As a result, the plant produced "off-spec material" all that autumn and into the following year. In December 1991, I was asked to travel to Calgary to be interviewed by Paul Augustin (Husky Engineering) for a three-month position as superintendent of the construction and process revisions to the Sulchem Plant. *(He must have been impressed as I got the job.)* What Husky had planned was to make a number of mechanical improvements and change the process as the plant continued to produce too much off-spec material which was totally lost and therefore making the operation unprofitable. This was a three-month contract in Calgary where I would supervise the construction workmen and plant personnel. *(Yes, Augustin placed everyone at the plant, including Duncan McRae, in my charge. Lee and McRae were not that jubilant about this arrangement but nevertheless that's the way it was.)* Arriving in Calgary on Sunday Janurary 5th, 1992 and staying at the Bow Delta Hotel that night, I then moved into the Executive Suite Hotel which would be home for the next three months.

The process changes were under the guidance of Prodecor, a firm based in Saskatoon, that had considerable experience in pan granulation of potash.

This firm revised the flowsheet and devised a method of feeding and mixing the turbulator and pan granulator evenly and steadily. This required installing a feed hopper and vibra-screw feeder below the dust-collector and above the turbulator. A bit of structural work was required here but it was possible. They also recommended the introduction of a feeder to dispense fine seed particles of sulphur to the pan granulator from which the fine sulphur dust would accumulate. Other changes were to the baffles and spray nozzles on the pan of the granulator as well as new mixing tanks for the reagents required as a binder. Among the improvements were the addition of a dust collector for the bagging units and a drum dedusting unit. Some safety improvements were also engineered such as a safer stairway to the pan granulator operators' mezzanine, as well as to the roof where the dust collector for the granulator was located.

At the same time as the process changes were being made, some management systems were introduced as well. The plant never had a requisition system nor purchase order system in the past. A number of ready-made requisition booklets were purchased from a stationery store and issued to Mike Haug, George Ayres, Duncan McRae and his side-kick Walter Kiceluk who acted as production superintendent. *(There was some reluctance on the part of Kiceluk as he, in the past, was used to just picking up the telephone and ordering anything he wanted. Some of the stuff he ordered was for his own personal use.)* Also, to keep control of costs, a cost control system was introduced, called JOCS (job order cost system). *(Roger Harris, when he worked in Timmins, put together this system of cost control which worked very well and gave one instant control of construction costs.)* A log book was started where the shift foremen on shift would record the highlights of the shift so that the on-coming foreman would have a record of the previous shift's activities as well as the present condition of the plant. Safety was not neglected. In the time spent superintending the changes, a safety check system was introduced where the workmen were issued a safety card each shift to be filled out and returned at shift's end. Through the Compensation Board, we were able to have an instructor come out to the plant and conduct sessions on fire-fighting and rescue for operating sulphur plants. Prior to this there were no set procedures for fire nor any of the above listed management and administrative systems.

After certain improvements had been made in the plant, the operation was

Why Mining?

started first by attempting to produce sulphur particles by the "agglomeration" process which was the method that the plant had tried in the past but had failed. The idea of giving this process one more trial was based on the fact that improvements in the plant had been made (particularly in feeding the pan) at a more steady rate. As before, this process again failed. The next attempt was by the "granulation" method where a sulphur "seed" particle would be introduced onto the pan granulator just before the sulphur powder was fed. This process right from the start showed some success; however, another old problem of explosions reared its head which took some of the jubilation out of the situation.

Sulphur dust at 44 microns can be highly explosive as the Crossfield Fertilizer Plant can confirm. When the plant was built, the persons in charge decided that a suppression system should be built into the plant rather than a preventative system such as inerting the atmosphere within the tanks and dust collectors. That initial decision cost the plant considerably in initial cost and was the cause of much downtime in operations. The idea behind the decision of a suppression system over a prevention system was the thought that inerting was more expensive to operate; whereas, the reverse was true. The capital cost of the suppression system was $231,000 in 1990 dollars. An inerting system, using the gas company's equipment (which was supplied free with the gas purchase), would have amounted to only the cost for the installation of a liquid air tank and some ducting. The savings in downtime would have more than made up for the cost of carbon dioxide or nitrogen gas used. After the initial trial of the granulation process, management decided to install ducting to inert the dust collectors. This job was added to the construction list and was completed in a week or so.

Before the construction contract was completed, the plant was started again. The product made was on specification but this time an explosion occurred in the dryers. Someone (D McRae) had installed a small fan at the head of the dryers to cool the first of the three dryers in series as it was overheating. This fan was pushing dusty air directly into the dryer. That caused the explosion in the first dryer. What had happened in this particular instance was that a new operator, who had never been in a startup mode before allowed too much powder to be added onto the pan granular before turning on the spray nozzles. What was thought to be a good idea (cooling fan) turned out to be a disaster.

Sulphur & Potash

After the changes to the plant had been made, the three-month contract came to a conclusion and I returned home to Kingston. But that was not the end of the association with the Crossfield Fertilizer Plant as there was a call from Husky Oil requesting an evaluation of the plant in order for them to put the operation up for sale. In December 1992, after moving to Vancouver Island, I was asked to evaluate the whole operation. With the help of my son Jimmy, who did all the adding and subtracting of numbers from reams of computer print-outs of capital costs, a replacement cost was calculated at $5.27 million (in 1992 dollars) with a discounted cash flow of 36% to 41%. *(All the original construction costs were collected under one account number as there was no attempt to assign expenses to any section or activity during construction.)* As a result of this evaluation (and the tax credits for the past operating losses) Husky accorded a lease/purchase agreement and leased the plant to Hans Schmidt who immediately renamed the company from Sulchem Products (1989) Limited to Solterra Mineral Inc. and retained Mike Haug as a vice-president and general manger of this new company. *(As Mike found out afterward, he was general manager in name only.)*

The plant operated trouble-free during the first quarter of 1993 producing bagged sulphur prills and powder. In June of that year, Solterra asked that the operation be evaluated for the purpose of issuing a prospectus as Schmidt wanted to go public and raise capital in order to purchase the plant outright from Husky. On the basis of the first quarter operation, the replacement cost was determined at $5.24 million in June 1993 which was about the same as was calculated six months earlier. The rate of return was well over 100%, which in anyone's language is extremely good. Schmidt had me do some other small jobs for him but refused to pay for these. In all, these extras amounted to about $1300. At the same time, he owed George Ayres, formerly the accountant at Sulchem, over $5000. After writing a number of letters, I more or less gave up ever collecting the $1300. It was at this time that Mike Haug left the employment of Solterra because of Schmidt's lack of integrity. *(While this was going on, McKay Mining Engineering was in a legal battle with Armistice Resources to get them to pay $12,000 owing for a report that was written in 1991.)* As part of going public, Schmidt's accountant for Solterra telephoned that autumn requesting the notes and backup calculations for the evaluation report made the previous June. Hans Schmidt even had the general foreman Norm Grantholm phone, saying that I was still their consultant and please send the notes. Answering him, I said

Why Mining?
the notes would be sent providing Solterra paid the $1300 owing. The cheque for $1300 was received and the notes were sent as my part of the bargain. A few months later, Solterra's lawyer called saying they were compiling the data for the prospectus and requested that they have my written permission to submit (as part of the prospectus) the June 1993 report "Description & Evaluation of the Crossfield Fertilizer Plant Solterra Minerals Inc." I said, "Yes, providing that I receive a cheque for $500 first." *($500 was small compensation for all the trouble Hans Schmidt had caused.)* The lawyer sent the cheque and I sent my permission to use the report as part of the prospectus. Again a few months later, the same lawyer called requesting written permission to submit my report to the Alberta Securities Commission for the purpose of listing Solterra Minerals Inc on the Alberta Stock Exchange. I again said, "Yes, providing he first sends a $500 to cover my trouble". The lawyer was taken aback, but I said, "I'm sure you don't work for nothing and I am certainly not going to go out of my way for Hans Schmidt for nothing". A cheque for $500 was received and a letter giving Solterra Mineral permission to again use the Description & Evaluation Report was sent in return. *(Score two for the good guys.)*

This was not McKay Mining Engineering's last dealings with Solterra, or one Hans Schmidt, as Husky Oil in March 1995 (who held a security document on the plant and equipment) were accepting offers on their security interest. MMEI's terms of reference were to confirm the existence and status of the assets, describe the condition of the equipment and plant circuits, confirm the status of the operating licences, estimate the volumes of product and inventory, and the condition of the plant regarding operability and state of maintenance. I met Hans when visiting the plant, although he didn't stick around and gab like he did in the past. *(Neither one of us mentioned our past difficulties but it sure was on my mind and I'm sure it was on his.)* This time I was working for Husky Oil and was paid right on time.

Something should be said about this last payment from Husky Oil. When making out the invoice, I decided to include it along with my Status Report which would be sent Priority Post. On calling Canada Post, they quoted an estimated price of roughly $6.50 which was included in my expense account. When mailing the report, the postage came to $6.33 and not the $6.50 used in the expense report. It must have made one of Husky Oil's bean-counter's day, as the cheque payment was shorted by 17 cents. Nothing was considered

Sulphur & Potash

for the time taken in mailing nor auto mileage to the post office.

SulFer Works

In the Autumn of 1993 Mike Haug had a falling out with Hans Schmidt over policy at the method that Hans wanted Solterra to be operated. When Mike told me of his plan to leave Solterra, I advised him not to leave at this time, or until he had another job but that advice fell on deaf ears. Mike left Solterra shortly after our phone conversation; as a consequence, Mike and his family had some very lean times in the next eighteen or twenty months that he was unemployed. In fact, Mike and his wife Jan had to sell their house and move in order to make ends meet. All the time Mike was unemployed, he never lost faith in starting his own company for the manufacturing and processing of sulphur fertilizer. *(Mike has more fortitude than most people. He also had a lot of integrity because what Hans Schmidt had proposed was illegal. The full story regarding Hans was learned a couple of years later.)*

That November, Mike phoned to say that he was very serious about starting his own company and wondered if I would go along with him as he needed some engineering done but didn't have the means to pay. I told him I would take my chances and to let me know what he needed in the way of estimates or drawings. In February 1994, the first of many flowsheets for Mike Haug's new company "Newco" and an estimate to build a sulphur fertilizer plant was started. These estimates were sent by Fax and the AutoCadd drawings and disks were sent by mail to Mike in Calgary. In March, he asked for revisions to the estimate based on a simple one-line (one-train) operation showing two scenarios with annual rates of 30,000 and 60,000 tonnes as Mike had an investor with limited finances; however, nothing came of that proposal. Mike and I did have a few conversations over the phone that year regarding his project but he seemed no closer to his goal of obtaining financing.

The following spring (1995), Mike contacted Fernz, a fertilizer company headquartered at that time in New Zealand (but now in Australia under the name of NuFarm) who apparently were interested in Newco and its sulphur process. In June of that year, Mike Haug sent me some bidding documents and specifications for review. These were reviewed and returned with comments regarding the scope of the project which now had blossomed into a semi-automated two-train processing plant. That summer, Monenco-Agra were contracted to prepare a preliminary feasibility design and estimate for

Why Mining?

the Newco sulphur fertilizer plant. In August 1995, Mike and I traveled to Saskatoon to review the plans and estimates prepared by Monenco-Agra. The folks at Saskatoon were the remnants of the old Cambrian Engineering Group (Gordon McMahon, Stuart Middleton and Ed Hinz) with whom I had previous dealings when working for Cominco Potash in Saskatoon. *(I knew Ed Hinz's brother Wilf very well as we worked together on his first consulting project with Cominco Potash.)* The Newco project had reached a cost of $14 million as it now had changed from a semi-automated to a fully-automated plant. Mike had hoped that the cost would be less but this is never the case. He was somewhat taken aback because one of the stipulations in the scoping documents was for the consultant to review other methods, processes and equipment which in this case, they had not done. This and the fact that the estimate was higher than expected should have been a warning.

Mike made the trip on a shoestring as he had no money. His plane ticket was bought on his Visa Card and I paid for his meals and hotel room which in turn I invoiced to Fernz. For these expenses, Fernz paid promptly as was Mike's I guess, but all the same, Mike was very short of spare cash at that particular time. The trip to Saskatoon was not all business even though Mike was broke. We were taken to the University Faculty Club for lunch on Monenco-Agra one day and on one afternoon Mike and I had a drink on Jim Bubnick, vice-president of PCS Mining at the Ramada Inn. That same evening, I had dinner with Dave Porteous, the best man at our wedding who was now living in Saskatoon.

After reviewing and digesting the preliminary estimate, the next step in the process was for Monenco-Agra to prepare a more comprehensive design and estimate incorporating some of Fernz's ideas. That November (1995) I traveled to Calgary where I met Mike Haug and Earl Stevens, vice-president of Fernz. Earl had just arrived from New Zealand and, after the meetings in Calgary, he was slated to go to New York and then on to London before returning home to New Zealand. *(He was a very busy traveler indeed.)* The purpose of this day-long meeting with Monenco-Agra was to hammer out the details as to how the project should proceed and generally review its scope. The Monenco-Agra Group consisted of Doug Annable, Stuart Middleton, Murray Holmlund and Bob Young while the Newco Group consisted of Earl Stevens, Mike Haug, Dave Whelan (lawyer) as well as myself. This was a very friendly meeting, somewhat in contrast to those that were yet to come.

Anyway that winter and the following spring the Monenco-Agra Group along with Mike Haug as reviewer (or monitor) began their comprehensive design and estimate for the Newco Sulphur Fertilizer Plant or Elemental Plant Nutrient Sulphur (EPNS) Facility.

In June 1996 a explosion of expletives was issued by Mike Haug when the comprehensive estimate completed by Monenco-Agra came to $30,247,762, more than twice as much as the preliminary feasibility study made by the folks in Saskatoon. Mike sent a copy of that report and estimate for my opinion. Mike, (who said he could kick himself for not getting me involved the past spring when Monenco-Agra were preparing the final designs and estimates), apologized and asked me to come to Calgary and prepare an audit for Fernz SulFer Works Inc the new name for Newco. When in Calgary, the people of Monenco-Agra were met, starting with Dave Scoulding (senior specialist), Eric Nielson (senior mechanical engineer), Eugene Freese (electrical engineer) and Don Davies (material handling specialist). The questioning lasted all of one day and a good part of another. Without going into detail, Dave Scoulding said that the estimated cost of the plant was increased because 1) the site was changed, 2) suppliers increased the cost of their equipment, 3) the client had made a number of requests for change and 4) other changes in design and development. After completing the discussions with Monenco-Agra, I attended a meeting at the Canterra Towers with Randle Black, Steve Wilson and Dave Whelan, lawyers for SulFer Works; Ed Dyck, the new project manager for SulFer Works and Mike Haug. I stated that the estimate was high and I would put together an estimate based on this latest design. I also stated that there appeared to be some gold plating on the office facilities which was not included in the original scoping documents.

The audit report was completed in August and showed the Preliminary Feasibility Study (FS) at $13.3 million with no contingency; the Interim Service Agreement (ISA) $30.2 million which included $600,000 for unforeseen items, $2.3 million profit and a 10% contingency; and McKay Mining Engineering Re-estimate at $18.1 million which included a 25% contingency. It should be mentioned that the FS and MMEI estimates are about equal when a contingency is added to the FS estimate. But the pudding is in the eating. The base plant when built cost $18 million but extras, when added later, brought the cost up to $20 million. This still is $10 million less than the plant estimated in Monenco-Agra Interim Service Agreement.

Why Mining?
After my August 1996 Audit Report was issued and from that time onward, no more money was paid to Monenco-Agra and of course legal action was started by Monenco-Agra with a counter-suit started by SulFer Works. On the engineering side, SulFer Works hired Stanley Engineering of Calgary to start all over again on the basis of the MMEI estimate and scope. I met the group (Bob Ramsey, Paul Foss, Roger Dingman, Dean Luborsky, Kathy Butler and Anthony Thorpe all of Stanley and Ed Dyck of SulFer Works) at the Stanley Calgary office in late August 1996. My job at that time was to educate these people in the process and to generally bring them up-to-speed regarding scope and estimate. That was the end of my involvement for some unknown reason. The plant was designed and built in the next 15 months and started tuning up in November 1997. During the design and construction period, I was not involved whatsoever. I had dinner with Mike Haug and his marketing people one night in January 1998 at Vancouver but that is the last time I saw or had anything to do with him and his company for a while. SulFer Works did take legal action against Monenco-Agra and vice versa but, as I mentioned, Mike Haug dropped me as a consultant so I didn't know at the time how that legal action was resolved. *(I felt badly that I didn't hear from Mike as I felt we had a good working relationship which spanned over eleven years from Canterra, Saskterra, Sulchem, Solterra, Newco and SulFer Works.)*

Wow! Things did change! Later in January 1998, after meeting Mike Haug and his marketing group in Vancouver, there was a fire at SulFer Works and the plant was shut down. It wasn't until the following November (after considerable repairs) that the plant was operational again. The plant operated for all of 1999 more or less under a cloud as the parent company Fernz Inc had lost faith in the Canadian operators. After starting up in late 1998, nothing seemed to be improving; in fact, things began to get worse as, not only was production under budget but safety and housekeeping were falling behind as well. This must have been a very frustrating time for Mike Haug because even though he was CEO, he was told by Earl Stevens (Vice-President of Fernz) to keep his hands off the operation and concentrate on marketing as Ed Dyck, an engineer, would handle the operations. Besides losing faith with the SulFer Works group, Fernz lost faith in Earl Stevens and fired him in early June 1999 for overspending on a number of projects, one being SulFer Works. After the departure of Stevens in mid 1999, Fernz began sending trouble-shooters out to the SulFer Works plant to see what could be done to bring that operation up to speed. Also, with Earl Stevens gone, Mike Haug was able to get more involved in the operation but things didn't change much. I guess finally in desperation they asked Mike Haug if he knew of anyone who could help them.

This is where MMEI got involved again. Mike called me in early January 2000 to ask if I would be available to come out to the plant for a week a month and see what could be done to improve both safety and production ... in that order. I was only too happy to oblige. *(After two years of not hearing from Mike, it would be exciting to be working together again.)*

In mid January 2000, I went out to Irracana Alberta, a town about 50 km northeast from Calgary where the SulFer Works plant was located, and stayed there collecting data for two weeks. On the second week, I met with Mike Haug and Dr Peter Bibby (NuFarm & Fernz trouble-shooter) at dinner one evening where we could talk about the problem areas of the SulFer Works operation. Three problem areas were brought to the table, 1) management, 2) production and 3) personnel, but the main one was management. I recommended that Ed Dyck the vice-president operations *(fancy title for plant manager)* be moved to one side or removed altogether and that Mike Haug replace him as manager of the plant. Dr Bibby thought that moving Ed Dyke to the side was what should be done. Mike thought otherwise and said that wouldn't work. He said, "Ed Dyck should go". I agreed and finally after some discussion, so did Peter Bibby.

The second problem area was production where I suggested a number of improvements could be made such as air-blasters in the dust collectors, more reliable feeders for the pans and a hammer-type mill for the oversize material. These were money items which could have easily been incorporated in the next capital budget. As far as safety goes, (which was the third problem area), the appointment of a safety/training officer and the adoption of a safety system was suggested. All these suggestions were adopted and to say the least... they worked. Production increased after two months by 46% and safety incidents decreased to practically nil.

Up until the end of 2001, I was travelling periodically to SulFer Works to monitor the operation for NuFarm, the new name for Fernz. Although production had increased, the plant still had not reached satisfactory levels

Why Mining?
either for plant utilization nor product yield. In the year 2000, a number of improvements such as a bulk storage system, bulk loadout system, air-blasters, screw feeders to the grinders and some drag conveyor to belt conveyor conversions were made. In 2001 there were other improvements in the works such as an off-spec reclaim system, conversion to belt conveyors from drags in the product storage bin as well as other minor projects. It was also planned to operate, in the spring of 2001, both production trains simultaneously, something that was never done before as the plant operated only as a single train entity.

As an update, it should be mentioned that, because of a downturn in the agricultural market and because fertilizer was not NuFarm's core business, NuFarm decided to sell the assets of SulFer Works. As of May 2002, there were no buyers who would meet the asking price.

Fernz SulFer Works Fertilizer Plant Irricana Alberta 1997

Fernz SulFer Works is the name of the Canadian subsidiary company. **Fernz,** the name of the parent, is derived from **"Fertilizer New Zealand"**. The name,**"SulFer"** is derived from **"Sulphur and Fertilizer"**. **"Works"** was added as it sounded very dynamic and full of drive.

Chapter 11

Belmoral

Broulan Reef

Back in April 1986 when starting to look for other clients besides CDC Canterra, Colin Chapman, an associate from Kidd Creek, advised me to give Malcolm Slack of Belmoral Gold Mines a phone call. As a result of that call and after a very friendly conversation with Malcolm, he said he would get back to me. I found out afterwards that he had called Colin Chapman and asked about me. Colin was in the same boat as myself; that is, he too was starting a consulting business because of the Falconbridge take-over. Colin gave me a good recommendation. Slack also called Terry Walton of Lafarge Canada Inc who also gave me a good rating. According to Terry, Slack said, "...he remembered me from Timmins when he managed the Noranda properties there, as being a gentleman". *(I never thought I gave that impression but I guess it didn't hurt in this case.)*

Slack, true to his word, did get back to me and said that he was busy right then but asked me to come to his office the next day where he discussed his plans regarding his company and what consulting services would be needed. He asked me if I would be going to the Annual General Meeting of the Canadian Institute of Mining and Metallurgy (CIM). *("Petroleum" was not at that time included in the name.)* As Irene and I had already planned to attend the 1986 CIM Convention in Montreal, meeting Slack at this gathering worked into our agenda perfectly. While in Montreal, I met with Slack and he suggested that I should work on the Broulan Reef project in which Belmoral was interested. The project, as proposed by Slack, was to explore and possibly mine the deep high grade veins that dipped and raked onto the Broulan Reef property from the adjacent Hallnor mine, by either intersecting and gaining access to these veins via the Hallnor shaft or rehabilitating the capped Broulan Reef shaft. Belmoral wanted to utilize the governmental tax incentive known as "Flow-Through Shares" to explore this mineralization. While in Montreal, Slack said he would gather all those involved with that project, including myself, for a meeting the following Sunday at the Belmoral office in Toronto. At the Sunday meeting, besides Malcolm Slack and myself, were Bill Longworth of Timmins who would manage the project and Don Lavigne, manager of the Belmoral Val d'Or properties.

Why Mining?
At the meeting, it was agreed the three of us (Longworth, Lavigne and myself) would spend the next two weeks in Timmins investigating the most suitable access to the deep mineralization raking onto the Broulan Reef property from the Hallnor. The Broulan Reef mine plans had been stored, since closing down in 1965, at the Gorf Construction shop at Porcupine, Ontario. These were subsequently moved to the office space at McIntyre mines rented by Belmoral, and were reviewed. As a result of this investigation, it was proposed in my report that there would be three methods of accessing the mineralization, namely, 1) drifting west onto the Broulan Reef property from the lowest Hallnor level, 2) deepening the Reef shaft from the 2500-foot level by 1000 feet to the 3500-foot horizon and drifting east towards the Hallnor property, and 3) accessing the lowest level (2500 level) of the Broulan Reef property via the Reef shaft and ramping down, and eastward towards the Hallnor property. The first proposal (accessing the deep mineralization from the Hallnor property) was recommended as it accomplished the task in the least possible time at the most economical cost. Unfortunately, like all governmental programmes, Revenue Canada commented that accessing from an existing mine (even though it was under "care and maintenance") was not acceptable exploration for the "Flow-Through Share Programme" yet accessing from an abandoned shaft was quite acceptable. *(How did they justify that?)*

Belmoral explored the possibility of purchasing the Hallnor property and driving west from the lowest level using its own funds instead of financing this through the Flow-Through Share Programme. As a result, a number of cases and reports on this scenario were prepared. In September, a pre-feasibility report on "Reactivating the Broulan Reef Shaft and Mine" was written. Later that month, I was asked to prepare a second report, "Reactivating the Broulan and Hallnor Mines Utilizing the Hallnor Mine Plant" (for hoisting). The third report was requested and written, enlarging the whole project to mine the remnant ore from both former producers at a 1000-ton daily rate. Negotiations for the purchase of the Hallnor broke down when the president of Pamour, Don McLeod backed out at the last minute. By not having access through the Hallnor mine, the project was doomed but Belmoral pursued the reactivation and exploration of the Broulan Reef property through the Reef shaft anyway. What should have been an exploration project, Malcolm Slack decided to squeeze it in as a full-scale mining project, hoping that the Flow-Through would pay for the exploration

Belmoral

as well as for a mining plant for production. It just didn't happen.

The original report, "Reactivating the Broulan Reef Shaft and Mine" was updated, expanded, presented and accepted in January 1987. This report included surface exploration drilling as well as reconstructing the surface mine plant, rehabilitating and extending the Reef shaft 1000 feet, advancing the 2500 level east to the Hallnor and developing three new levels below the 2500 level and connecting these with 4200 feet of ramp. The surface drilling was under the direction of Ed Van Hees, a consulting geologist. Ed proposed that $2 million be spent on surface drilling while the report recommended only $$1/2$ million be spent as it was felt that the surface had been drilled to death as it was. In hindsight, what should have been done was to bring in a Navi drill and determine if there was any mineralization at depth, then decide if it was worthwhile deepening the shaft. This was not done and it was decided to rehabilitate and deepen the shaft. As a result, a secondhand hoist was purchased in Ogden Utah with the help of Tiley and Associates and then refurbished at Peacock Limited shops in Mississauga Ontario. The DC electrics for this hoist were acquired from a former Steep Rock friction hoist that I was involved in purchasing some 30 years earlier when working at Steep Rock. An old headframe located at Quebec Lithium was visited along with Don Lavigne (manager Belmoral Val d'Or) and Wilf Coutu (Pamo Construction Noranda) in the dead of winter. *(It must have been 30^0 below that day we inspected the headframe.)* Anyway the headframe and bins looked like a good fit for the Ogden hoist and as a consequence were purchased and removed from its Barraute, Quebec location and erected on the Broulan site. The compressors were secondhand also and came from Inco. The only new item was the building complex that housed the hoists, compressors electrical switch gear, shops and dry. The surface plant took shape early 1987 and dewatering began that summer. This was a fine looking plant but definitely not an exploration facility as defined in the Flow-Through Share Programme. To make a long story short, dewatering and reactivating the various levels took place for the next two years. The project continued down to the 2500 level, the lowest existing level in the Reef shaft. Some exploration and development was done on the 500 level to explore what was known as the "C" zone, first drilled from surface, but nothing of any consequence came of the underground exploration. Up to and including 1989, a sum of $14 million was spent on the project which produced little in the way of results. Had Revenue Canada allowed the exploration programme to access

Why Mining?
through the Hallnor shaft, the project would have cost only $9.5 million with a good possibility of being successful. Three mistakes were made here, 1) Revenue Canada negated exploration through an existing shaft but sanctioned access through an abandoned shaft, 2) Belmoral should have backed away when the Hallnor proposal was turned down, and 3) Belmoral should not have spent money on a permanent plant but rather they should have rented temporary facilities for exploration. *("Twenty-twenty" is great after a project is completed. But at the time, I was quite contented to go along with the Belmoral management's decision as it kept me employed. Looking back now, was I really being forthright?)*

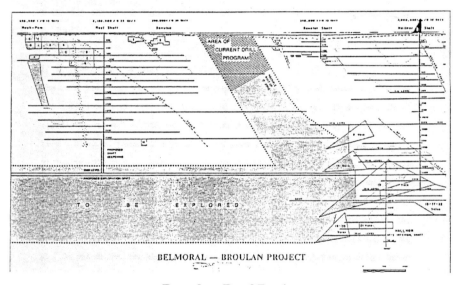

Broulan Reef Project
The plan was to deepen the Reef shaft (left) by 1000 feet and drive east to intersect what was thought to be the Hallnor veins that plunged into the Broulan Reef property. The two vertical lines (right) represent the Hallnor main shaft and the internal shaft (winze). The dark section (top center) represents the area diamond drilled under the direction of Ed Van Hees. The lighter shading (lower section) was be explored when the shaft and level development reached the lower horizon. Some level development was done on the 500 level but little was found. The Reef shaft was dewatered to its original depth (2500 feet) but the project was abandoned before further work was done.

While working on the Broulan project, Ed Van Hees and myself were requested to travel to Noranda to review the development (due-diligence) of the Eldrich-Flavel project. This property was owned and managed by Mines Sullivan Inc. of Montreal. How Ed and I got involved was through a request by Mike Ross of Orocon Inc of Vancouver B.C., who had been approached to joint venture with Mines Sullivan Inc to supply and erect a 500-tonne mill. We travelled to Noranda and visited the office to view plans etc. the first day, and toured the surface and underground plants the following day. What was seen on touring the surface plant was very good, but the reserves underground, from reviewing the drilling were very spread out and therefore required much lateral development in order to access them. Also the fact that the orebody dipped 37^0 made mining difficult as once the ore was broken it had to be scraped instead of carrying downward on its own by gravity. As a consequence, this project was not recommended. Ed Van Hees was a good team player and I enjoyed my association with him on this project and on the Broulan project as well.

Vedron

The Broulan project led to the Vedron project which was another Belmoral assignment in the Timmins area. The Vedron property consisted of the single Fuller claim adjacent to the abandoned Buffalo Ankerite mine and 57 other claims formerly held by the Buffalo Ankerite Group. Leon LaPrairie who, at the time, was president of Vedron had high hopes for this property. Belmoral got involved in this joint venture with Vedron on a fifty-fifty basis where Vedron would put up the property and Belmoral would supply the cash and management. This project started in late summer of 1986 via ramp access on the Fuller claim. My involvement did not really start until the following spring when I was requested to write a proposal to reactivate the Buffalo Ankerite No 5 shaft and connect the Buffalo Ankerite workings to the ramp that was in progress on the Fuller claim. This original request was enlarged to include mining and milling the 8.0 to 9.1 million tons of potential mineralization (determined by Ed Van Hees) on the Fuller claim and the other 57 claims controlled by Vedron. The key word here is "potential" as nothing was proven.

As one can see, Belmoral and Vedron were very optimistic regarding this venture. Leon LaPrairie, a very charming fellow, was adamant that Vedron had a mine on the Fuller claim alone. His consulting geologist, Alexander Po,

Why Mining?
was equally convinced that the Fuller claim would be a mine. *(I have seen it many times, in the course of my experience with geologists, that they often get so involved with a property that it becomes an obsession, that they actually fall in love with the property where they cannot let go.)* In any case, the ramp was extended down 500 feet below surface where three levels (275, 375 and 500) were explored by drifting and diamond drilling.

After Belmoral had spent $14 million on the Vedron property, Steve Harapiak, who had taken Malcolm Slack's place as president of Belmoral, asked me to evaluate the total project. *(This is the same Steve Harapiak who had been president of Potash Company of Saskatchewan but was let go by Premier Devine after the 1987 provincial election in Saskatchewan.)* The $14 million spent was mostly on the Fuller claim showed a reserve of 118,000 tons at 0.152 ounces per ton (opt) proven, 21,000 tons at 0.169 opt probable and 162,000 tons at 0.172 opt possible for a total of all classifications as 301,000 tons at 0.164 opt. This amounted to an exploration cost of nearly $5 per ton of reserve, a very high cost indeed. The report was very pessimistic regarding any more work on the Fuller claim and again reminded Belmoral that my original recommendation favoured more work on the Buffalo Ankerite claims rather than those on the lone Fuller claim. The report confirmed Harapiak's opinion but LaPrairie was most upset and went so far as to hire another consultant to give him what he wanted to hear. Without mentioning any names, LaPrairie did get a report that made out that the Fuller claim would make a mine. Belmoral pulled out of the project and left the joint venture soon afterward. *(There is a good possibility that Leon LaPrairie may be right but what was proven at that time would not in any circumstances be enough to make a mine.)*

The Buffalo Ankerite Headframe as seen from the Vedron Property.

Belmoral

Armistice

The association with Belmoral led MMEI to conduct work on the Armistice property at Virginiatown in Ontario. Steve Harapiak had just been appointed president of Armistice Resources Limited and asked for, during the Annual General Meeting of the Canadian Institute of Mining and Metallurgy held in Toronto, the preparation of an outline of an exploration programme that Armistice was considering in the summer of 1987. Being extremely busy working on the Broulan project at that time, I turned this assignment over to Wyatt Hegler (my former boss at Steep Rock Iron Mines) to conduct this assignment for Steve Harapiak. This was not a good idea, as Wyatt being Wyatt, liked a bit of humour, he then proceeded to insert little jokes into the report as well. For instance, he described the weather in Northern Ontario as 'eleven months winter and one month tough sledding". None of these little comments went well with the client and as a result the report had to be revised.

One interesting thing that transpired regarding the Armistice project was the enlarging of the compartment size in the shaft without further slashing of the walls. What Dynatec Mining Contractors came up with was the removal of the wall and end plates and filling in of the void between the walls and the outside edge of the plates with concrete, thus enlarging the compartments the thickness of the plate timbers. The Armistice shaft, instead of its hoisting compartments being 4ft x 5ft size, were increased to 5ft x 6ft instead. All this work (forming and pouring concrete and removing the timber) was done while the shaft was being dewatered. *(It was a young draftsman at Dynatec Engineering who thought of this idea of enlarging the compartments. I often wish I had come up with that idea for the Broulan Reef shaft as it would have assisted in the shaft rehabilitation considerably.)*

Ferderber & Dumont

Considerable amount of work was done for Belmoral in Val d'Or at the Ferderber and Dumont mines. Belmoral held significant land holdings (22,888 acres) on the Bourlamaque Batholith north-east of the city of Val d'Or, Quebec. This was like going back home as I had worked in Malartic (which is a distance of only 17 miles from Val d'Or) some 35 years before. How things had changed since then. When working at Malartic Goldfields in the 50's, the mining towns, Noranda, Malartic and Val d'Or, were relatively bilingual. The towns in those days, seemed to be proud of the fact that both

Why Mining?

languages were spoken. The management at the mines at that time was all English speaking while the language of the workforce was French. One could notice some changes in the 60's when travelling this same area for Rockiron that some of the mine management was French speaking but then in the 80's it was totally French. The fleur d'lis flag was very much in evidence as well. Only on the few federal government buildings would there be the maple leaf flag. Regardless of this change in culture, I seemed to be accepted in my travels throughout Northwestern Quebec. Nowhere was I ever insulted but in fact I was more than accommodated for my lack of fluent French. One thing that did not change in this part of Quebec (and probably never will) is the fact that the young women in the business offices and in the shops always dressed very nicely. The French girls love fashionable clothes and dress smartly at work even when the office is away out in the bush like the Belmoral office was at 5 miles off the main highway.

While in Val d'Or I always tried to stay at the Auberge du Gouverneur at the east end of Val d'Or, although at times, it was necessary to stay at the Pic d'Oré Inn in Bourlamaque when the Auberge was full. One of the things that disturbed me at the Auberge du Gouverneur (and also when Irene and I went out to dinner at the Senator Hotel when we lived in Timmins) was the number of Indians that seemed to be always wining and dining at these top flight hotels. At the evening meal at the Auberge, there would be a number of Indians having drinks and dinner at the same time. Then in the morning at breakfast there would be only white people at that early hour meal. When returning at lunchtime for a meeting in town, one would invariably see the same Indians there for lunch as were there the night before for dinner. Obviously they didn't get up for an early breakfast as they had no pressing business matters to attend. *(Where did they get all this money? The answer is: the Federal Government, the Department of Northern Development and Indian Affairs. The nation's tax dollars are being wasted and if one complains too loudly as in this case, one will be branded a racist.)*

The original assignment was to evaluate the mining methods used at the two Belmoral underground mines. The Dumont mine was the smaller of the two mines producing 70,000 tons of ore annually and was accessed by a three-compartment shaft that was sunk to 1678 feet. The Dumont mine had pretty well exhausted its ore reserve in 1987 but for some reason it was able to keep producing for another three years. The mining method was by blast hole and

Belmoral

shrinkage which seemed to work well in the ground encountered at this mine. The other mine, Ferderber, was somewhat larger producing 170,000 tons annually, also from a three-compartment shaft which bottomed at 1250 feet. Cut and fill mining was employed at this mine which, in most cases, seemed to work well except that, if filling was delayed, then the walls would squeeze into the open stope and cause considerable stope rehabilitation expense. The original MMEI report, written in July 1987 recommended the development of more stoping areas for both mines which was somewhat followed to the limit of the company's funds. What had happened to Belmoral was that it landed into receivership to the Continental Illinois Bank (Canada) as a result of a disastrous cave-in in 1980 where eight underground miners were killed. As a result, the bank would not advance any funds for new development but demanded that the operators concentrate on extracting the existing stope reserves to pay back the outstanding debt. This led to a shortage of working places; therefore, it was necessary to "play catchup" on development. *(The Flow-Through Share Programme didn't help in this case. Belmoral seemed to find all kinds of funds to chase wild schemes where they should have spent money deepening the shafts of its producing Val d'Or mines.)*

In the spring of 1989, another report was written on the two Val d'Or mines so that Belmoral could issue a prospectus in order to raise $20 million for the development of the mines and increase the capacity of the mill to 2000 tons daily. After the report was written and submitted in February 1989, Belmoral management chose to delay the issuing of the proposed prospectus because of the decrease in the price of gold at that time. Belmoral management did ask me to travel along with their chief geologist, Victor Popoff to Washington D.C. that spring to submit and discuss the ore reserves with the United States Securities Commission for the purpose of listing on NASDAQ.

The discussions with the United States Securities Commission was a full day affair at its office in Washington. The impression projected by the gentleman with whom we had our conference was that he had little to do and therefore used that meeting to fill out his day and look busy. Victor Popoff, the chief geologist, was an interesting man to travel with and, like most geologists, was very honest, thorough and dedicated to his work. *(It is often said that the type of person who becomes a geologist could just as easily have taken up theology or vice versa.)* Victor was an immigrant from Poland having arrived in Canada after World War II. He had spent all his mining career in gold

Why Mining?

mining in the Val d'Or - Malartic area. He had worked at Canadian Malartic during the same time that I had worked at Malartic Goldfields. Although we had not met prior to my present association with Belmoral, we did know many of the same people in the Malartic area.

A year later, Belmoral requested again a full report on their Val d'Or mines reviewing the Five-Year Plan and determining the problems and outlining solutions for increased profitability of the Val d'Or Operations. This required another extensive visit to the property in order to see firsthand the operations and to collect the latest cost data and ore reserves. Again, it was recommended that the two shafts be extended in order to access the reserves on the lower levels. As the operation stood in 1990, other than more development in the lower reaches of both mines, there was little to recommend as the operator had improved the ventilation throughout, had tightened up on the grade control in the stopes, improved the ground control, especially in the stopes by better filling practices, and had corrected the filling deficit with the addition of another sandfill line.

The Belmoral management kept MMEI very busy not only writing reports for a Prospectus and evaluating the Five-Year Plan, but they wanted alternatives for future development at Val d'Or.

The alternatives investigated: Deepen Ferderber Shaft
Deepen Dumont Shaft
Sink a New Shaft Between the Two Mines
Drive a West Zone Ramp at Ferderber
Excavate a Hoisting Raise at Ferderber
Drive an East End Ramp at Dumont
Sink a New Shaft to the New Zone
Do Nothing

It was recommended that the Ferderber West End Ramp be started immediately as there were ample reserves outlined to amortize this undertaking. The next best was the deepening of both shafts providing that the deep drilling exploration programme from underground showed ample reserves to justify the shaft-sinking expense. As a result, Belmoral did nothing which was the last alternative. It was a pity that none of the recommendations (except "do nothing") were acted upon; instead, Belmoral financed such ventures as Vedron, Broulan and Wrightbar.

Belmoral
Vertical Sections of the Dumont, Ferderber & Wrightbar Mines

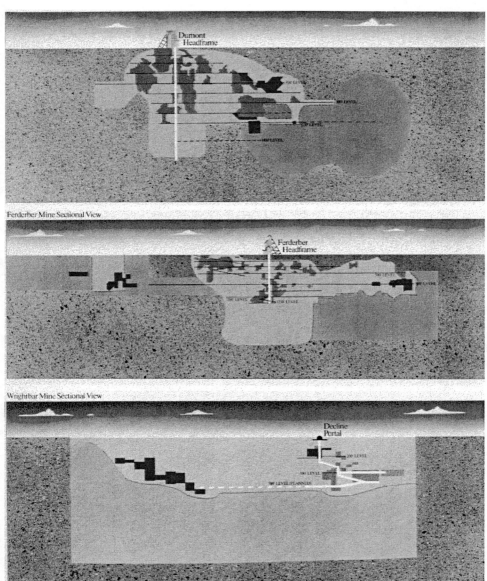

The light shaded areas adjacent to the workings were explored zones while the darker areas at depth and at the extreme right were areas to be explored.

Why Mining?

Wrightbar

In the case of Wrightbar, that was another Flow-Through Share Programme fiasco. In the spring of 1988, MMEI was asked to evaluate the mining method at the Wrightbar mine in which Belmoral controlled 33% of the Wrightbar shares. The Wrightbar project was a ramp development to a depth of 500 feet below surface where three levels (200, 400 and 500) were advanced. The Wrightbar geology was similar to that of the Ferderber and Dumont mine except that the veins dipped at 40^0 which required slushing down dip instead of having gravity convey ore to millholes below. After reviewing similar dipping orebodies at such mines as the Eldrich-Flavel in Quebec, Pamour No 1 in Ontario, Puffy Lake and Namew Lake mines in Manitoba, it was recommended a modification of a horizontal longhole method where the stopes are 30 feet in breadth with pillars 20 feet. Horizontal holes would be drilled from a raise within the pillar and would be blasted so that the muck would be thrown down the dip of the stope thus requiring a minimum of slushing. Belmoral's local manager, Serge Gagnon, didn't like the idea and therefore it was never tried.

Later that year in June, again Belmoral and Wrightbar management requested a review of the Wrightbar project as to possible profitability, grade estimate, and cost to mine and mill. Using a cutting grade of 0.5 opt, the calculated grade was only 0.127 opt and therefore the project would not be viable at the then $500 (Can) per ounce of gold price. At a cutting grade of 1.0 opt, the grade increases to 0.19 opt and, therefore, could just break even. To say the least, there were a few disappointed people such as Bill Cluff, the president of Wrightbar, when he read the report. At the time $13.4 million had been spent in Flow-Through Funding which would not be recovered unless more ore was found. To find more ore would require more money to be spent which was a vicious circle; that is, to be profitable requires finding more ore but this requires spending more money which again requires that more ore be found requiring once again spending more money on exploration. A Progress Report was written in October '88 which came to the same conclusion as the previous one written in June. As a result the project was shut down and the plant was placed on "care and maintenance".

Louvem

Belmoral in the autumn of 1987 was very aggressive in expanding its operations in the Val d'Or region where it made advances to acquire, or at least integrate, its operations with those of "La Societe Miniere Louvem Inc". I attended a meeting with executives of Louvem (A Lacroix and C Marti) and Belmoral (M Slack and S Gagnon) at the executive and administrative offices of Louvem in Val d'Or. At that meeting, it was decided that a plan be prepared for the purpose of amalgamating the Louvem and Belmoral operations into one operating entity. Louvem had one operating mine, the Chimo, which produced 26,000 ounces of gold in 1987; the Pascalis Nord mine which was under care and maintenance; the Manitou mill and four interesting prospects (Simone West, Monique, Courvan and Akasaba). Belmoral, on the other hand, had the Ferderber and Dumont mines which produced 53,000 ounces of gold in 1987; the 1500-ton Belmoral mill; and the Wrightbar project. The report showed that there could be savings in manpower by processing all ore through the Belmoral mill and closing down the Manitou mill. Also, some positions such as administration, warehousing and maintenance would become redundant through amalgamation. In any case, the two companies kept talking about such a move for the balance of 1987. The following year, Belmoral management requested an update on the report written on the Belmoral-Louvem integration of operations showing the savings that might be made through acquiring the Louvem assets. As a result of this report, Belmoral approached St Genevieve Resources Limited who owned 56.6% of Louvem with the purpose in mind to acquire the assets of Louvem. This transaction was at a point of being completed when, at the eleventh hour, the Quebec Government stepped in and acquired the Louvem assets for Soquem, a crown company of the provincial government. It was the opinion of those involved at Belmoral that the Quebec government (under Levesque) was not about to allow an Ontario based company to gain control of these mining assets in that province.

Yorbeau

Belmoral was not only active in the Val d'Or area of Quebec but was interested in properties in the Noranda area as well. That autumn Steve Harapiak replaced Malcolm Slack as president of Belmoral *(Steve, who was president of the Potash Company of Saskatchewan was squeezed out when the Devine government was re-elected in Saskatchewan that year. That must have been a tremendous blow to Steve's pride as the PCS presidency was one of the top mining positions in Canada.)* After the failure of the Louvem amalgamation, Steve Harapiak called a meeting (November 1987) at the Belmoral office in Toronto where Doug Stirling and Victor Popoff would

Why Mining?

collaborate with me in writing a report on the Astoria and Ellison properties held by Yorbeau Resources Inc in the Rouyn-Noranda District in Quebec. Victor Popoff would supply the geology and reserves, I would estimate the construction and development required and Doug Stirling would calculate the financial data. From Toronto, I travelled to Val d'Or and met Serge Gagnon (manager of Ferderber and Dumont mines) and drove to Noranda to visit the Yorbeau properties. We met Pierre Tessier, the Yorbeau chief engineer and his assistant, Yves Gregoire. Both of these guys were very co-operative. Presumably they were told that Belmoral had money and, by showing us everything, perhaps Belmoral would buy into their project and therefore it would be a chance to get one of their properties underway.

The report was written within a week and the following week we gathered at the Belmoral office in Toronto to discuss the possible Yorbeau acquisition. While working on plans to develop and mine the Astoria property, I was also working on a report for Wrightbar as well as making inspections in Timmins on the Vedron and Broulan properties and in Saskatoon for Saskterra. This was indeed a very busy time and Christmas was coming up as well. What was finally recommended was the enlargement of the shaft compartment size similar to what was done at the Armistice shaft in Virginiatown. Also, this would require modifications to the headframe which also changed the fleet angle on the hoist. Serge Gagnon contacted Roger Demers in Quebec City and got his opinion regarding the fleet angle where he said it would not be more than $1^{0}18"$ which was quite acceptable. Pamo Construction in Noranda was then contacted where they gave suggestions on the changes to the headframe tower. These changes were checked out with Ken Short of Short Engineering in Toronto. To make a long story short, these plans were presented to Pierre Tessier, Yorbeau's engineer, who was not at all pleased. Pierre, who worked on the Astoria Project for a long time was not happy to see his plans changed. *(I don't blame him. It would upset anyone if a stranger came in and changed everything that had been previously set up.)* To win over Pierre and his crew, arrangements were made for them to travel to Virginiatown and visit the Armistice shaft where we had previously enlarged the shaft compartment size from 4 x 5 feet to 5 x 6 feet without blasting or excavating the shaft walls.

Our delegation to the Armistice property at Virginiatown was composed of Serge Gagnon, Pierre Tessier, Yves Gregoire, Larry Colin and myself. Merdy

Armstrong, owner's representative for Armistice Resources on the property, welcomed us graciously and guided the delegation around the project and down the shaft where Dynatec were removing the wall and end-plates and replacing these with concrete. The delegation were duly impressed. The procedure even impressed Pierre Tessier and his assistant Yves Gregoire, who were previously against any changes to the shaft size or configuration. As a result, the changes that I recommended for the Astoria shaft were completed. A few months later, Bill Longworth, who was in charge of the Broulan Reef shaft dewatering and rehabilitation project at that time, was shown how the walls were stabilized in the Armistice shaft. He too was impressed. The idea of increasing the compartment size by removing the timber and replacing it with concrete was certainly an idea that was quite revolutionary. Fred Edwards of Dynatec, although not the originator of this idea, did write a paper on enlarging the Armistice shaft using this method and received many favourable comments.

Ketza River

January 1989 started out with a bang. Belmoral management, under Steve Harapiak, were very bullish in expanding its producing mining properties. Ken Dalton, chairman of Belmoral, was also a director of Canamax Resources Inc. When Canamax decided to divest itself of the Ketza River mine in the Yukon *(which is it, "The Yukon" or just plain "Yukon"?)* Belmoral, because of Ken Dalton's association with Canamax, had an excellent opportunity in this acquisition. Anyway, a number of us (Steve Harapiak, Jim Johnstone, Jim Fortin, Victor Popoff, Lawrence Melis and myself) flew from Toronto to Vancouver, where we had a four-hour stopover and then on to Whitehorse where we stayed overnight. While waiting in the Vancouver Airport, Steve Harapiak took us into the Maple Leaf Club, something I had never experienced before. By coincidence, Mike Stewart of Canterra was waiting for his flight to Calgary in the Maple Leaf Club also. *(That synched it, I just had to find out how one becomes a member of that facility and, when getting back to Kingston, I applied to join because of considerable travelling done on MMEI business in those days.)* After spending the night in Whitehorse, the next morning we flew by charter aircraft to Ross River, Yukon, and then travelled by van about 50 miles to the mine. *(The Town of Ross River was named after the first commissioner of the Yukon, James Hamilton Ross, who was married to my Great Aunt Barbara Elizabeth McKay. Barbara Ross was drowned along with 65 others in 1901*

Why Mining?
when the Steamship Islander crashed into an iceberg off Douglas Island near Juneau, Alaska.)

The Ketza mine, located at 1370 meters elevation and 200 km by air northeast of Whitehorse, consisted of an underground mine, a 400 tonne cyanidation mill and a 150-man camp. On our arrival there, the temperature was about -30° Celsius. *(At that temperature, it doesn't make much difference what thermometer you use.)* On the way out a week later, there was a wait in Ross River for our aircraft when it was -40^0. The pilot didn't dare turn off the plane's engines while we boarded for fear that the engines would not start again because of the cold.

This was my first experience with a "fly-in fly-out" camp and to say the least I was impressed, especially with the meals. Since that time, I have experienced living in a number of other similar camps and can honestly say that the meals at the Ketza River mine camp were the best of all camps visited. The rooms consisted of Atco Buildings where each person had a room to his or herself but had to share common bath and shower room facilities which seemed to be adequate for everyone. As we were visitors, we could use the staff lounge where the mine manager Bob McCombe kept a bottle or two of Scotch. Yes, Ketza River mine operated a wet camp. The rule was, "any rowdiness due to alcohol, the party involved would be terminated and the camp would revert to a dry camp." *(Apparently there was never any incident related to alcohol, which speaks well for the people employed at this mine.)*

Wayne David, who worked for me as a shift boss and later as a safety supervisor at Kidd Creek, was at the Ketza River mine. Wayne was at this time employed by Canamax as a safety instructor and travelled from one Canamax mine to another. At that particular time (1989) Canamax, as well as operating the Ketza River mine in the Yukon, operated three gold mines in Ontario, namely the Bell Creek mine in Timmins, Kemzar mine in Wawa and the Matheson mine in Matheson. Also, working at Ketza River as the mine superintendent was Elmer Olafson (formerly of Steep Rock Iron Mines) who was first a surveyor, then a shift boss at Steep Rock back in 1953.

The mill was a standard cyanidation/carbon-in-pulp gold mill operating at 300 to 350 metric tonnes per day but could operate comfortably at 400 tonnes

Belmoral

daily. The mill, according to Lawrence Melis, was quite adequate and exhibited very few problem areas. I toured the underground with Elmer Olafson on three occasions while on site. The mine was accessed via four portals between 1430 and 1550 metres elevation. The major development and facilities such as the materials handling, ramp and ventilation systems in the mine were very good. Unfortunately, the mining method (drift and fill) chosen to exploit the orebody was only half complete. They started out stoping by crosscutting the orebody from hanging to footwall and then filling each crosscut lift with unconsolidated rock material obtained from a talus slope on surface. As a consequence, when they tried to mine the pillars between crosscuts, it resulted in much dilution. Management had plans to use consolidated fill but the slurry mixing hopper and pump had not been delivered at the time of our visit; therefore, in the meantime placed unconsolidated fill instead.

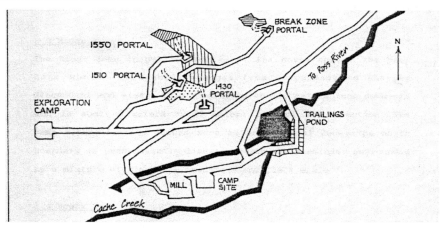

The Surface Plan of the Ketza River Mine

Another mark against the Ketza River mine was the shortfall in the ore reserve. The specific gravity of the mineral was grossly over-estimated as was the grade and, as a result, there was a shortfall in gold content. *(Estimating ore reserves using various computer techniques such as geostatistics and other computer estimating programmes can be misleading if used by the wrong people who rely more on the computer instead of first ensuring proper sampling, understanding of the deposit's geology and confirming the accuracy of the data base.)* The recommendation after doing this due diligence study was, " ... the purchase of the Ketza River mine would **not** be a viable enterprise to acquire."

Why Mining?
Chapter 12

Sweden

World Mining Congress

The spring of 1987 was an extremely busy time for McKay Mining Engineering Inc as there were a number of jobs not only with Belmoral on projects at Broulan and Vedron but with Saskterra Potash as well. To top this off, Irene and I had made plans to attend the 13th World Mining Congress 1987 in Stockholm Sweden that June. We had been to Stockholm before and were anxious to return as our plans were to rent a car while there and travel to the southern part of that country and visit the area from where my maternal grandparents were born and raised.

The Swedish Trade Commission was organizing a group tour where the group would not only attend the congress but would tour a number of manufacturing plants catering to the mining industry as well. This looked like an interesting adventure so Irene and I went along. A very pleasant event occurring during the tour was the trip over and back. Prior to going on this tour, I phoned Doug Bertoria, my friend with KLM and told him about flying to Amsterdam on KLM and the switching to SAS for the balance of our journey from Amsterdam to Stockholm. Doug asked for our flight information and said that he would arrange a little surprise for us. We joined the tour group at the Pearson Airport in Toronto and to our surprise at the check-in desk, the flight clerk said there were two first-class tickets there for us, compliments of the Deputy General Manager of KLM Canada. Irene and I had never been treated so well on an overseas flight as this one with KLM. Certainly that airline is world class as my later experience, flying with them, would confirm. When Irene and I arrived at Skippel Airport in Amsterdam, we met some of the members of our group tour, one being Mike Amsden the general manager at Kidd Creek, who wondered why we were chosen to travel in the first-class section. We didn't say but played dumb. When we got to the designated hotel, the Karelia Hotel, Irene and I were not impressed and immediately made our own arrangements and checked into the Sheraton Hotel where we had stayed before in 1981. The tour group, not knowing what the city had to offer, accepted the shoddy arrangements made for them by the Swedish Trade Commission (STC) *(Shame on the STC)*. We did join the group for the social events and did tour with the group outside of Stockholm but there was no way we were going to stay at the Karelia Hotel.

Sweden

One event that made quite an impression was the opening ceremonies for the World Mining Congress. There were translation headsets (French, Spanish, German and Russian) given out to those who were not familiar with the English language as all communication was in English even though the Congress was held in a country where the national language was Swedish. The King of Sweden was in attendance and was asked to perform the duty of opening the Congress, which he did, **all in English**. *(This is contrasted with two other events: one was the Montreal Summer Olympics and the other was the Calgary Winter Olympics. In the former, Queen Elizabeth II opened the Olympics in Montreal, a French speaking city, in French which was only right. But, the Governor General, at the time, Madame Sauve opened the Winter Olympics in Calgary, an English speaking city, in French. It shows that royalty in the case of the King of Sweden and Elizabeth II exhibited a lot class while in the case of Madame Sauve, a commoner, one couldn't say the same for her performance.)*

Production Drilling at the Kirunavaara Mine in Kiruna Sweden.

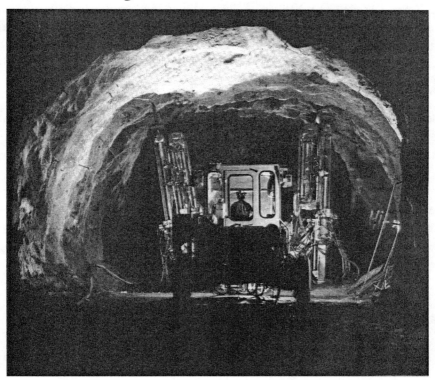

Why Mining?
Chapter 13

Bond Gold

Northern B.C.
The next venture in the consulting world was with Bond Gold after having interviewed in the autumn of '88 with Desmond Kearns in Toronto but not hearing a word from him or Bond Gold until much later. Apparently, Jack Kokanan their accountant who, incidently had worked at one time in the Toronto Office of Kidd Creek, gave Kearns a negative report about my ability as he judged me on my handling of the position of manager of marketing services. Fortunately, Desmond Kearns, who knew Colin Chapman, asked him about me. Colin gave me a topnotch report saying that I was thrown into a very difficult job at the head office of Kidd Creek and was in the process of sorting out the situation when the company was taken over by Falconbridge. *(I was wary of Kokanan after Chapman told me about this. Kokanan, who was only a junior at Kidd Creek, had no idea of what was going on in marketing where I wore five hats which involved: office management, sales control, contracts, traffic and potash.)* Upon returning from visiting the Ketza River mine in the Yukon, Kearns called and requested we meet in Toronto. Kearns was an interesting study. The guy was a "fuss budget" and very hypercritical, and not that well organized either. He seemed to be worried and fussed about everything to a point of interfering. *(Anyway, more about Kearns later.)*

At the Toronto meeting with Kearns, MMEI was awarded a 12-month contract *(thanks to Colin Chapman and Nat Scott)* at $5000 a month (10 days per month) and expenses and then was given an overview of the current Bond Gold projects. It was at this time I met David Molloy, Bond's head of exploration. After our meeting, Dave and I flew out to Vancouver, but not before calling home to tell Irene that I was headed for B.C. and then to Wrangel, Alaska. On arrival in Vancouver, Molloy hailed a limousine for downtown and registered at the Vancouver Hotel. *(Molloy liked to do things up right.)* The next morning we met with Vic Tanaka for breakfast at the hotel. Later at Tanaka's office we reviewed various properties Tanaka had in his portfolio in which Bond, at the time, had some interest. Not much came of this meeting as Bond was just fishing.

In the afternoon, we went to Bond's Vancouver office where Dave Kennedy, a very pleasant and friendly guy, was met. So there we were: Dave Molloy, Dave Kennedy and Dave McKay all with the same first name. *(When the three of us went to the various mining exploration offices in Vancouver, everyone had a chuckle when we were introduced... all these "Davids".)* Kennedy showed us what he had compiled on various properties that were of interest to Bond. There was geological data on such properties as the East Arm; Willoughby; and Westmin located in the Stewart Area just south of the Alaska Panhandle.

On the following day, while in Vancouver, the three of us reviewed the Skyline documents of the Johnny Mountain property after signing confidential declarations. That mine was in deep trouble and Skyline Explorations were trying to unload the property which reportedly had a good mine plant but had only limited reserves. A preliminary due diligence report, as a result of this visit, was written which recommended that no further effort be made in acquiring that property.

The following day an aircraft was chartered in Vancouver for a flight to Wrangel, Alaska but had to be aborted due to icing of the motors when half-way into the flight. As a consequence, the plane turned back. The rest of the day was spent in Vancouver visiting Jim McLeod who gave a presentation on the Tenajon property in Stewart B.C. *(McLeod was a likeable guy even though what he was telling us was BS.)* The following day, Friday, we flew back to Toronto on Air Canada. A lot of writing and capital calculations were done on the plane home so that the due diligence report on the Skyline property could be completed over the weekend as Bond required me back in Toronto Monday to review the East Amphi property in Malartic, Quebec.

East Amphi

Peter Huxhold (Bond's geologist overseeing the East Amphi Joint Venture with Breakwater Resources) and I flew to Noranda via Vogageur Airline the weekend after the aborted flight to Wrangel, Alaska. Peter was a very good person, as most geologists are, but his failing point was that he was in love with the orebody at East Amphi. *(As mentioned before, often times a geologist falls in love with a property and will not face reality when there is really nothing of great value in the property. This was seen in Alexander Po at the Vedron also. Was this not the case with Michael de Guzman of Bre-X fame when he salted the drill core the first time so as to keep the*

Why Mining?

exploration of the property going and then couldn't stop as he got into this scam too deeply?) After arriving in Noranda we travelled to Cadillac, Quebec first, where the joint venture had its field office, and then on to the mine at Malartic. The underground was accessed via an alimack instead of a hoist as the lowest level was only 475 feet below surface. *(The use of an alimack was a quick and cheap access method down a shaft.)* The ground conditions were the same as those found at Malartic Goldfields as both these properties are located in the "Break" while the other mines in the area are off the "Break" and are related to the Sladen Fault as in the case of Canadian Malartic; Barnet; and East Malartic mines.

That evening, when phoning Dave Molloy, I stated that the East Amphi would not be a mine according to the present results at hand. Dave was very upset and somewhat angry that such a statement was made. *(Often times its the messenger who is condemned.)* In the report later, Bond was given three options and a recommendation.

>**Option 1.** Terminate the project altogether.
>**Option 2.** Shut down the underground segment and explore the west-end by surface drilling.
>**Option 3.** Continue the exploration from underground at an added cost of over $1.8 million over and above the present budgeted amount.

Recommendation:
>**Option 2.** Explore the west end of Amphi East by surface drilling.

The recommendation proved me right as, 10 years later, when drilling the west end, enough ore was found where a viable open-pit mine could be developed and worked by the McWaters Company, the same company who purchased the Sigma mine from Placer Dome.

Golden Patricia

The next job with Bond was a trip to the Golden Patricia mine in the Pickle Lake District of Ontario. We, which included Desmond Kearns of Bond Gold, Geoff Pearse of the Northern Miner and myself, flew to the mine from Thunder Bay by charter plane and landed at the small airstrip at the mine. The property included a three-level underground mine accessed by two ramps, a 250-tonne mobile mill purchased from Orocon Ltd, surface shops,

Bond Gold

an office and dry facilities and a 100-man camp. Almost immediately on landing we changed into our mining clothes and were guided underground by the manager, Pat Moore, an Irishman, who was a mill man and not too acquainted with underground mining. The reason for me being along on this trip was that Kearns wanted some comfort knowing that the mining was proceeding well and not deviating from good mining practice.

The Golden Patricia was a little jewel box of a mine. The quartz vein was only 12 inches wide with a ribbon of visible gold about $1/4$ to $1/2$ inch in width running down the center of the vein. The mining method was shrinkage stoping where they attempted to keep the stoping width to one metre. Prior to taking over the mining in January 1989, Bond contracted the mining to Redpath of North Bay who were somewhat more concerned with tonnage than grade and, therefore, often times had considerable over-break rather than adhering to a stoping width of one metre. After touring the underground, I typed out my report on site and faxed a draft home to Kingston for Irene to edit. *(Irene did a good job in editing my reports. Nothing was sent out without first having her critique the work.)*

The camp at Golden Patricia was not that great as the cabins were better than a kilometre or so from the cookhouse. As for the cookhouse, it was very small and overcrowded and not a bit roomy as what we found at Ketza River. The meals at Golden Patricia were not as good as those at Ketza either. Of all the camps (Ketza River in Yukon, Golden Patricia in Ontario, Huckleberry in B.C. and Hope Brook in Newfoundland), the poorest camp all round was that of Golden Patricia. The best meals were at Ketza River and the best room facilities were at Hope Brook. Golden Patricia did have satellite TV where we were able to watch the 1989 NHL All-Star Game that evening. *(Incidently Gretzky scored a goal and had two assists and won the most valuable player award that night.)*

The next day being February 8, 1989 in the dead of winter in Northwestern Ontario, we drove to Pickle Lake via a winter ice road which was a new experience for me. From there we took Bearskin Airways to Thunder Bay and then flew to Toronto via Air Canada. Leon LaPrairie was on the flight to Toronto and he gave me a ride to the Chelsea Inn where Irene was staying that weekend. *(When travelling often, one is always meeting a familiar face in the airport or on the plane. It happened to me many times.)*

Why Mining?
On one occasion when visiting the Golden Patricia mine in the Pickle Lake Area with a Northern Miner rep, I asked the personnel manager how many employees were on site at any one time. Kearns took me aside afterward and said not to ask those sort of questions in front of the Northern Miner reporter. If the number of employees was a deep dark secret, it wouldn't have taken much detective work to determine the numbers for the work force from the size of the camp. *(Anyway, the question was asked... too bad.)*

To show how uncompromising and disorganized Kearns was, is to relate an incident that happened a few months later at the Golden Patricia property. Generally, at most isolated properties, the mine manager is allowed to have certain leeway regarding personal favours, sometimes called "government jobs", involving work done about the manager's house, (or whatever), by workmen from the mine. In this particular case, the manager, Pat Moore, had two workmen re-shingle the roof of his house in Dryden and he charged the labour to the mine. When Kearns found out about it, he fired Moore on the spot; then after firing him, he asked Moore to stick around until a replacement could be found. Moore said, "stick it". Kearns asked me if I knew of anyone who was immediately available. I suggested he contact Jim Johnstone who was the manager at that time in Timmins for Belmoral as that company was in the process of shutting down its Timmins operations. Johnstone met Kearns and liked the mine and the salary offered but could see that he couldn't work for Kearns as he said he thought Kearns would be too interfering and therefore turned down the job offer.

Two trips to the Golden Patricia were made that year, one in February as described and the other in August. By August on the second trip to Golden Patricia, Bond had another manager in place in the person of Frank Hamonic, another mill man. They did have a very capable young mining engineer (who acted as mine superintendent) by the name of Doug Hayes. Hayes acted as the guide on two underground tours of the mine during the August visit thus I got to know him better. *(When going underground, you get to know a person quite well as you have a captive audience and therefore his full attention.)* On this particular trip, the new development was toured on the lower west end of the mine, where the bottleneck was the haulage from that section of the mine. This also impacted adversely on the whole underground operation. The recommendation was for track haulage on the lowest level and a hoisting raise to surface. A couple of years later, Dynatec Engineering were given the contract to design a hoisting facility somewhat similar to what I suggested. *(I don't know if Bond ever developed that facility as Bond Gold was taken over by Lac Minerals at about the same time.)*

Shoal Lake

In April 1989, Kevin Leonard, Bond's geologist for the Shoal Lake project and I flew via Air Canada and Bearskin Airlines to Kenora Ontario. On arrival at Kenora, a Beaver aircraft was chartered and flown to the property on Shoal Lake which is west of Lake of the Woods. What makes this property environmentally sensitive is that the City of Winnipeg obtains its water supply from Shoal Lake. While walking over the terrain at the property, Kevin and I located a good spot where a ramp could be collared. We also, while inspecting the shoreline near the old Mikado mine, found a good location for a camp. The abandoned Cedar Island shaft near this area was examined as well. The next day, we walked to the property over an old three-kilometre trail. The "wood-ticks" were very bad and we had to stop a number of times to clear them off our shirts and trousers. Afterward, when getting back to the hotel, it was necessary to take a good shower and thoroughly shampoo one's head and scalp to make sure none of these little insects were attached.

The Shoal Lake property was a joint venture between Bond Gold and Kenora Prospectors and Miners Limited, the latter company controlled by Sue Dobson and her husband Eric. I have never met Sue Dobson but everything we heard about her from others, (who had dealings with her), was all bad. Apparently everyone who had business dealings with her always got the short end of the stick. Some people said she and her husband were downright dishonest. Others said she was crazy.

The report on the property investigated three alternatives for access and for exploiting the orebody which had a drill-inferred resource of 775,000 tonnes at a grade of 12.04 grams using a cutoff of 5 grams located in a long, narrow, shear-type structure. The recommended access was a ramp from surface on the mainland, 1200 metres east of Cedar Island rather than rehabilitating the existing Cedar Island shaft or sinking a new shaft 600 metres east of Cedar Island on the mainland. The overall cost for mine and mill was $27 million with a cash cost per ounce at $250 Canadian. Even at today's low price of gold this would be a viable project but because of its size, it was too small for

Why Mining?

a large company like Bond; therefore, they decided to walk away and leave the property to some junior mining company to exploit.

Eskay Creek

Back in B.C., Bond was interested in the Eskay Creek property *(Who wouldn't be?)* which, at that time, was owned by Calpine Resources Inc. At least Bond was interested enough to fly a group of us to Smithers B.C. and then into the property in late May 1989. Our group was composed of Dave Molloy, Dave Kennedy, myself and a young Austrian geologist who worked for Bond by the name of Anders Vogt. *(Anders was also a mountain climber and therefore was a natural for exploration in that mountainous part of B.C.. However, a sad occurrence happened the following year while Anders was exploring for Bond in Northern B.C. when he was stung by a wasp. He died before the helicopter could get him out for medical treatment. Ander's death left a young woman a widow and young child without a father. I often wondered ...do mining companies do enough to look after those who are fatally injured while in their employ? I know in the case of my wife Irene's former husband, the late Tex Clark, who was drowned in a thickener tank at the Adams mine in Kirkland Lake while under contract to Jones and Loughland, his family received nothing. I also know that the families of those killed in the Belmoral Val d'Or cave-in never so much as received a Christmas card nor any acknowledgment of their loss, at least to my knowledge, during the period I consulted for them.)*

The flight to Smithers was uneventful but the one to the Eskay Creek property did have its moments. At Smithers an aircraft was chartered and flown to Cominco's Bronson Airstrip located about 85 kilometres east of Stewart just off the Iskut River. Cominco at the time was developing the Snip mine and, while waiting to change aircraft from fixed wing to helicopter, our group looked around the Cominco camp. Cominco seemed to go all out in constructing a very comfortable camp as the bunkhouse appeared to have private rooms for each employee and comfortable mattresses for the beds. *(While there, workmen were assembling the beds and unpacking the mattresses.)* One ominous circumstance which brought us down to earth was the remains of a crashed aircraft that had gone up in flames and was still sitting along one side of the Bronson Airstrip. *(That brought on some sober thoughts.)*

Bond Gold

After a short stay at the Bronson Strip, our helicopter attempted the flight over the pass to the Eskay Creek Property. *(This was my first experience flying in a helicopter so I was quite excited.)* On reaching the pass, my excitement changed to fright as the clouds in the pass were very low and the helicopter was completely shrouded in the clouds ...like a "whiteout". With the mountains on both sides of the pass so close, is it any wonder we were nervous? Anyway, the pilot didn't take any chances and lowered the aircraft back to where he had good visibility and landed on a snow-covered area to wait for the clouds to lift. After waiting a short time, he tried again to no avail. After the third attempt, he aborted the flight to Eskay Creek that day and returned to the Bronson Airstrip.

At the Bronson Airstrip, our fixed-wing chartered aircraft was boarded for a flight to Stewart which was also aborted when a thick fogbank in Bear Pass on the way to Stewart prevented flying further. *(When flying down the valley to Stewart, I wondered why the pilot seemed to insist on flying to one side of the valley and not travel down the center which, to my thinking would be safer and not so close to the steep walls of the mountains which formed the valley. I found out when we came to Bear Pass when the fogbank barred our way and we had to turn right around in that narrow valley. Had we been flying down the center, we would not have had enough room to turn the aircraft around, so the pilot knew what he was doing. Since that incident I have much respect for bush pilots.)* Returning to Smithers, we rented a van and drove 160 miles along Highway 37 to Stewart where we stayed overnight.

While in Stewart, I looked up an old friend from Timmins...the one and only Elmer Borneman who had been personnel man for Texasgulf-Kidd Creek. Elmer was always a very friendly guy and very quick-witted. Elmer, true to form, was in a very jovial mood bringing me up to date on things going on in the Stewart area. He said, "...you haven't lived unless you've made a visit to Hyder, Alaska, while in Stewart". Nothing would suffice but we three: Elmer, his wife Louise and I travelled the few miles (less than 5) to Hyder. There is no customs or immigration to go through as it is an open border up there. *(There is a sign at the border when re-entering Canada from the Alaska side which reads, "Purchases made in the USA can be declared by phoning "such n' such" number in Prince Rupert." It was not even an 800 number either. Now, how many people would call that number and declare their*

Why Mining?

purchases?) Hyder, Alaska is probably the next thing to "Dogpatch" as there is no law or order and the general layout of the town is very haphazard. Also, many of the people carry guns.

Elmer tells the story of Westmin, when bringing hydro-electric power through to the mine, the powerlines had to traverse from Stewart B.C. through Hyder Alaska and then back out into B.C. again. In order to build good will with the citizens of Hyder, the company offered to erect five electric light poles where the line crossed through the village as there was no street lighting in the village at that time. The citizens of Hyder were also given the say at a town meeting as to which of the poles were to have lights installed. After much discussion, they voted on an arrangement as to where the lights would start and where they would terminate. Some wanted the first pole to start at the south end of the village and terminate at Mr Whoknows' house near the north end of the village and others wanted it to go one past Mr Whoknows house and start 500 feet from the south end. Anyway it was brought to a vote. The vote was something like 23-for and 6-against for the start at the south end and terminate at Mr Whoknows' house. But before the chairman could declare the motion carried, someone at the back of the room hollered that Mrs Godknowswho didn't have a chance to vote as she was outside nursing her baby. Again there was much heated discussion. The mining company supplying the powerline in exasperation said it would add another light standard so that both options would be satisfied. *(I think that's what the people of Hyder wanted in the first place. Maybe they weren't so dumb.)*

On one's first trip to Hyder one has to go through the ritual of becoming "Hyderized": that is, where one has to drink a full tumbler of whiskey in one gulp or else buy the house a round. Elmer warned me about this so, when asked, I could answer that I had been Hyderized before and paid my dues. *(Anyway, I didn't have enough money on me on this trip to buy the house a free round as I knew I couldn't have downed a whole glass of whiskey in one shot.)* When we got out of Elmer's car at Hyder, I naturally just closed the door. Elmer said, "Lock the door". I asked why, as Hyder was only a small town not the city. Elmer said, "You have to beware of the "Hyder Hobo" who doesn't like Canadians and when he sees a car with a B.C. licence, he opens the door and pisses all over the inside." *(Now that's something to know... beware of the Hyder Hobo when vacationing in Hyder, Alaska.)*

Bond Gold

The next morning, the Bond group chartered a helicopter at Stewart and attempted to fly again into the Eskay Creek property. The first trial was unsuccessful as a cloud bank prevented further travel and the pilot had to abort that flight. The pilot landed on an island on the Unuk River and waited there for the clouds to lift. I wondered at the time why we landed on this rocky island as there seemed to be a nice grass meadow on both sides of the river. The pilot explained that he was cautious of grizzly bears and, by being on this island, he could see them if they tried to cross the river rather than being on the meadow and having them come out at you from nowhere. On the second attempt the helicopter was successful in getting to the property where the Calpine Resources had its camp. Its exact location is 56° 38'N and 130°28'W or roughly 84 km east of Stewart between Tom Mackay Lake and the Unuk River with an elevation ranging between 900 and 1300 metres above sea level. *(The exact location is given for the sake of good order as the Germans say.)* The camp was composed of a cookhouse, an office and a half-a-dozen other small buildings which housed sleeping quarters, drill shop, core shack and storage unit.

Apparently the property was first discovered by a party headed by Tom Mackay in 1932. In the span of 57 years such mining exploration firms as Unuk Valley Gold Syndicate, Premier Gold Mines, American Standard Mines, Western Resources, Canex Aerial Exploration, Stikine Silver, Kalco Valley Mines, Texasgulf, May Ralph Industries, Ryan Exploration, Kerrisdale Resources and Consolidated Stikine Silver all had a hand at exploring this property but only Calpine Resources seemed to put it all together and show interest in making this property a mine. *(It goes to show what effort it takes to get a mine started.)* At the time of the visit, the drilling showed 2.7 million tonnes at a grade of 0.92 ounces per tonne with the possibility of increasing this to 3 million tonnes. *(It's not good practice to mix imperial and metric measure but that's how the people there at that time recorded its resources.)* The preliminary capital cost to bring the property into production was estimated at $60 million but what we could see at the time, it was more like $100 million.

Getting out of the Eskay Creek property was uneventful as the sun was out which burnt off the fog and cloud bank so the pilot could fly back to Stewart without difficulty where we picked up our van and drove back to Smithers. As we had a few hours to wait in Smithers for our commercial flight to

```
Why Mining?
```
Vancouver, I visited the home of Dave and Sharon Porteous who lived there. I had lost touch with Dave and Sharon as they had moved a number of times since Dave and I worked together in Timmins. It was a coincidence that I met Sharon Porteous at the airport on arrival in Smithers (as she was one of the security personnel at the Smithers airport) and invited me to their home on my way back. Dave Porteous was my best man at my wedding in 1972 and was now the mine inspector in the Smithers District. We had a good, but short visit at their Smither's home as I had to rush off to catch my plane for Vancouver.

After we over-nighted in Vancouver, our Bond group visited the office of Ron Netolitzky, a geologist and major shareholder in the Calpine Resources whose major asset was the Eskay Creek Property. As it turned out, Calpine Resources wanted more than Bond was able or willing to pay and therefore Bond's fascination and interest declined in that property. Of interest though was Ron Netolitzky. This man, in anybody's language, was a very successful exploration geologist. He had considerable success money-wise in the oil patch in Alberta before arriving on the scene in British Columbia. He had the ability to see a property's promise before actual diamond drilling confirmed its potential. With his brains and talent, he was able to become very wealthy indeed. He lived in a very large home in Victoria and commuted to Vancouver by flying from harbour to harbour two or three times a week. He was also able to stake his wife to five very fashionable specialty stores featuring antiques and fine household furnishings. *(Irene and I have often shopped at Chintz & Company in Victoria and have come away drooling at the wonderful things they had for sale.)*

As an aside, while at the Calpine office, Robbin Porter ex-Cominco of Kimberley was there doing consulting work for Cominco. It was good to see Robbin again. Prior to that, I saw him when he toured Kidd Creek while I was superintendent there. The last time we met was at the CIM Annual General Meeting in 1997 in Vancouver.

The one year's contract with Bond expired in October 1989 just at the time Mr Bond of Perth Australia of racing yacht fame was finding it difficult to cope with the high interest rates on the money borrowed to build the Bond Empire of breweries and gold mines. He was forced to sell his company to Lac Minerals.

Chapter 14

Dynatec

Montanore

The next consulting venture (subcontracting) was with Dynatec Engineering of Richmond Hill, Ontario. Dynatec Engineering was at that time a subsidiary of Dynatec Corporation, a firm headed by Bob Dengler with Fred Edwards in charge of its engineering subsidiary. Fred, at No 1 mine and I, at No 2 mine of Kidd Creek, were old friends. This would not be the one and only occasion that Fred would call for help in performing consulting services for him and his company. On this particular occasion, it was a job first in Montana, then later another in Wyoming.

The first job was at the Montanore Project in Libby Montana managed by the Noranda Minerals Corporation, a joint venture of Noranda and Sunshine Mining Company. The purpose of this investigation was to review a feasibility study completed by Redpath Engineering Inc, and to prepare revised capital and operating cost estimates and schedules for Sunshine Mining.

After flying to Cranbrook B.C. and renting a car, I drove to Libby Montana through Kingsgate and Eastport on the border travelling through to Bonners Ferry and then along Highway No 2 to Libby Montana. That highway was under construction and was in terrible shape so, on the way back, another route, Highway No 37 through Roosville to Cranbrook was taken instead. Something that one couldn't help but notice was the difference between the Canadian and American cultures. When driving towards the border from Libby to Roosville, one notices, on the American side, the many honky tonks along the highway beckoning people to gamble or drink, while on the Canadian side of the border, there are no such things, and everything is well-kept and orderly. *(Irene and I noticed this when crossing into the States on the Ambassador Bridge between Windsor and Detroit. On the Canadian side there were very well-kept homes with flower gardens while on the American side, there was a slum area to travel through before seeing the real U.S.A.)*

On arrival at Libby late that afternoon, I checked into a hotel. At the Noranda office the next morning I met Joe Scheuring, project manager for the Montanore Project, and then began to read the Redpath Report on the project. Fred Edwards arrived later in the afternoon. After Fred arrived, we

Why Mining?

looked over the core of a couple of the drill holes that penetrated the ore zone. The ground, from what was seen from the cores, looked very strong and competent. The next day, Bob Sadler, project superintendent for Gilbert/Dynatec (who were developing a very long access tunnel for the Montanore Project) took us underground. The Montanore orebody was within a Wilderness Area and therefore all mine construction had to be outside of this area. As a consequence, no access adits nor ventilation raises could break through to surface within the Wilderness Area. This meant four very long (anywhere between $2^1/_2$ to 3 miles) tunnels for access and ventilation had to be driven from outside the Wilderness Area in order to gain entrance to the orebody. The orebody, which consisted of two sub-parallel horizons dipping at 15^0 and averaging 32 feet in thickness, held a geological reserve of 142 million tons at 2.1 ounces of silver and 0.78% copper per ton. With good ground, a large reserve and fairly good grade, it was an orebody that would interest any mining company, except for the environmental considerations. *(Montana, a State, which attained its status as a "Mining State", has indeed had some strange citizens who had recently voted (Initiative I-137) to ban the use of cyanide in the processing of gold. I have read that this proposition was voted on by the citizens who were prevented from being given the complete facts on the case until four days before voting day. That's hardly democracy.)*

Montanore Project Libby Montana

Establishing a mine portal for one of the long access tunnels.

This was a very busy time for MMEI. After spending a few days in Libby, it was necessary to fly to Saskatoon to attend the Allan Potash Participants' Meeting with Mike Haug. Going to Saskatoon was a pleasure as there were many very good restaurants and, on this particular trip, Mike Haug of

Dynatec

Saskterra, Jim Bubnick and Bill Eatock of PCS Mining and I had dinner at John's Steakhouse. The next day, after touring the underground and surface plant at Allan Potash in the morning, all those from the dinner the night before then attended the Participants' Meeting in the afternoon. The next day being Friday, Mike and I flew to Calgary where the report on the Participants' Meeting was written at the Husky Oil Company's office. *(Before the report was presented to Husky, it was faxed to Irene in Kingston for her editing.)*

After spending the night in Calgary and then travelling to Denver, Colorado the next day, I met John Wearing, a former colleague from Kidd Creek, at the Stapleton Plaza Hotel. John Wearing was the design engineer for the underground development at No 2 mine. John, at the time, was having women trouble. In Timmins, he and his first wife separated and then divorced. John left Kidd Creek just after the completion of No 2 shaft to join Cementation Company, a mining contracting firm that had a shaft-sinking project in New Mexico. While he was working in the Southwest States, he married a younger woman than himself and, at this time in Denver, he was going through a divorce from this second wife. After some time between contracting jobs, he left Cementation and was holidaying in England when I, at the time, was looking for someone to act as project engineer on the Broulan Reef Rehabilitation Project. Having lost touch with John and after calling a number of his old cronies, he was found at his parents' place in England. *(Every time John and I met, he would remind me of the time he was tracked down in England.)* Anyway, John didn't take the Broulan offer as he had an offer in hand in Denver with Pincock, Allen & Holt, a geological consulting firm, where John was their one and only mining engineer at that office. It was a coincidence that this same firm were the geological consultants that Dynatec were using to review the ore reserves on the Montanore Project. Our paths crossed again, a few years later, and he was still having women trouble. He met a woman his own age while working for Stone and Webster in New Jersey. This one, he didn't marry. The problem in New Jersey appeared to be this woman's teenage daughter, who was rebelling against her mother and John. The last time we talked, he was back with this woman in New Jersey as her daughter had grown up and moved out.

The work on the Montanore Project was finally concluded while in Denver on December 11, 1989 where John Merrington and Al Walsh of Dynatec Denver, Fred Edwards of Dynatec Richmond Hill, Bob Reuss independent consultant

Why Mining?

from Tucson and myself worked together at the Denver Dynatec office for the next two days on details regarding what should be included in the final report. MMEI was given the task of estimating the capital required, as well as elaborating on the notes gathered at the property.

Bob Reuss, the little guy from Tucson, was an interesting study. He could put numbers on everything. Where many of us would use experience to determine the number of machines or men required, Bob always had a factor he used to determine what he thought should be the correct quantity. *(I think he was more right than wrong.)*

Something should be said about the Dynatec Denver office. Apparently the office was built by Orotec, a company headed up by Alex Scott, who was a year ahead of me in mining engineering at UBC. Alex worked as the shaft superintendent on the original K-1 shaft at IMC in Esterhazy and from there specialized in shaft work. He did very well for himself and made many millions of dollars in consulting and construction in shaft work. He sold his Denver office to Dynatec for $2 million when he liquidated his company in the late '80s and then retired doing the odd consulting job when it pleased him.

After supplying my input (which was really rushed) for the Montanore Report, it was left up to Fred Edwards to write the final report as Irene and I flew out to Victoria that December 19th for Christmas.

General Chemicals

Now getting back to cases, the next day after meeting John Wearing, was a Sunday (November 19, 1989), when I flew to Rock Springs, Wyoming and waited there for Fred Edwards to arrive from San Francisco. While waiting for Fred, Roger Harris, another old colleague who now was managing the Texasgulf Trona Mine in Green River Wyoming, was called and informed that Fred Edwards and I were down in his part of the world for the next couple of weeks at the Holiday Inn and suggested we meet. This we did at the golf course club house there in Rock Springs where we enjoyed a very fine dinner together. We also toured the Texasgulf Trona Operation. At the mine, our escort underground was Herb Price, who Fred and I had met before in Saskatoon and in Timmins. Herb was the owner's representative for Homestake at Allan Potash when Homestake owned what was then the Texasgulf's 40% ownership of Allan Potash Operators. *(Allan Potash*

Operators (APO) was a joint venture company where Homestake and US Borax owned 40% of the shares each, while Swift Canadian owned the remaining 20%. Later Texasgulf purchased the 40% held by Homestake, and the Potash Company of Saskatchewan (PCS) bought out Swift's 20% and USA Borax's 40%. Finally after some time, all the shares of Texasgulf's potash interest were transferred to Saskterra. Later PCS purchased that firm's 40% interest to hold 100% of APO.) Herb had toured the underground at Kidd Creek during the planning period of the Texasgulf Trona mine so now he was just paying us back by showing off his mine. Texasgulf ran a very good operation in Wyoming, much better than the General Chemical Corporation's mine in the same area.

The purpose of the trip to Wyoming was to review the underground operation of General Chemical Corporation (GCC) and offer suggestions for improvement. The GCC mine was the dustiest and dirtiest mine I had ever toured. *(I have been underground in over 120 mines in my career.)* It was surprising to see men working under conditions such as those found underground at this mine. Labourers were shovelling dust spillage from under the conveyors, and every time they threw a shovel full of this spillage onto the top of the conveyor, a cloud of dust would rise up and engulf the whole area where it was next to impossible to breathe. There were no atomizers used either along the roadways nor along the conveyor galleries to control the dust. Fred and I found a number of things that could be done easily without much cost to improve the operation. For instance, the loadout storage area was designed for 3300 tons storage but could accommodate only half that amount because a manlift had been installed inside the storage bin instead of enlarging the existing raise at the east end of the loadout area and installing the manlift there. As a result there was considerable starting and stopping of the main conveyors that fed the storage because that facility was too small. As mentioned there was no dust suppression at the transfer points on the conveyors nor, for that matter, along the travellways. The backs in the mining section were very ragged and should have been trimmed with an undercutter which had considerable standby time as it was used only for shearing the center cut; whereas, it could have been used to trim the back after excavating the center cut. The rockbolts used were the expensive resin-grouted type where the cheaper split-set type could have been employed in the temporary sections. The loader operator was not encouraged to clean up the heading but only loaded the easy material in the center of the heading, thus leaving much rubble behind.

Why Mining?

The ventilation in the mining area used a negative system to draw dusty air away from the face instead of a more positive system where the dusty air is pushed away from the face. The cutting pattern in the panels and along the main entries had to be improved as there was much slabbing in all areas. We suggested GCC contact Dr Shosi Serata at the University of California at Berkley to determine a safe pillar-to-room size. We found that the mine workforce was organized by function rather than by shift as there was a superintendent each for continuous mining, bore mining, conventional mining, services, maintenance and engineering. No one person was in charge of overall production on the off-shifts, nor for that matter on day shift either except for the mining manager who spent much of his time on surface. The crews, by working five 8-hour shifts per week, lost considerable time in travelling to the work-place at shift change and when the maintenance personnel performed inspections on the equipment. We suggested that less time would be lost if the shift schedule was changed to four 10-hour shifts with the maintenance men arriving two hours earlier than the production crew and, in that way, all equipment would be available when the production crew began their shift.

At the end of this two-week study period at GCC, the findings were presented to Jim Wilkinson, vice-president; Rick Casey, general manager; Ron Hughes, mining manager, and a representative of the Steelworkers Union whose name escapes me. The presentation was made in Casey's office. Wilkinson was a no-nonsense guy with little in the way of a sense of humour. Casey, on the other hand was quite friendly but one could see, when in the presence of Wilkinson, he was very quiet and noncommittal. Ron Hughes was a somewhat know-it-all but knew we were right about the matters we brought to the table and, like Casey, was very quiet at the meeting. The union representative was non-committal on everything. Fred Edwards led off with the recommendations and suggestions which are summarized as follows: the operation could be improved by reorganizing the crews into teams, reorganizing the shift schedules, automating the loadout, controlling the loading on the conveyors, suppressing the dust along the roadways, improving the face ventilation and revising the cutting patterns which are things that require very little capital to inaugurate. The meeting seemed to go well as the suggestions were well received but it is strange that Dynatec Engineering did not secure any subsequent work.

Regarding Casey, he got into trouble a month later at GCC's Christmas party when the union set him up by getting one of the girls in the bargaining unit to seduce him into an uncompromising situation where sexual harassment was claimed. As a result, Casey's association with GCC was terminated. As for Hughes, I escorted him and two of his associates (Dean Wear and Randy Pitts) underground at both PCS Allan and Cominco Potash in March 1990. They were interested in seeing how the borers with extendable conveyors were operated. *(I was hoping to get some extra consulting work by organizing this tour for them but no such luck.)*

The Holiday Inn was home for the full two weeks while in Rock Springs. We had our breakfast at the Inn but found the dinners there were second rate so, as a consequence, we frequented other establishments mostly at "The Outlaws" up the road a piece. Because there is little likelihood of spoiling breakfast, we continued having breakfast at the Holiday Inn with some improvement thanks to Fred's initiative. For the first couple of days, we both requested All-Bran cereal for breakfast but the waitress said that they didn't have any but would order some that day. After the second day, and getting the same answer, Fred went out and bought two boxes of All-Bran and gave them to the waitress on the third morning with the request that these be kept for us only. After that, we had our healthy breakfast of All-Bran cereal. At lunchtime, Little America, right on Interstate I-80, not very far from the mine, was the preferred place for the noonday meal. At Little America, the largest gas station in the world with over 150 pumps, they really knew how to fill you up as the food portions were very large. Also, they advertised the "25¢ Ice Cream Cone", except a 2-cent state tax was added making it 27 cents.

On November 25th the US Thanksgiving, we were still working in the Rock Springs/Green River area. The waitress at the Holiday Inn was very concerned that Fred and I were not home with our families for Thanksgiving. We had to explain to her that the Canadian Thanksgiving was a month earlier and always on the second Monday of October. We told her that what Canada has instead, (for the weekend that the US has its Thanksgiving), is our Grey Cup Day . She thought that was very strange and somehow couldn't comprehend having Thanksgiving so early and not on the same day as in America. When having lunch at Little America on Thanksgiving Day, it was intriguing to see families who had reserved tables being served whole turkeys where the head of the table would carve the bird there just like he would if he were at home. It was a good touch as it gave a homelike atmosphere and the womenfolk a break from cooking on this "family day".

Why Mining?
Chapter 15

Newfoundland

Hope Brook

Out of the blue so to speak, John Auston, president BP Resources called in December 1989 just before I was about to fly to Denver to wind up the work on the Montanore project. After returning from Denver, I met with John Auston in Toronto at his BP Canada office where he outlined the assignment at the Hope Brook mine in Newfoundland. It was agreed that the mine would be inspected and evaluated sometime early in the New Year but first he would talk with those at Corner Brook, Newfoundland at the mine before making arrangements. *(It was just as well that the Newfoundland trip was not pressing as, with Christmas coming on Irene and I had planned to be in Victoria. January was to be a full month with a trip to Saskatoon to investigate the No 2 shaft leak at Allan Potash. Also there was another trip to Victoria to attend a funeral of a friend. Finally there was the writing of a report on the five-year plan regarding the Belmoral mine in Val d'Or. It was a very busy month indeed).*

Prior to going to Corner Brook, it was necessary to meet the senior staff at the Toronto Office. Those that were met besides John Auston were, Dave Crowley, vice-president; Bill Fotheringham, mine manager of Hope Brook mine and Norm Dufresne, mining engineer. A lot of information was received from Norm Dufresne including the "availability statistics" of the two Kiruna trucks that were used at Hope Brook for hauling from the production level underground to the mill surface bins. Also, a report was received, written by John Smith, a rock mechanics consultant (who also lived in Kingston). Dufresne, without saying so, as much as said that the mine at Hope Brook was doomed to failure. The assignment given to McKay Mining Engineering by Hope Brook was to appraise the capability of the mine to achieve consistent levels of production at or above the rated capacity.

When flying into St John's on our way to Corner Brook, the pilot announced over the intercom that the plane would be landing shortly on the "granite planet". The granite planet was quite appropriate, for most of the Island was rock except for the area around Corner Brook where there was an abundance of trees which supplied the pulp and paper mill located in that town. The Sunday night February 4th, 1990, when we landed in Corner Brook (actually Deer Lake), it was the coldest night of the winter as the temperature went to

Newfoundland

-35⁰ C. It was freezing at the Holiday Inn as the heating system couldn't keep up. The hotel wasn't that well built either as the windows were only double glazing at best and not triple as they should have been for weather such as was experienced that night. Breakfast the next morning was with Bill Fotheringham who gave an update on the mine as well as covered the arrangements for travelling to the property. *(Bill was intrigued with my Tilleys outfit I was wearing at the time and asked where it was purchased. This was the same Bill Fotheringham, whose father was formerly president of Steep Rock.)*

After the breakfast meeting, we went to the Hope Brook office in Corner Brook to meet Errol Ladner, mine superintendent, and Dick Arndt, chief engineer who gave a briefing on the operation. Hope Brook was one of those "fly-in fly-out" operations except that the crews "sailed-in and sailed-out" from Port aux Basques. I expected to travel to Port aux Basques by automobile and then sail to Hope Brook on the south coast of the island on a small steamer that was chartered by the mine to transport the crews and supplies: instead I was given the VIP treatment as Fotheringham chartered a helicopter for the journey to the mine.

From the airport at Deer Lake, the helicopter flight was overland south to Grand Lake, then it followed Highway 480 to Burgeo for part of the way before heading west to Hope Brook which is on the south coast about half way between Burgeo and Port aux Basques. Except for the first few miles out of Deer Lake, the landscape was completely devoid of trees. As the 'copter got closer to the mine, there were some small scrub trees in the gullies and low- lying areas as well as many caribou. Around Deer Lake the moose were plentiful while down near the south coast, it was home for the caribou.

The mine camp at Hope Brook was the best anyone could encounter. For instance, after experiencing amenities in a number of different camps: three as a miner, one as a visitor and four as a consultant, definitely the Hope Brook camp was the best. It was like a Holiday Inn away off in the wilderness where the main diningroom had laminated wooden beams all nicely varnished in their natural state, each employee had his/her own private room complete with bathroom. Supplied were laundry facilities on each floor wing as well as lounges with TVs on each floor wing also. There were game rooms and snack facilities for those relaxing on their off hours. All told, it was a very nice

Why Mining?

layout to be sure. The meals were good but not up to the standard set by the cooks at the Ketza River mine in the Yukon. *(While at Hope Brook, I didn't care for the coffee so instead I drank the fresh milk that was supplied. After a few days of drinking nothing but milk and sometimes as many as two glasses with each meal, I developed a bad case of diarrhea which I couldn't shake for a month.)*

The ore grade at Hope Brook averaged between 4 and 5 grams per tonne (0.12 to 0.15 troy ounces per ton); thus, in order to mine at a profit it was necessary to mine in quantity. This was difficult due to the shape of the orebody because it did not have much in the way of strike length and, as a consequence, all activity was concentrated in one location rather than spread over the strike distance. The mine was developed under an open pit which supplied ore during the initial stage of operation. Three levels: an overcut level; a stope drilling level; and a mucking-undercut level were developed from a very well maintained 12% decline ramp. The mining method chosen was blasthole open stoping with delayed fill to recover the pillars. Due to the close proximity of the active stopes and pillars, it was necessary to adhere to strict sequencing in mining and filling operations; otherwise, production would be adversely affected, especially on the mucking level as the congestion was easily aggravated thus slowing the mucking operation. Another thing that slowed production was the choice of Kiruna electric trucks for ore-hauling to surface which at the time were somewhat in their experimental stage. On the initial visit in February 1990, the availability of the two trucks on hand was only 49.5%; whereas, the manufacturer quoted a 75% availability when those units were purchased.

On this first visit, it was the concern of management to increase the availability of those trucks as this was the bottleneck in the operation preventing the mine from achieving its budget tonnage. A number of recommendations were made regarding correcting the freezing conditions in the ramp that affected the availability of the trucks. Also, it was suggested that marshalling the traffic was required. But mostly, the mine had to improve its maintenance procedures on the trucks so that, when problems arose, the trucks would be out of service for only a minimum amount of time.

That same spring the mine ordered and had delivered a third Kiruna truck which alleviated the truck shortage problem but what did improve was the maintenance crews' knowledge of the electronic controls which were the bugbearer to increasing the availability in the first place. With the availability and truck shortage concern improved, now the problem of low ore production shifted to other factors.

That October, I returned to Hope Brook to critique the 3500-tonne per day mining plan, to review the mining and maintenance procedures and to study the backfill system. Again the VIP treatment was given, in that I was flown by helicopter from Deer Lake to Hope Brook. *(The coffee had improved since February which was a big plus.)*

Since the earlier trip in February, the mine operators, for the sake of expediency had mined a stope out of sequence and had fallen short of broken tonnage to be mined later. *(They were warned about this practice in the earlier report.)* The mine was still excavating the mucking headings too large, thus accelerating the time when remote control mucking had to be performed. *(Remote control mucking is slower than that controlled directly by an operator with hands-on the controls.)* The "effective shift time" was very low, caused by servicing the equipment underground. The excess water was entering the backfill thus weakening the fill. The mine still needed to improve its housekeeping. All the above problem situations could be improved by adding one more supervisor per shift which was one of the recommendations made in February. It was also stated that unless there was improvement in the areas discussed, the mine would not achieve the 3500 tonne rate of production. It was not long before the Hope Brook mine was placed on the block for sale and Peggy Witte of Royal Oak Resources purchased the mine a year or so later but even Peggy couldn't make the Hope Brook mine into a viable operation.

An Isometric Sketch of the Hope Brook Mine

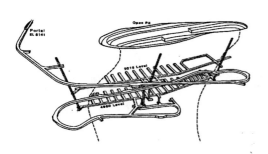

The sketch shows the open pit at the top, the main ramp, 4960 (undercut) level very bottom and 5015 (drill) level. Not shown is 5060 (overcut) level below the pit and above 5015 level.

Why Mining?
Chapter 16
Revenue Canada

As mentioned, the association with Revenue Canada came unexpectedly. What this involved was an auditing of mining research projects selected by Revenue Canada to determine whether or not the tax payer conducting the research was eligible for a federal grant. To determine eligibility the project had to comply with the following guidelines: "...each project must show that there is a technological advancement. It also must show that there is technological uncertainty and, thirdly, it must show that qualified people apply a systematic approach in dealing with the project. All this is in the context of the particular company's environment." This all sounds simple but the truth is, it was sometimes quite difficult.

Well after passing the security clearance, I was assigned to the Ontario Division of Revenue Canada, headquartered in Hamilton and reported to Dennis Marcon, scientific officer, who was very fair and patient with those just starting and learning the ropes. Dennis had graduated as a metallurgical engineer but he had not worked very long in metallurgy prior to joining Revenue Canada. We got along fine and continued our association even after moving out to British Columbia. As a consulting engineer, one would never get rich working exclusively for Revenue Canada as the per diem rate it allowed was only $600 which was supposed to be exceptionally high and even so, they would limit the total time that was thought the job would take for completion and, as a consequence, even though more time was spent to complete the task, one had to justify why the charges were more than that estimated. *(It's nice to know that the taxpayers' money is being watched closely.)* Meals and incidental expenses were limited to $42.25 per day in 1990. *(Being used to the high living that Canterra and Saskterra would allow, my living costs had to be toned down when travelling for Revenue Canada.)*

Drillex
The first job with Revenue Canada was the auditing of four projects of Drillex International of Canada Inc located in Sudbury. Upon arrival in Sudbury and after calling Bob Lipic (who was the manager at that time of Drillex) arrangements were made for a visit to his plant. The chief engineer for Drillex was Allan Guse who described and displayed the four projects that

they had underway.

The four projects were:
1) Swivel Mount Shock Sub Element Molds
2) In-Hole Lubrication System
3) Hard-Facing For Ribbed Stabilizers
4) Rod Handler & Carousal

My draft report on these projects was not very complementary as these were rather "mickey mouse", especially the last one because Drillex was "reinventing the wheel". The Tamrock Company of Finland and Atlas Copco of Sweden both had a much better unit which was proven and was working in the field. Anyway, the Revenue Canada people in Hamilton thought the report was being too harsh and asked that it be revised and less critical. The report was rewritten thereby passing the whole lot. *(To hell with being objective.)*

Mintronics

The next assignment was to audit Mintronics Systems Corporation who were located in North Bay. This company had developed a laser guidance and computer control concept for vehicles in a mine. The man behind this project (and who developed the concept) was Glenn Brophey, the principal owner. His "Opti-Trak" vehicle guidance system was designed (and tested) to automate control of all operations of an underground mining vehicle. The system converts the trucks into hands-off robots for moving material where the truck would travel from a set starting point to a dumping area, using its own guidance system. Inco came up with something similar a few years later but not half as good as the unit developed by Mintronics. *(Inco is a strange company. Because they have their own automated mining research and development department, they believe, "...if they didn't invent it, it can't be any good." That's what you call the NIH factor. "Not Invented Here".)*

Meta-Probe

Another project was the auditing of Meta-Probe Inc, a company situated in Picton, very close to home in Kingston. This company had developed a drilling machine for sampling liquid-bearing materials. Its claim to fame was its drill or sampler which could produce an undisturbed sample in unconsolidated material. It was the initiative of John Porritt along with Jean-Charles Potvin, both geologists who prompted the development of this

Why Mining?

machine. Anyone interested in this project was shown a video of their machine working and then shown, firsthand, the machine in the shop. The company was well organized with a general manager, sales force, manufacturing group and office staff. It had sold a number of these machines at the time of this audit inspection and was well on its way to success. *(The project was endorsed wholeheartedly.)*

Phoenix

The Phoenix Belt Scrapers Limited, a company which was developing a conveyor cleaning device was next audited. This company was originally located in Val Caron, (Sudbury) but was now located in Parry Sound at the Shaw-Milex plant. The principal, Norm Morin, a young man who was formerly a maintenance supervisor at Inco, recognized the need for a better conveyor belt cleaner which would give longer service and do a more thorough job. His idea was, when the scraping blade for cleaning the return side of a conveyor belt wears, it could be replaced from a roll of urethane. The roll of urethane is placed within a pressure vessel and is fed by the differential pressure between vessel and atmosphere so that there would always be a fresh blade of material to act on the surface of the belting. The project had a lot of merit. *(Norm Morin was rather interesting: prior to my departing for home in Kingston, Norm handed me a copy of the New Testament as he was also a lay preacher. He probably thought consultants needed saving.)*

Auburn

The next assignment was an odd one from Revenue Canada. One of these was the metallurgical dust agglomeration project of the Auburn Equipment Leasing Inc where the company had contracts to clean up blast furnace dust that accumulated and recycle it back into the system. The company had contracts with various metallurgical plants in Sorel, Quebec; Sudbury and Burlington, Ontario; Cleveland, Ohio and Salt Lake City, Utah. The process, patented as the Produ-Kake Process, was similar to another process that was familiar to me; that is, it used agglomeration with emulsified resin as a binder. When visiting their headquarters in Burlington, Ontario in October 1990 and meeting with Messrs Ounpuu and Parker (who were the controller and the technical consultant respectively), Don Parker showed off his laboratory facilities as well as pictures and flowsheets of the portable agglomeration plant. He then went on to explain how he was able to emulsify

Revenue Canada

the resin in order to reduce the cost as well as the quantity of resin in the dust so that, when recycled, it would not cause excessive heat in the furnace. It was a technical advancement and it was recommended that it be approved.

Drillnamic

Two other audits were conducted for Revenue Canada, one of them was very timely. Irene and I had sold our townhouse in Kingston and were about to leave for Vancouver Island when a call was received to investigate an automated diamond drilling machine in Sudbury at the Falconbridge Strathcona Mine. This drill was being tested by Falconbridge for the Drillnamic Inc., the company which developed this drill. *(We had to travel through Sudbury anyway because Irene wished to say goodbye to her aunt who lived there. This assignment was a real bonus.)*

The principals of Drillnamic Inc, Dennis Martin, president and Steven Korski, manager were met personally, as well as Harry Budden, accountant via telephone conference call who, at this meeting, explained the project and its financing. Then, the next day we all went underground to see the drill working. The drill had a hardware arrangement and a software control programme which sets and maintains machine parameters through ever-changing load conditions. The hardware components control machine functions (settings from hydraulic cylinders and pumps) were adjusted with electrical signals generated from a Programmable Logic Controller. The software programme, developed for the drill, provided the ability to adjust operating parameters through ever-changing rock formations without being touched by the operator. *(This drill was certainly an advancement from the diamond drill used when drilling on the Mineral King claims with Bob Mitchell 40 years before.)*

After viewing the drill underground and completing the discussions with Martin and Korski, Irene and I travelled to Timmins and spent the night at the Venture Inn where a draft report was written on an old 286 lap-top computer. Upon arrival in Calgary, three days later, the report was finished but required a printer to generate a hard copy for Revenue Canada. On a whim, Mike Haug of Husky Oil was phoned as a courtesy call and also perhaps to see if his printer was available. It was a lucky call as Mike had a job for me at the Sulchem plant in Crossfield, Alberta. At the same time, his secretary printed and bound the Drillnamic report which was mailed to Revenue Canada from

Why Mining?

Calgary. *(The job with Revenue Canada and Husky, not only paid for our personal travelling expenses for our move west, it also covered some of the expense in shipping our furniture.)*

Tomco

The move west did not terminate the technical audits assigned from the Hamilton office. A new audit was for the study of a mesh gripper and domed compression indicator developed by the Tomco Steel Company located in Stoney Creek, Ontario. The gripper involved the development of a device to grip and secure mine support screening to the back (roof) and walls (sides) of excavations. The domed compression indicator was a development from observing plate washers fail due to heavy loading. It was rationalized that a rockbolt washer could be manufactured where it would fail at a predetermined load, thus it could be an indicator of a potential dangerous condition in the mine. Both these items were of value and merited acceptance. This was the last assignment with Revenue Canada as, shortly after that, the research programme was somewhat curtailed.

The Mount Hope Waterpower Project (see Chapter 17)

The picture shows the four proposed shafts. At the top left is shown the surface reservoir which supplies water through the shaft at the far left down to the power station at the bottom left for the generation of peak power. This water is then sent to the lower reservoirs at the bottom right where it is stored and then pumped back to surface using off-peak power. The three other shafts are for servicing and ventilation.

Chapter 17

New Jersey

Mount Hope Hydro Project

We had just arrived at our new place of residence on Vancouver Island in September '92 when a call was received from an old friend, John Wearing, asking me to come to Dover, New Jersey and meet with Frank Fisher of Mount Hope Hydro, Inc, Paul Sheel and Fred Harty of Stone & Webster Engineering Corporation along with John Wearing himself who was now working for Golder Associates, Inc. *(Yeah, this is the same John Wearing who was previously mentioned as having considerable women trouble.)* After having completed a ventilation report in May of '92 on Mount Hope, based on information given by Wearing, the principals involved wanted a meeting to discuss and perhaps update that original ventilation report.

The Mount Hope Waterpower Project was without a doubt the most interesting project in which one could be involved. It's a project that is now lying dormant but, as time goes by, it will undoubtedly be resurrected and completed as planned because there is a need for extra power in that particular area. The project consisted of excavating a large reservoir on surface; the sinking of four large circular shafts to a depth of 2500 feet; the excavation of an equally large-size reservoir underground and the construction of an underground powerhouse and electrical galleries. The purpose of these shafts and reservoirs was to generate "peak power" for the City of New York at times of high usage; that is, the gates of the surface reservoir would be opened so that the water from the surface reservoir would flow down the shafts to an underground electrical generating station and then be stored in the underground reservoir. Afterward, the water that was stored in the underground reservoir would be pumped back to the surface reservoir using "off-peak power" when the power demand was low so that the system could be ready to generate peak power again when needed. The scope of my involvement was to determine the shaft sizes, the underground equipment requirements and the method of ventilation during development. In the previous report, the method of excavating the underground reservoir and powerhouse was determined. It also showed the sequence as to how these excavations would be developed before the size and pieces of equipment could be determined. After determining the size and quantity of equipment, it was possible to calculate the ventilation air requirements. Assisting in this project was Maurice White, a ventilation consultant from North Bay, Ontario,

Why Mining?

who completed the ventilation calculations and critiqued the method of ventilation chosen. White calculated the ventilation air quantities and pressures for each fan and its location. MMEI's involvement was to determine the sequence of events, the type and numbers of equipment and then tie this altogether into one report. As my preliminary report had been written some months previously, it was a matter of updating the plans to include the changes that had been made in the project since that previous report was written. The report was well received and, therefore, it was my hope of getting more involved in the project had it gone through, but that was not the case. As stated, in time, this project will be resurrected.

Regarding the trip to New Jersey, Wearing said to fly to Newark, New Jersey and to stay at the hotel across from the airport terminal where we would meet for dinner. That evening, John and I enjoyed a very good dinner and had a very congenial evening where he gave an update on the project and on his latest affairs of heart. The next morning, he did the driving to Dover, NJ which was great as the New Jersey Turnpike is not something one would want to try without knowing first the ins-and-outs of that area of New Jersey. The Mount Hope property, outside of Dover, was a site of an abandoned iron mine. There was an open pit, where a party at the time, was excavating aggregate rock for use locally which, with some modifications, Mount Hope planned to use for the surface reservoir. The shafts and underground reservoir of course would be part of the construction work that had yet to be started. The meeting with Messrs Fisher, Sheel and Harty went well. Fred Harty tried to challenge some points in the previous report, but having been warned about Harty by John Wearing the previous evening, it was no great challenge to rebuff any of Harty's pointed questions or remarks.

The meeting lasted for only the morning. Returning to the hotel with Wearing, we had lunch and bade farewell. Having all the afternoon to kill, as the flight to Toronto wasn't scheduled until the next morning, I decided to take the bus from the Newark Airport to Manhattan via the Holland Tunnel. The airport bus was a very cheap way to get downtown New York. It only cost $7 one-way. Getting off at the World Trade Centre and going up to the top of both buildings was exciting. Ordering a glass of wine at the restaurant on top of one of the towers and viewing the city from that height, was very impressive. *(It's a must-see for anyone travelling to New York.)* This visit to the World Trade Centre was only weeks before terrorists first attempted to

New Jersey

blow up the buildings in the fall of 1992. *(An odd thing in the States, is the fact that one will see the sublime and the ridiculous side-by-side. There, in Lower Manhattan, are these two lovely towers and right beside them are some very derelict buildings such as one would find in a slum.)*

Anyway, on returning to the Newark hotel and calling home, Irene said there was a telephone call from Mike Haug (of Sulchem and Husky), who requested some help on a problem. Mike knew from a previous conversation that I was travelling through Toronto and therefore, when returning his call, he asked that I visit the Hosakowa Plant in Brampton, Ontario and discuss inerting a micronizer (grinder) for the Sulchem plant. The Hosakowa people marketed machines for grinding prills into powder. They also marketed and built dust collectors as well as other units for dust handling. *(The trip east was indeed a very profitable trip with two clients paying the shot.)*

As a postscript to September 11, 2001, the tragedy of that date destroyed what were probably the two most impressive buildings one could visit in the western hemisphere. That particular act was a crime against our culture and our civilization as a whole. *(What a shame. What devastation.)*

Why Mining?
Chapter 18
Simons & Peru

Coroccohuayco

In the summer of 1994, Bryan Nethery of Simons Mining & Metals in Vancouver, called regarding a job for his client Metall Mining. Nethery and I had never met before but apparently he got my name through Ken Midan, ex-Wright Engineers of Vancouver, who was now out on his own as an independent consultant. Bill Gilmore of Wright Engineers introduced us when both he and Ken were with that company. *(Always, when passing through Vancouver, I made a network call to Wright Engineers.)* The Simons job consisted of designing an open pit and underground mine to exploit the reserves of the Coroccohuayco copper orebody in Peru. The Peruvian Government were in the process of privatizing some of its resources and had the Tintaya Project (which consisted of Coroccohuayco orebody as well as nine other deposits in the Tintaya District) up for option.

Simons had a number of people at work evaluating these various deposits. The task given on the Coroccohuayco deposit was to take raw geological data and resources of the Coroccohuayco orebody and prepare a mining plan where these could be mined at a profit. The orebody was in the final stages of being drilled. All the drill data, such as core logs and assays were sent from the property to Tucson, Arizona where these were compiled on maps and sections to be analyzed so that resource quantities and qualities could be calculated. This data was then shipped by disk to the Simons Vancouver Office for printing on its computer. *(Note: this was before e-mail had become popular.)* The resulting plans and sections were then either sent by mail or by faxsimile to the MMEI office. Initially, the job required working out of the Metall's office in Vancouver's Gas Town above the corner where the steam clock is situated.*(A very interesting place for an office.)*

Gas Town's Steam Clock

274

Regarding the stay in Vancouver, Simons had an executive suite at the Pacific Pallisades Hotel complete with kitchen, living room and bedroom where they accommodated visiting consultants. This was quite handy as it was a nice walk, (about eight blocks), to work each morning and back at night. Breakfast was enjoyed in the suite but lunch was generally with one of the Simons men or with Ray Dujardin of Metall. For dinner in the evening, the hotel was very close to where my Aunt Marguerite McKay lived, so often we had dinner out together. *(Consulting does have its perks.)*

Incidently, while working on the Coroccohuayco project in July 1994, Ray Dujardin received word from Metall informing him that he was to be transferred to Chile. He said he and his wife were happy about this move as he had only two more years left before he was about to retire and that this would be an interesting way to end his active working career.

For the Coroccohuayco ore zone (which had a mineable resource of roughly 20 million tonnes at 2% copper), open pit mining was proposed for the first $3^1/_2$ million tonnes near surface and blasthole underground mining with consolidated backfill to mine the remaining $16^1/_2$ million tonnes. For both of these operations, it was necessary to draw rough sketches and forward these to Metall's draftsman so he could prepare the final drawings and sections. For the surface mining, it was necessary to design a small pit, choose the equipment required and estimate the operating and capital costs. In the case of the underground mine, the means of access had to be chosen which, in this case, was a ramp for men and material and a hoisting facility for the ore and waste. Next came the level layouts, stope design, major development, including service shop with fueling stations and lunchrooms etc., ventilation and pumping facilities, backfill system, electrical distribution and finally the materials handling arrangement such as ore passes, chain-gates, crusher station and loading and spill pockets. A ten-year production plan was prepared showing the annual grade mined. Again the equipment for the underground mine was chosen; also operating and capital costs estimated. The production plan and operating cost as well as the ten-year capital cost were then sent to the financial analyst who compiled all the mining plans and cost estimates from the various team members to consolidate these into one calculation to determine the overall value that Metall should bid for the Tintaya Project. *(Rio Tinto bid was $350 million for the project, out-bid Metall by $4 million, a difference of 1%. As a consequence, Rio Tinto was the successful bidder.)*

Why Mining?
Chapter 19
Taku & Tulsequah

Polaris Taku
In June 1996, John Tulley of Fluor Daniel Wright (FDW) of Vancouver called, asking if I was interested in writing a technical and economic review for Canarc Resources Corp who wished to use that report for issuing a prospectus in order to raise money for their project. Apparently, FDW had been contracted to write this report but refused to issue it when they were told that it was for a prospectus. For some reason it was their policy not to get involved with anything that endorses an investment venture. Having done this type of report before for Belmoral (on the Broulan Reef Project), I jumped at the opportunity as it meant work in an area of British Columbia, the Taku and Tulsequah River Area, that was new to me.

A few days after that phone conversation with John Tulley, I met the Canarc people, namely: Bard Cooke, president; Cliff Davis, vice-president finance; and Peter Karelse, chief geologist in the Vancouver office of Canarc Resources where it was decided, after reviewing the existing data, that a visit to the Polaris Taku property would be in order. A report on these would be written for the use in a prospectus that this firm planned to issue in order to raise capital. The trip to the property was exciting as this was all new country to me. It meant flying to Whitehorse and chartering an aircraft to fly into the Polaris Taku Property which was located 60 miles south of Atlin B.C. and 40 miles northeast of Juneau, Alaska, but on the Canadian side of the border. There is no road into the property and access could only be made either by aircraft to the property's airstrip or by barge up to the confluence of Tulsequah and Taku Rivers and then overland for five miles to the property.

The Polaris Taku property was a producing gold mine off-and-on between the years of 1937 to 1951. Cominco did operate the concentrator in order to mill its base metal ore from its neighboring mines, the Big Bull and Tulsequah Chief. All supplies for these mines had to either be flown in or barged up the Taku River from Juneau Alaska and then transported five miles overland from the barge landing to the mine. It was very interesting reading the old reports for these operations. There were a few rowdy characters that worked at these mines and it was interesting to read how these people were handled.

Taku & Tulsequah

On June 26, 1996, I flew from Victoria to Vancouver and met Peter Karelse and Stewart Lockwood (accountant) at the airport, then we all proceeded to Whitehorse. At Whitehorse, Peter Karelse chartered an aircraft for the flight to the Polaris Taku property. It was the same charter company I remember flying with from Whitehorse to Ross River a few years before (Jan'89) when conducting a due diligence study on the Ketza River mine for Belmoral. The flight from Whitehorse was uneventful except for the scenery which was out of this world. On the way, there were a number of gossans along the side of the hills south of Atlin and I wondered if any of these had ever been thoroughly explored. The glaciers in the Tulsequah Valley were something to behold. *(Anyway, I'm getting ahead of myself.)* From Whitehorse we flew directly to Atlin B.C. and landed there in order to pick up some supplies and a lady cook. This lady cook, along with a lady geologist, were the only two women in a camp of a half-dozen diamond drillers at that time. But as time went by that summer, the numbers of people got out of hand as the camp was enlarged to over 30 persons.

The plane landed at the property safely. This was the first time I had ever landed on an airstrip that had a dog leg in the middle. Apparently, all the smaller aircraft used this airstrip which was built on the former townsite next to the mine, while the larger aircraft such as DC-3 planes used a much larger strip some six miles to the south. After a sandwich and coffee at the cookhouse, we inspected the surface facilities and equipment on site that had been unused for some years and were at the time languishing in deterioration. Most of the buildings, except for the former manager's home and staff house, showed signs of roofs collapsing or falling over. The drill crew were using the former manager's home as their sleeping quarters, and the staff house as the cookhouse. All former surface equipment (such as generators, compressors etc.) was in a high state of disrepair. As anyone with half an eye could see, there was very little of this equipment that could be renovated. Some of the underground workings were accessible via adits and it was via the AJ adit where we accessed AJ level and inspected the hoistroom and shaft (winze) collar. I was interested to see if the shaft could be enlarged *(Yes, it could)* using the Dynatec Method.

In my prospectus report, a two-stage programme was proposed where, in Stage I, $5 million would be spent on further exploration which included dewatering the existing shaft. From 1988 to 1995, there had been 129 diamond

Why Mining?

drill holes drilled totalling 104,380 feet of surface drilling and what the proposed exploration covered was 10,000 feet of surface fill-in drilling and a 50,000-foot drilling programme to the deep zones from the lower levels of the mine. The underground drilling required that the shaft be dewatered which could be done quite easily using an alimack raise climber for shaft transport. If the results of this initial exploration stage were successful, it was estimated that $10 million would be required for mine development in Stage II as well as $7 million for infra-structure and $40 million for a mill. Using the tonnage and grade of the existing resources, this expenditure would result in an 18% return on investment at the then price of $385 US for gold.

The report was presented to Canarc who in turn submitted it to the Security Commission where it was accepted and a prospectus was issued. That summer, with the funds derived from the issuance of new shares and from other sources, Canarc began its exploration programme. Instead of holding Stage I to a basic exploration programme, the persons in charge began to work on projects and structures that were not necessarily exploration but more like projects for Stage II construction. What was visualized in Stage I was perhaps a six-man drill crew, half a dozen miners part-time for mine rehab and shaft dewatering, a camp cook, two geologists and a project leader. What they ended up with was a site crew of thirty persons, building facilities that were not required at the exploration stage such as a complete fuel receiving and distribution system; a large office, change house and core logging complex; and at head office, others were conducting due diligence studies on other venues and facilities for ore treatment, as well as transportation studies. Also, much equipment was purchased in lieu of renting. As a result of completing considerable out-of-scope work, the costs climbed to $7.2 million, a good 44% more than what was estimated. But worst of all was that not all exploration targets were met. Of the 50,000-foot underground drilling programme, which was the most important part, only 40,000 feet was completed. *(If Canarc had put a geologist, such as Peter Karelse, in charge of the Stage I programme (which was supposed to be exploration), instead of a mining engineer like Dave Watkinson, I dare say that the drilling would have been completed; whereas, Dave Watkinson was attempting to get a jump on the mine development and plant construction and somehow he neglected the task at hand which was drilling and exploring for ore.)*

One of the studies conducted, (which should not have been charged to the Stage I Exploration) was the Environmental Impact Study. This reminds me of an environmental problem that was a result of working on out-of-scope work. Canarc borrowed a small dozer to do some road work and to level some ground for the oil storage tanks. The dozer had to cross the Tulsequah River which flows over a flat flood plane of gravel and sand. The Tulsequah River is not a salmon spawning river because of the flooding caused by the phenomena where the Tulsequah glacier dams the river and periodically this dam breaks (Jokülhlaup) causing flooding of the valley. Anyway, before Canarc was allowed to drive the dozer across the Tulsequah River, which was almost dry at the time, the government environmental people got involved and demanded that the dozer be thoroughly washed clean and a representative of the provincial government to be on site to monitor the crossing of the flood plane. On the return trip, the same procedure had to followed. *(This added expense of a representative travelling from Smithers B.C., to witness a D-6 Cat crossing an almost dry river bed was asinine.)*

I was called back to the Canarc office in October 1997, eighteen months later, where Dave Watkinson gave me a verbal update on the Polaris Taku Project. First of all he said that, since the time of the last report, the project had gone through a name change and was now called the "New Polaris Project". Dave also handed over copies of the monthly progress reports for the Polaris Taku project. *(One could see from these reports that many out-of-scope activities had been preformed. For the life of me, it was hard to figure out why senior management also couldn't see the same thing and ask a few questions because that "out-of-scope" work was costly and added nothing to the mineral inventory of the property.)*

It was agreed that another visit to the property was in order to inspect firsthand what had been accomplished in the eighteen months interval since the last visit. On October 23, 1997, Peter Karelse and I flew to Whitehorse via Canadian and from there Peter again chartered an aircraft for the flight to the property. A rather interesting coincidence happened as we were walking over to the air charter office. I told Peter about a high school buddy, Doug Williston who, as a teenager, was very enthusiastic about flying, even to a point where he would play hooky from school if he knew that an aircraft had landed at the local airport in Cranbrook. In those days during wartime, it was a rarity for an

Why Mining?

aircraft to land at Cranbrook. I went on to tell Peter that when Doug began to earn money after high school, every nickel went into flying lessons. He finally got his commercial pilot's licence, then landed a flying job with a small west coast company and finally progressed to flying for CP Airlines before he was forced to retire because of health reasons. *(It was unfortunate that Doug died quite young as a result of a brain tumour.)*

Peter made the arrangements to charter the aircraft. Our pilot was a young man, no more than 21 years of age at the most. We asked him his name.
He said, "Mark".
I said, "Mark who"?
He said, "Mark Williston".
"My God", I said, "You're Doug Williston's son Mark. I haven't seen you since you were nine years old, just before your father died".
I said to Peter, "The friend I was telling you about (while walking across the tarmac to the charter office) who wanted to be a pilot was this boy's father". I had a lot of questions to ask about Mark's mother, his grandmother and aunts as I had lost touch with the Williston family after Doug died.

Anyway, Mark turned out to be an excellent pilot. *(It was in his genes.)* At one point in the flight, he tried to take the direct route to Tulsequah Valley but was forced to turn back because heavy snow blinded the flight path when he tried to fly over a small range of mountains. He then flew the long way there, by travelling southeast to the Taku River Valley then north up the Tulsequah Valley to the short landing strip at the New Polaris property that doglegged in the middle. *(I felt safe flying with Mark Williston for some reason, maybe because Doug was part of him.)*

Because we took the long way to reach the property which consumed precious time, no time could be wasted if we wanted to return to Whitehorse and catch the last plane to Vancouver that same day. We immediately changed clothes for underground and accessed the mine first via the escapeway where new ladders were installed to the 300 level. After climbing part way to the 300 level we came back to surface and entered the mine on the AJ and Polaris levels to check on the conditions of the shaft and the newly installed winch. It was surprising to see all the construction work that had been done since the last visit to the property in June 1996. A large office, changehouse and core-logging complex had been constructed. Two 90,000 gallon oil-storage tanks from the previous operation had been restored. As well as these, two 8000-gallon transfer tanks and pumping facility had been installed, a one mile 8-inch diameter PVC discharge pipeline had been installed to the river; sleeping quarters for a 35-person camp had been set up; the former shop had been refurbished which presently housed: a D-6 Cat dozer; two Toyota jeeps; two 5-ton dump trucks; four Land Rovers; and three scooptrams. There were three 205-kilowatt diesel generators on site plus two 200-cubic foot per minute air compressors which had been installed to supply air to the underground. All remaining buildings (which were formerly in a poor state of repair) had been removed. All this work was out-of-scope which added to the cost of Stage I Exploration and also had interfered with the progress of exploration drilling. After talking to the person on site (whose name I have forgotten) who was in charge of the site activities during the past summer, it was obvious he was more concerned with mine development and construction than exploration drilling; in fact, he had a ramp already laid-out on paper for development during the coming season. This was not the rustic drill camp that I had envisaged but a full scale mine construction complex. This was similar to what one would see at Broulan Reef. As you will recall, the Broulan Reef project was supposed to be an exploration activity funded by the Flow-Through Share Programme but it turned out to be mine construction under another name. *(In the Broulan Reef project, truthfully I do share some responsibility in what one would call a disaster as $14 million was spent on more mine construction than on exploration. But in the Canarc case, I do not share any responsibility. Thank God that I learned from the Broulan fiasco.)*

An updated report was written and issued where the rate of return was somewhat decreased to 7% from 18% due to the falling price of gold at that time from $385 US to $325. By increasing the plant size to 1000 tons daily, this would almost double the rate of return to 13%. As it was, with the fallout from Bre-X and on-set of Asian Flu, there was little hope of raising capital to finance this project. Canarc did work on its geological model in order to understand and get a better handle on the geology. Perhaps, when the conditions are right, this property will again be among the gold producers in the country.

Why Mining?
Chapter 20

Davy-McKee

Kennsington

When I first started networking with Kvaerner, the firm went under the name of Davy-McKee. Then the name was changed to Davy International and from there to Kvaerner Davy and finally to plain Kvaerner. Steve Harapiak first joined Davy-McKee after he left Belmoral and it was through him that I became acquainted with Ian McColl, executive vice-president of this firm of consulting engineers. Steve left Davy-McKee after a short stay and then went to Eastern Siberia where he brought a large gold operation into production. After Steve left Davy-McKee, there were still networking calls made to Davy-McKee, first through Ian McColl and then through Fred Edwards who joined that firm after leaving Dynatec Engineering.

Fred Edwards

In May 1996, Fred Edwards called asking if I would travel to Juneau, Alaska to review the mining plan made by Redpath of Arizona for Coeur Alaska Inc on Coeur's Kensington gold property. This meant obtaining a Free Trade Agreement temporary "Green Card" to work in the States. Having obtained these before, it was almost routine. John Howard of Coeur Alaska was contacted and sent me a letter requesting my services, length of time and the method of payment. Upon receiving this letter, it (along with a passport and university diploma) was taken to the American Immigration Office in Victoria on Belleville Street where the Americans at that time had an office. As these documents were in order, I was told to board the plane in Victoria and a temporary card would be issued in Port Angeles, Washington as the plane cleared Customs there before flying on to Seattle.

The flight to Port Angeles on Horizon Airways went fine, but the temporary card at the Immigration Office wasn't ready. As a consequence, the departure of the plane was delayed until the officials there could confirm my credentials and prepare my card. I felt a bit sheepish getting on the plane as its load of passengers looked at me as being the cause of delaying the flight to Seattle and perhaps forcing some of the passengers to miss their connections. Anyway after a stopover in Seattle, the flight to Juneau with a stop in Sitka Alaska on Alaska Airlines was uneventful. *(On U.S. airlines, the meals served were very meagre and nothing compared to what was served on comparable Canadian air carriers at that period of time.)* John Howard met the plane and gave me the grand tour of the Capitol *(note the spelling)* of Alaska. There is some controversy about Juneau being the capitol where it has no road connecting it with the rest of the State. There is a group who are lobbying for the capitol to move to Anchorage instead. Whether or not Juneau gets a road or Anchorage gets to be the capitol, one will have to wait and see.

After meeting the staff at the office in downtown Juneau, John Howard gave me various reports to read on the Kensington property prepared by previous consulting firms, namely Redpath, Bechtal and also by Davy-McKee. *(The staff made a bit of a joke about David McKay working for Davy-McKee.)* The Kensington gold property, owned by Coeur Alaska, a subsidiary of Coeur D'Alene Mines Corporation, is located on the Lynn Canal some 45 miles northeast of Juneau. The project at that time was in the final phase of permitting and was moving towards developing a 4000-ton underground mining operation. It had a resource of $13\frac{1}{2}$ million tons at a grade of 0.151 ounces of gold per ton which would sustain operations for ten years. Roughly, my report on the property could be summarized in the next sentence. "Although the gold ore is not necessarily low grade, the operation will have to be careful in its mining practices as grade is critical." In that report, I suggested three types of stopes based on mining width that could be developed as well as a method of attack where the higher grade of ore could be mined at the commencement of operations without jeopardizing the integrity of the orebody.

During a lunch-hour break while in Juneau, I made a tour of the very modest capitol building. While following a tour with its tour guide, one had to chuckle when the tour guide said, "I guess you folks are very surprised that we don't live in igloos here in Alaska. Well, the only people who live in igloos are the Eskimos who live around Hudson Bay, you know, just a little north of New York." On another occasion, I visited a very well equipped museum and library located there. At the library, they had a copy of "The San Francisco Call" dated August 19, 1901 which headline read, "Steamship Islander Crashes Into a Northern Iceberg in the Night and Sixty-five Lives Are Lost in the Sea Off the Alaskan Coast." Those mentioned on the list of passengers drowned were Mrs J Ross, Governor Ross' wife, baby and niece. The Mrs Ross, mentioned in the

Why Mining?

newspaper article, was my great aunt Barbara Elizabeth McKay the wife of James Hamilton Ross who was at the time Commissioner of the Yukon Territory and not governor as stated in The San Francisco Call.

In the summer, Juneau is a very popular stop for the cruise ships. Each day there were at least three and sometimes four large cruise ships tied up at the quay. I'm sure most tourists found Juneau a bit disappointing as it rains at least 365 days a year and besides, there isn't much to see except for a glacier that comes right down to the north end of town, and an old Russian Church at the other end. *(The town has a population of only 35,000 people.)* One of the more quaint places is the Red Dog Saloon with sawdust floor and all. It has a sign at the door that says, "No Dogs Allowed." One time when a Canadian Navy destroyer docked at Juneau, my son Richard (who was a member of its crew) said the saloon changed the sign to read, "No Dogs Allowed Nor Canadian Sailors."

So far (May '02) the Kensington Project is still in the project stage. With the price of gold at around $300 US and financing very tight, this property is sitting dormant waiting for more favourable times.

Bakyrchik

The next assignment with Kvaerner was in Kazakhstan. In February 1997, Fred Edwards called again asking if I was available for a six-week assignment for the purpose of preparing a "Conceptual Study", showing a production rate of 3.5 million tonnes per annum from the various zones in the area granted to the Bakyrchik Mining Venture (BMV) in Kazakhstan. This would require travelling to Kazakhstan to tour the property then work in the Kvaerner office in Toronto for a couple of weeks and finally finishing up at home writing a report. I jumped at the opportunity as that part of the world was closed to Westerners while the Soviets were in charge because the area was their nuclear and rocket testing range... very high security area indeed.

In order to apply for a Visa to Kazakhstan it was necessary to send my passport to the administrative personnel of Kvaerner who then forwarded it, along with details regarding me, to the Kazakhstan Embassy in Washington DC rather than to the embassy in Ottawa. Apparently, the Russian Embassy handles business for Kazakhstan in Canada and are notably very slow so Kvaerner, who had done business with the Ottawa group before, chose to by-pass Ottawa in favour of Washington. *(It's rather strange, when the Visa arrived at my home in Cobble Hill B.C., it was noted the Kazaks spelled Kazakhstan without an "H" like Kazakstan. Yet, one always sees it spelled with the "H".)* With all the visa papers in order and the list of **"what you need to know"** sent by Kvaerner, I flew on to Toronto where Fred Edwards met me and gave an update of the assignment and said that three other fellows would be travelling to Kazakhstan as well. *(The "what you need to know" listed the various contacts in Kazakhstan; Canadian, British and American Embassies; flight information; accommodation; currency exchange rates... like 76 tenge to each US dollar; as well as medical information. This was a very informative three pages of information.)*

Fred introduced a fellow traveller, Ian Chisholm, who would be one of the geologists on this venture. Ian was a world traveller who had been all over. His passport was nearly full of stamp marks from the various countries where he had completed assignments. *(I shouldn't use the word "completed" as, even though I have great admiration for Ian, who is a very charming fellow, he never seemed to complete anything and procrastinated on everything.)* Another travel mate was Gary Giroux of Giroux Consultants Limited of Vancouver. Having read some of Gary's reports from my Canarc assignment of the Polaris Taku property, I was therefore familiar with his ability. *(Gary was very capable. I wished he, rather than Ian, had been assigned to supply me with geological data.)*

The last of our travel mates was Peter Tiley of G.L.Tiley Associates, who was a son of the late Gerald Tiley, a very old acquaintance of mine. Peter was the hoisting expert of our team who was contracted to look into the hoisting facilities for the project. It wasn't until we were in Frankfurt Germany that we caught up with Peter as he was on a different flight from the rest of our party. After a three-hour wait in Frankfurt, we boarded a Lufthansa flight to Almaty (formerly Alma Ada) Kazakhstan. *(Why is it so many Islamic countries' names end with "**stan**" as in Afghani**stan**, Kyrgyz**stan**, Uzbeki**stan**, Turkmeni**stan**, etc.? This is a good question which our group could not answer; however, since then the question was resolved in the February 2002 edition of the National Geographic magazine.* **Stan** *means "**place**" or "**land**".)* The client stipulated that we fly tourist class and not executive class for our flight to Kazakhstan. Having already flown from Victoria to Toronto and then from

Why Mining?

Toronto to Frankfurt tourist class, I was not looking forward to another long eight-hour flight to Almaty. As it turned out, the one and only Lufthansa flight to Almaty was only about a third full at best; therefore, our group was able to stretch out on the four seats in the centre row and sleep the whole trip. An hour out of Almaty, I had a shave and generally freshened up. The service on this Lufthansa flight was excellent. One can't say enough about those German flight attendants in the way that they served the meals and looked after the passengers' comfort.

Maurice McKee, Gary Giroux, Peter Tiley and Ian Chisholm in front of the TV Broadcasting Building in Almaty, Kazakhstan in February 1997

On landing in Almaty, I thought we had landed on a gravel runway as it was extremely bumpy. Later we found out, that's the way all asphalt feels in Kazakhstan as the roads were full of potholes too. We got off next to the very dimly-lit terminal building and were escorted across the tarmac to the immigration and customs room and were met by the mine's interpreter, Ada Tereschenko. Ada had spent some time in the Toronto Kvaerner office helping to translate some of the Bakyrchik mine documents and to become familiar with the ways of Canadians. Ada said that clearing immigration may take some time so she suggested that we all pay $50 US for the privilege of sitting and waiting in the VIP lounge which we all four did. Now the VIP lounge was a room measuring perhaps 8m by 5m (25 feet by 15 feet) with very old-fashioned furniture similar to what my grandmother had on her farm back in Golden

B.C. in the 1930's. The room was very hot as it was heated by steam radiators, again the old fashioned kind. Ada was right, it took two hours to clear customs and immigration. But to be honest, both Gary Giroux and Ian Chisholm *(our two geologists wouldn't you know it)* failed to obtain Visas to gain access into the country and therefore it took a bit of time to get them sorted out. In Peter Tiley's and my case, there was some sort of a delay too but not to the extent of Giroux's and Chisholm's.

After gaining clearance, we followed Ada, our guide, to the parking lot where a van with chauffeur were waiting to drive us to the hotel. The drive was about five miles over paved but bumpy roadways and streets. Although the roadways were clear, there was still some snow remaining in banks along the sides as this was mid-March 1997. *(As a guess, one would say Almaty has about the same climate as Toronto.)* After the bumpy ride, our party arrived at the Ankara Hotel (a hotel financed by Turkish money) which was a very modern five star facility that had just opened the previous autumn. The Ankara Hotel was indeed the best hotel in which I have ever stayed and the most secure as well. To go through the front door one had to pass a metal detector. There was also a security guard *(not in uniform though)* whose purpose was to discourage the prostitutes from entering. *(Because of the difficult financial times in this former soviet republic, many of the citizens, even the normally good hardworking and law-bidding folk, have had to resort to various methods to survive.)* Everyone in the hotel spoke English, which was surprising. When questioning the girls behind the desk where they had learned their very perfect English, they said it was at the university there in Almaty. I said, "That university must be a very good one and you girls must be very good students". Everything in the hotel was first-class, even the 30 beautiful Russian models who had arrived shortly before us to put on a fashion show the next day. The rooms in the hotel had the latest in the way of furniture and bathroom fixtures... it was unbelievable. The TV had about 30 channels, twelve in English, which included CNN, British TV and a number from India. There were about six channels from Turkey and a number of local ones but only one from Russia. *(I have forgotten the call letters of the Moscow channel as I had seen that same channel when visiting Finland a number of years before.)* As a curiosity I dialled long distance for home in Cobble Hill B.C. and to my surprise my wife Irene, when answering, sounded as if she was only next door. *(We live in a very advanced technological age.)* The meals at this hotel and the way they were presented, left everyone with a good impression of its management and staff.

Why Mining?
The Ankara Hotel in Almaty Kazakhstan 1997

This was a five-star hotel which had opened in the autumn of 1996 where all the staff spoke perfect English. The hotel was very modern with superb facilities. The cuisine was excellent.

Our party met next morning at breakfast with Maurice McKee who had just arrived from Auezov, the mine site, who filled us in with the latest information on the project. After breakfast, in order to put in time waiting for our flight to Ust Kamenogorst and the drive to Auezov, the group went for a long walk to the Almaty Market, which was an experience. This market featured cultures of the Moslem Kazakhs and the Orthodox Christian Russians. Regarding cultures, it is interesting that, up until a few years ago, the Kazakhs were a minority in their own country, where the majority of the population was composed of 45% Russian, 10% Turkish, 5% Chinese and only 40% Kazakh. The language of commerce is Russian but the government, which is very nationalistic, has declared Kazakh the official language of the country and has adopted the Roman alphabet instead of the Cyrillic. The government, fearing that the northeastern part of the country (which is heavily populated with Russians) may break away and join Russian Siberia; moved the capital in 1997 to Astana (formerly Aqmola) which is more central; whereas, Almaty is located in the very southeastern corner of the republic. *(Kazakhstan reminded me of Quebec 50 years ago when all commence and communication in the work-place was in*

English and the language in the home was French. All the positions of authority in Kazakhstan (except some governmental jobs) are occupied by Russians but the labouring jobs are performed by the Kazakhs similar to former times in Quebec before the "quiet revolution". Also similar is the fact that most Kazakhs are bilingual, speaking Kazakh and Russian, while the Russians speak only Russian.)

As mentioned, the streets in Almaty, although well laid out, were in need of much maintenance. *(Almaty would be a pretty city in the summertime as there were many trees lining the streets.)* It was also noted that the apartment buildings which are all very similar in design *(it seems one design covered all buildings)* and were in great need of maintenance as the stucco (or cladding) was falling off or was badly cracked. The local buses were badly rusted too; in fact some of the doors were completely rusted at the bottom, leaving a gaping hole between the bottom of the door and floor of the bus. *(No comment on the conditions inside the buses as I didn't use that mode of transportation.)* After the site-seeing walk and lunch at the hotel, we checked out and rode again in our chauffeured van to the airport to catch our flight to Ust Kamenogorst now called Öskemen which is located in the northeast corner of Kazakhstan about 100 kilometres south from Siberia and nearly 1000 kilometres north of Almaty. Ust Kamenogorst was and still is a badly polluted city of 600,000 population where two copper smelters, without smoke control, operate. It was in this area that the Semipalatinsk Nuclear Test Range of the Soviet Union was located. The United Nations and World Bank spent considerable sums of money helping to cleanup the pollution in the area.

The airport terminal in Almaty is something to witness, especially the system of checking in. There doesn't seem to be any organization to the method of taking the tickets or checking the luggage. Our party milled around the terminal (as we didn't have Ada with us) trying to find out at which wicket (like the old fashioned train wickets) we were to line up in order to obtain a boarding-pass. *(Line-up is the norm in Kazakhstan.)* Somehow by reading the names of destinations (in Cyrillic alphabet) and the flight numbers (in Arabic numbers), we found the right wicket. But that was only the wicket to hand in your ticket and show your passport. Behind, in the little room, we had to go through a door, carrying our bags and hand them over to the baggage attendant for his inspection and then lift them onto the baggage cart. *(No convenience whatsoever.)*After having done that, we returned to the wicket where the ticket and passport were surrendered to obtain a boarding pass. But, no! We were directed to a crowded wicket where everyone was reaching for their tickets and boarding passes all at the same time. Fortunately being tall, I was able to see a guy being handed my ticket and passport which I immediately grabbed from his hands. *(There was no organization, no convenience nor any smiling face to welcome one for taking Kazakhstan Airlines. We, in the West, were terrified of the Soviets during the cold war thinking that they could defeat us. What I have seen of their organization and the maintenance of their facilities, there is no way they could ever have defeated us even with their atomic bombs which would have undoubtedly failed to explode. I'm glad though we didn't have to give them that chance.)* After obtaining boarding passes, our party went to the security area where soldiers *(maybe security guards in very military looking uniforms)* looked at the boarding passes and passports and then passed their metal detector wands over us very thoroughly. *(Not a very friendly experience.)* After passing that hurdle, we waited outside (under a lean-to) on benches. The benches were the only convenience given to the travelling public that we could see. When the plane was ready for boarding, everyone rushed out onto the tarmac and milled about the portable stairway that was pushed up to the door of the plane. I was one of the first in line, but somehow within seconds, I was pushed to the back of the line. *(The people of that part of the world have had to line up all their lives and are able to manoeuvre their way in order to secure a good place in front of the line.)* Fortunately we were saved by the stewardess who could see we were foreigners and held the crowd back and let us board first. I sat down at an aisle seat until one big Russian came to my place and said something in Russian and motioned me to move. Sitting there and ignoring him, he got a bit angry. My buddies told me not to cause trouble and move to another seat which I did. By this time, all the good seats were gone and I had to share a seat with a Russian "babushka" who had done considerable shopping for vegetables in Almaty and was on her way to the northern part of the country. She had bags of carrots, onions and other veggies tucked about her, under her seat and mine, and also under the seats in front of her. Besides that she was a big woman and wore a heavy fur coat. *(Man, it was crowded!)*

Something should be said about the plane. It was a Russian make, of what type I wouldn't know, but it was old and it was not well maintained. *(Now that's a comforting feeling.)* What was noticed when I first boarded (when walking up the aisle way) was that the carpeting in the aisle-way was worn and was replaced with plywood in places. My seatbelt wouldn't fasten either nor would the seat backs stay in place when adjusted. *(At this point I began to be concerned about the maintenance on plane's air frame and engines, wondering if it was air worthy or not.)* It was an uncomfortable two-hour flight without a secure seat and a seat mate who commandeered all available space. To make the situation worse, the flight attendants served "borsch" as the in-flight meal. Passing on having any borsch, I was afraid of spilling the entire bowl onto my lap and then wouldn't be able to clean the red beet juice from my clothing. It was the most uncomfortable flight one could experience.

Upon arrival in Ust, we deplaned to very winter conditions as the tarmac was extremely icy and with snow banks along the sides. It was also incredibly cold. The baggage was dumped *(yes dumped)* off outside the terminal building onto a little platform similar to what is seen in movies depicting conditions at railway stations in the 1920s. After retrieving our bags, we waited outside the terminal as we thought it was closed as there didn't seem to be any lights on inside. Anyhow other people were waiting in the same area catching rides from private cars and taxis, so we waited there for our chauffeur to arrive and transport us to the Bakyrchik mine which was some two hours by road from Ust Kamenogorsk. We waited and waited and still no one showed up, until we were the last group waiting there. *(Here we were in the middle of the cold night (10 pm & -30° C), halfway around the world, unable to communicate and nobody shows up to take us in hand.)* After half-an-hour of standing around, I had to relieve myself, so walking up to the side of the terminal building where no one could see, I had my relief. *(The only ones around were my buddies.)* While having my sturr *(that's Scottish)* and at the same time noticing a dimly lit light inside the terminal building, I went around to the front and found the main door was unlocked. Returning and telling my buddies what I had found, we all went into the building where it was warm and lit by half a dozen low-wattage incandescent lights. Here was this very large main concourse of the terminal building, which was originally installed with multiple clusters of flourescent lighting, being lit by half a dozen incandescent low-voltage bulbs because all the flourescent light bulbs had burnt out and were probably too expensive to replace. *(This was indeed third world.)*

After going inside, we noticed that there was music, and people were coming and going at the other end of the concourse, so being inquisitive we walked down there to investigate. The comings and goings were revellers who were

Why Mining?

enjoying a party on the mezzanine floor at the far end of the building. There was a small concession stand which was open as well, where they sold bottled beer. I said to the lady in attendance, "Dieh me beer". I knew from working with doukabors back in British Columbia that "dieh me" meant give me. I just assumed that beer was international. The lady understood and by signalling with four fingers, she gave me four beers (one for each of my buddies and one for myself). Handing her a bunch of money and hoping she would give the proper change, she took what I assumed was the cost of four beers. *(I may be naive but I think people in business are generally honest.)* While finishing our beer, we discussed our plight and made plans as to how we were going to resolve this situation. It was decided that Peter Tiley would phone Ada (our interpreter in Almaty) and enquire as to why the mine's chauffeur had not arrived. *(Peter Tiley, being a very clever person, was chosen to phone as he seemed to be the most fluent in communications. Peter wasn't afraid to test his knowledge of the Russian language, even the little bit he picked up from the information booklet given us on our flight with Lufthansa.)* Peter and I went to the concession booth again where he asked for change for the phones. The Kazakh phones required a special token, a somewhat square coin. With a handful of tokens, Peter tried to phone out to no avail. The phones in the terminal were for local calls only and Ada was in Almaty which was 1000 kilometres (600 miles) away. He then got the bright idea to phone the mine but he didn't know the number so he called the operator and said " Dieh me Bakyrchik." Somehow the operator understood and put him through as the mine apparently was a local call. Our contact at the mine said that our driver was busy and that we should go to the Irtys Hotel *(named after the river that flows through the city)* in Ust Kamenogorsk for the night and we would be picked up in the morning. *(We found out later that our driver was out partying Saturday night and was unavailable in spite of the fact that, as a good Moslem Kazakh, he drank heavily just as all the Christian Russians do.)* A very helpful security guard at the airport called a taxi and directed the driver to take us to the Irtys Hotel.

After a rough ride to the hotel, (as the streets were covered with snow and not well-plowed), we entered the dimly-lit hotel and approached the desk. Fortunately the taxi driver stayed with us to help us in with our bags and did the talking for us with the desk clerk, but regardless, the lady at the desk called the police. After much conversation in Russian, she passed the phone to me and to my surprise the person on the other end could speak English. He, being a member of the local police, wanted to know who we were and what business we had in Ust Kamenogorsk. I gave him our names, spelling each one, as well as giving the numbers of each passport. I told him that our chauffeur from the Bakyrchik mine had not shown up and the authorities at the mine told us that we should stay at the Irtys Hotel for the night and we would be picked up in the morning. The voice on the other end of the phone accepted my story and said that we couldn't stay at that hotel but could stay at another one, its name I have since forgotten. He said that this other hotel charges $20 US a night and asked if we could afford that amount. At that price, it was OK by us. He then asked that the taxi driver be put on the phone where the driver was probably given instructions regarding our transportation.

After another bumpy ride, we arrived at the designated hotel. Again, the desk clerk phoned the police, who presumably gave us our security clearance as we didn't have any trouble getting registered at this establishment. When the lady at the desk was asked how much (skólko stóit) for the rooms she said it would be $10 US. This was half the cost that the policeman had quoted. We were told to go to the fifth floor and the Babushka *(old lady)* would give us the key. *(In Russian hotels, at every floor sits an old lady who watches the comings and goings of each guest. This lady gives out the keys and demands that you give her your key when you leave the floor.)* The room I got was the worst I have ever seen. The toilet, washbasin and bathtub were badly stained brown from the rusty water *(high in iron)* in the area. No way was I going to use those facilities except when nature called. *(There are times like these when I'm glad I was born a male.)* After getting settled, my buddies and I were wondering what to do. We decided that we would open one of the bottles we had purchased as gifts at the Duty Free Shop in Frankfurt and celebrate our good fortune in arriving at such a POSH (?) hotel. We joked, laughed and drank until the bottle was gone, then retired to our respective rooms. I didn't bother to climb into my bed; instead, I took off my shoes and laid on top as there was no way that I would dare attempt to sleep in that bug infested "sack".

The next morning we met downstairs and were directed to the second floor for our breakfast as this was part of the $10 per night deal. We were the only ones in the little breakfast area so we had the exclusive attention of one old Babushka who sliced cucumbers, very dark brown bread and cheese besides making us tea as a beverage. Halfway through our appetizing breakfast, our chauffeur, who had the features of a typical Kazakh, turned up, all smiles from ear to ear. How he found us is anyone's guess but he must have gone to our first hotel and not finding us there had perhaps called the police. Anyway, to our surprise when we went to check out, the room rate was only $7 US per night. *(Why the reduction? That's a mystery.)*

After a two-hour ride and through a couple of police checkpoints along the way, we arrived at the town of Auezov where a soviet gulag was once located but now those same buildings no longer billeted political prisoners but instead housed local criminals and convicts. The town was nothing to write home about. It was a typical run-down Soviet village consisting of a number of poorly maintained wooden houses, cow sheds (not barns) and other miscellaneous structures like outhouses. There were manure stacks on each lot as well as garbage piles. Every house seemed to have a mangy dog which was kept outside even in cold weather and slept on the manure piles, perhaps to keep warm. The place was disgusting and it was surprising that the mine authorities hadn't taken a frontend-loader and truck and cleaned up the village. We were housed in Atco buildings right in the middle of this shack town. To get to the mine office, it was necessary to take the path which wound around the garbage and manure piles: it was appalling but thank God it was freezing or otherwise the stench would have been unbearable.

We arrived just before lunchtime, met with the mine manager, Ken Trotter (an English expatriate) and were assigned our sleeping quarters. Our first assignment was to hand over our passports to be stamped by the local police. *(Seventy-five years of communism is hard to shake. The civil authorities are still suspicious of outsiders even though we were there to help.)* Ken gave us a number of reports to read and assigned a young geologist (a Czech who spoke English) to bring us up-to-speed on the local geology. Unfortunately, the two geologists in our party (Chisholm and Giroux) monopolized most of this fellow's limited time to a point where I couldn't get much assistance from him at all; therefore, I had to depend on reading those geological reports that were translated into English. I did have about an hour in discussion with Ken Trotter but even then there were many interruptions as he was preparing to drive into Ust Kamenogorsk that afternoon on business and had to get certain information for that purpose.

Ken Trotter's main concern was the actual Bakyrchik main orebody where the mine of which he was in charge, was being developed. The main Bakyrchik mine itself had been first started by open pit mining in 1956 and then by underground mining in 1974. When we visited the property, the deposit was being mined at a rate of 120,000 tonnes per annum with plans to increase production to 400,000 tonnes per year, immediately. After reaching that target, it was planned to mine the main Bakyrchik orebody at a rate of 1,400,000 per annum using undercut drift and fill with paste backfill. This rate would expend the known resources above the 600-metre horizon in a period of 15 years. We toured the Bakyrchik underground the next morning with Trotter and his Russian assistant who acted as mine superintendent. Trotter didn't speak much Russian and the assistant couldn't speak English but somehow they were able to run quite an efficient operation together. They worked in the same manner as the Navy; that is, the "commanding officer" being Trotter, would give orders to the "coxswain" being the Russian assistant, who handled the workmen.

Our stay at the mine spanned a whole weekend and a couple of days where we had a look at some old core in the core storage and had a trip underground as mentioned. The present company had spent a considerable sum of money in bringing the mine up to western standards. That was a pleasant surprise seeing the high standards of the underground. The mining method though was an old-fashioned top-slicing method which was very labour intensive. The present management had plans to convert from the present labour intensive top-slicing method to undercut drift-and-fill method which they could mechanize. They even considered the possibility of using Alpine Roadheaders for excavating within the orebody. One thing that they will have to do before going into full production is upgrade the hoisting plant as it leaves much to be desired. The hoist installation would never meet Canadian nor American standards as it was based on old German standards that had two brakes but only one independent brake engine.

The change facilities were better than what I had seen on one of my trips to the West Virginia coal mines. At least in this case, the workmen had a place to shower and there were hooks for their clothes, as well as a sauna. We were given mine clothes but having my own, I wasn't in need of anything that the mine supplied except socks which hadn't been brought along on this trip. The miners in Kazakhstan don't use socks but instead wrap their feet in lengths of cloth about four inches wide and four feet long. The Russian mine superintendent showed how it was done but finally wound up wrapping my feet for me. He was really a nice guy. *(I wished at the time that I hadn't given our*

Why Mining?

gift bottle of whiskey to Ken Trotter, the expatriate manager, but should have saved it for the Russian mine superintendent.)

In the Conceptual Study that I was assigned, it was determined that a production rate of 3.5 million tonnes per year for the whole Baykyrchik area at an average grade 6.34 grams per metric tonne (0.185 ounces per short tons) is fair grade. In all, there were about 25 resource areas in the Bakyrchik region that had some geological investigation done on them in the past. It was my job to determine which of the 25 geological resource zones could be mined and then prepare an economic study of their potential. There was little in the way of geological information that could be found which was needed to prepare this study. I was completely on my own as both Ian Chisholm and Gary Giroux monopolized the attention of the young geologist assigned to us. The two geologists were more interested in the diamond drill database and what it contained. In the immediate Bakyrchik area, there were two other modest size deposits, the Bolshevik and the Promezhutochny which could be exploited, where the former would be mined by sublevel caving at a rate of 2000 tonnes per day while the latter would utilize the same mining method as that for the Bakyrchik orebody at a rate of 1800 tonnes per day. Four small satellite deposits (Globoki Log, Sarbas, Zagadka and Dalni) would first be mined by open pit mining to a depth of 30 metres and afterwards by sublevel caving via a decline to a depth of 90 metres below pit bottom. These deposits would be mined individually and would generate 650 tonnes per day each. The "overall blended cash cost per ounce" calculated at $215 US assuming Whole-Ore-Roasting treatment was used. The overall aggregate production would generate a total production of gold over 15 years of 8.94 million ounces or roughly 596,000 ounces of gold annually. The Internal Rate of Return (IRR) before taxes, using a gold price of $400 US, (the price at that time) was 15% with a six (6) year pay-back period.

While on site, there were crews from Kvaerner Davy overseeing work on the construction of a roasting plant for the concentrates as the ore was refractory. This was a huge expenditure. Although the present management had spent money on the underground, bringing the mine up to western standards, it didn't spend a dime on the mine office and dry. The Russian mine superintendent's office, which was located in the mine change house building, smelled like an outside latrine. There was no ventilation in the building whatsoever. When someone would use the washroom, the smell permeated throughout the building even to the office area. The expatriates and visitors didn't occupy these offices but used the main administration offices near the village. That was fortunate as I couldn't have worked in such a smelly atmosphere.

Our trip back to Ust Kamenogorsk was uneventful except for the usual checkpoints along the way. The airport terminal building on our trip back to Almaty was a beehive of activity compared to when we arrived a few days before, late at night. There were people milling about, waiting for their flights like any normal airport terminal building. Even the washrooms were open; whereas, they were locked when we arrived from Almaty. To my surprise, they charged for the use of the washrooms as there was a guy at the door who collected 5 Tenge (9¢ Can) from each customer as they entered the premises. As Ian Chisholm didn't have any Kazakh money, I had to pay his shot too. When we got inside there was the usual urinal trough which was very crude indeed and an oval hole in the floor for those who were in dire need. *(Having never seen anything like that before really shocked me.)*

The airport terminal organization at Ust was much better than at Almaty, perhaps because the air terminal wasn't as busy. At the Almaty terminal, when on our way to Ust, it was confusion at best. Not that Ust was in any way organized, it wasn't, but it was better relatively speaking than the confusion we experienced at Almaty. When arriving at Almaty on our flight from Ust, the pilot didn't park near the terminal building at all but taxied to where the plane would be parked for the night but, because the parking space was occupied, he just sat there until the plane that occupied the space moved. Finally when our plane was parked for the night, they let us all off at a spot some distance from the terminal building. *(Nothing is done for the convenience of the public in the former Soviet countries. They have a long way to go to catch up to the west in catering to the public.)*

We stayed again at the wonderful Ankara Hotel and enjoyed their lovely meals. One can't say enough about the good service of that hotel. It was the highlight of the whole trip. While in Almaty, we had time to spare waiting for our flight back to Frankfurt. Again, taken in hand by Ada in the chauffeured van, she had the driver take us to a winter athletic area about 20 kms from Almaty where there was a ski jump hill, an ice skating oval and other winter games facilities. It was all very interesting as it had been the plan to blast the mountain (to make these facilities) with an atomic bomb but the authorities thought better of it and

Why Mining?

used more conventional blasting methods instead. Some snapshots and pictures of this facility and of our group were taken. These were good memories.

The trip back to Frankfurt and Toronto was without incident perhaps because we slept most of the way. The next two weeks was spent in Toronto where the project regarding the Conceptual Study was discussed with Fred Edwards and Maurice McKee. The study was finally completed at home in Arbutus Ridge-Cobble Hill and forwarded on to the Kvaerner Toronto office.

Winter Sports Stadium March 1997

The Author at the Winter Sports Stadium located about 20 km to the southeast of Almaty, Kazakhstan near the Kyrgyzstan Border. This area of the country is very mountainous with a goodly number of trees in contrast with the area at the mine which is barren.

The landscape in the area of the Bakyrchik Mine in Kazakhstan.

As seen in the picture taken in March 1997, the landscape is deficient of trees. The development in the distant is the village of Auezov which was a gulag for political prisoners during Soviet times. At the present time, it is still used as a jail for those who run afoul of the Law.

Chapter 21

Wright Engineering

Sunshine

My first contact with Wright Engineers was through Bill Gilmore in 1988, an old buddy from Steep Rock in Atikokan Ontario, who was with Wright's mining group. *(Fluor Daniel purchased Wright Engineers some years later.)* Bill was always visited when passing through Vancouver so it was not unexpected that he would phone when there was an overload in that firm's business. Bill was in charge of the mining section for Wright Engineers and had two very able assistants namely: Dave Wortman for underground mining and Ken Midan for open pit mining. All three gentlemen were first-class.

At the end of November 1988, Bill phoned me in Kingston and asked if I would travel to Kellogg, Idaho to the Sunshine mine as they were bidding on a contract there. Of course his offer was accepted. After flying to Spokane and meeting Ken Boyd (Wright's conveyor and crushing specialist who would accompany me to the Sunshine mine) we rented a car at the Spokane Airport and drove on to Kellogg Idaho and stayed the night at the Silver Horn Inn, a place where Ken had stayed before. *(A few years later when Fluor Daniel bought out Wright Engineers, Ken Boyd and others left and joined Simons. People like Bill Gilmore and Dave Wortman took retirement while Ken Midan went consulting on his own. A new group, John Tulley, Jim Leader and Glenn O'Gorman (before Glenn joined Kilborn) replaced the old guard at Wright Engineers in Vancouver and it was through this new group that Canarc Resources became interested in me. This turned out to be a very profitable experience.)*

The next morning we went to the Sunshine mine offices and were met by Mark Hartman, the mine manager. Hartman escorted us into the conference room where three other groups of consultants, (Dynantec, Redpath and Stearns-Roger) had gathered. There we were, all four groups, all in the same room and all vying for the same contract. Hartman started in and began to give a good organized briefing of the problem when he was interrupted by his boss, Harry Cougher, general manager, who arrived late. *(I had met Cougher before when touring the Bunker Hill mine in Kellogg a few years earlier when he was employed as mine manager there.)* Anyway, Harry Cougher took over and rambled on giving a disorganized outline of the problem. We all could see

```
Why Mining?
```
Hartman was getting annoyed but he held himself in control and didn't say anything. *(After all Cougher was his boss.)* When Cougher was called out to answer a phone call, order was restored in the briefing and one could see what was required. Between Hartman and the briefing notes he handed everyone, all groups went away with a pretty good understanding of the problem. But, no thanks to friend Harry Cougher. *(One shouldn't be too hard on Harry as he is really a very nice guy.)*

After the meeting, we travelled back to Spokane where we outlined the stance that would be taken in the submission to Sunshine Mining which was due in ten days. Ken Boyd phoned his boss, Bill Gilmore in Vancouver, and told him I would be at his office the next day and would finalize the submission from there. When the next morning came, it was a very foggy day at the Spokane Airport and it was necessary for anyone flying out to take a bus to Seattle and fly on from there. *(I don't know why we had to drive the whole way to Seattle as the airport at Moses Lake (a few miles out of Spokane) was clear as a bell.)* As it was, it was very late at night before arriving in Vancouver and therefore a whole day was lost on account of the fog. On meeting with Bill Gilmore, I outlined the problem and the proposed stance that Boyd and I thought should be taken. Bill concurred and organized his group to prepare Wright's presentation. Unfortunately for Wrights, Dynatec got the job. But the Wright submission did come second on the list. *(Seconds don't count in this profession ... you either have the contract or you don't.)*

Aquas Tenidas

In October 1989, Dave Wortman of Wright Engineers phoned asking me to go to Spain on an assignment on the Aquas Tenidas property for a period of two weeks. With much disappointment I had to turn him down as I had committed my services to Fred Edwards of Dynatec on an assignment in Wyoming. As it turned out, although not going to Spain, I was sent the preliminary report to appraise and to criticize. It was possible to make a few suggestions to improve the report as outlined below:

- On the shaft layout - steel and compartment orientation should be changed to better resist stress from the shaft walls.
- On the mining method - access ramps into the stopes should be driven from unmined stopes thus saving on waste development.

Wright Engineering

McCreedy East

The next assignment with this group was at Sudbury in June 1991 after Wright Engineers had been taken over by Fluor Daniel. Ken Sinclair of Fluor Daniel, who was the manager at the Sudbury office, called requesting me for an assignment on the McCreedy East mine at Levack Ontario. After driving to Sudbury on June 20th, 1991 which was a Thursday, I met Ken Sinclair, who introduced me to Bill Deason, a Fluor Daniel employee from the San Francisco office, and the group leader Peter Hopper, who would assign those involved with the McCreedy East project. *(As an aside, this new Inco project was named after John McCreedy, the late Inco vice-president and ex-Maple Leaf hockey player. It was in 1962 that I was called by John McCreedy to come out to Copper Cliff and demonstrate Rockiron's new grouted rockbolt. I was all head-up at the time about going out to Copper Cliff and demonstrating before the "big man", which I did. Then, after the demonstration was over, McCreedy called me into his office and gave me a good dressing down for going out and demonstrating the bolt to the operating group at Creighton mine instead of first coming to Copper Cliff. I never cared for McCreedy then nor since. He was the typical Inco type of that era, heavy-handed and bullying.)* After getting squared around with the McCreedy East assignment, I drove home to Kingston early Saturday morning and, on the way, nearly hit a deer around Honey Harbour. It was a short stay at home as Fluor Daniel wanted to get started on the project first thing Monday morning.

Before getting started describing the McCreedy East job, it would be in order to say something about the people working on this assignment. Ken Sinclair, Fluor Daniel's manager at the Sudbury office, was a no-nonsense type of guy and pretty sharp as well. He was completely honest and definitely knew his stuff. *(A few years later Ken vouched for me regarding my credentials for renewing my consulting engineer designation.)* Peter Hopper, the project leader, was clever enough and for sure a good talker. Much of his talk was just that, as he seemed to expound on the obvious much of the time. Bill Deason, the American from San Francisco, was a likeable fellow but was unsure of what he was doing. He was always bothering someone about some particular point. He wanted to discuss everything he put on paper to a point where we were all ghost writing his part of the assignment. Another person was Ralph Adams who had worked for me as an electrical engineer at IMC and under contract at Texasgulf-Kidd Creek. Ralph was the strong silent type, who definitely knew

Why Mining?

his electrical engineering and was, at this period, working as a full-time employee out of Fluor Daniel's Vancouver office. While working on the McCreedy East assignment, I met Jim Leader from the Vancouver office. A few years later, Jim and I would be working very closely on assignments in China. A jovial fellow was Jim C Fagan who was the estimator on the job. Jim was ex-Kilborn and very anti-Jack Mitchell, a former president of Kilborn Engineering.

For this new assignment, we (those assigned to the project) drove out to Onaping where we went underground through the Coleman shaft and hiked the lower Coleman levels underground to the McCreedy East mine. The McCreedy East mine was a very strange setup. Falconbridge had the task of mining the top section of the orebody while Inco had the bottom half. It was the bottom half that comprised the McCreedy East mine. This trip underground was enlightening. Like all Inco mines, the housekeeping was very good and the equipment was first-class in quality. The trip underground started by going down the Coleman shaft to the 3300 level and then walking down the No 1 cable conveyor gallery to the transfer point where No 1 and No 2 cable conveyors intersected. This was the first time I had seen a cable conveyor; however, I had read and studied cable conveyors quite extensively when working for Steep Rock Iron Mines. The installation of conveyors and pumping system was of first-class quality construction. Even the galleries that had been excavated were supported by screening, making them look very safe indeed. It was very impressive. After viewing the transfer of No's 1 and 2 conveyors, our tour group went back up No 1 conveyor and to 3500 level to see the ramp being driven out to the McCreedy East ore zone.

The overall plan for the McCreedy East mine was to access the mine through the Coleman shaft and drive three ramps out to the McCreedy East ore zone. One of the ramps would be the 16° decline gallery for No's 1 and 2 conveyors as mentioned, which would carry crushed ore and waste to the Coleman shaft for hoisting. The other two ramps, somewhat flatter, would be an 18 x 16 x 4839-foot long access ramp for men and supplies and an 18 x 16 x 4300-foot long return air transfer drift from the McCreedy East to the Coleman shaft. Also included in this Inco scheme was a 12 x 12 x 1000-foot internal fresh air raise and an auxiliary 18 x 18 return air transfer drift. This was certainly considerable development to gain admission to the McCreedy East ore zone. Besides the long drives to access the ore zone, the travel time for the workmen

Wright Engineering

to reach their working areas would have been horrendous.

Inco Engineering was responsible for the initial planning of the McCreedy East mine. Apparently, the initial estimate was something in the order of $180 million but Inco had already spent $240 million when they requested Fluor Daniel to come in and review these figures. Inco Engineering did an excellent job at detailing the project but the flaw in the whole project was the initial concept...it was faulty. What Inco set out to do, as stated before, was to access the ore zone through the Coleman shaft as well as convey the ore to that shaft and hoist from it. That meant drives of nearly a mile out from this shaft to the vicinity of the McCreedy East ore zone. Besides the long drives, Inco had started two ventilation shafts in the vicinity. It was a very expensive concept. In short, Inco Engineers put together a weak conceptual plan but they did do an excellent job in detailing that plan's components.

Fluor Daniel made four proposals as a comparison to the Inco Engineering Plan and are outlined as follows:

- **Joint Venture Company** with Falconbridge and mine both sections of the McCreedy East ore zone jointly, instead of Falconbridge mining the top and Inco mining the lower sections of the ore zone.
- **Optimize the Base Case** of the Inco Engineering Plan showing where savings could be made in developing smaller headings, installing a single crusher, eliminating the proposed surface loadout, and proposing a track haulage system to eliminate excess trackless equipment.
- **Stand Alone Scheme** where the two shafts presently underway would be utilized, where one is rock hoisting and the other as men and materials hoisting. The work already completed involving the Coleman shaft would be abandoned. This would save the drives from the Coleman to the McCreedy East as well as the long travel distances for the crews. It would mean new hoisting and dry facilities installed at these two shafts.
- **Optimize Link Scheme** was my assignment. This would utilize the completed conveying system for the ore and waste handling and converting one of the two shafts being sunk as a service shaft. The savings would be in eliminating the long crosscut drives and the travel time for the crews.

Why Mining?

I don't know which scheme Inco used, if any of them. The operating people were very upset to think that one could operate a shaft system with as few people as the study proposed. When they were told that the numbers shown in the study were based on actual numbers employed at Kidd Creek (which was larger than the McCreedy East mine), they were surprised and unbelieving. *(Inco was not the most efficient operator in the country at that time. It didn't have to be because it had good reserves and fairly good grade.)*

Changma

In the spring of 1997, Jim Leader made a trip to Shandong Province in China where he gathered information on the Changma Diamond mine located at MengYin City which is the administrative centre for the nearby county. The operators were mining from open pit and were nearing the economic limit of the open pit mining reserves for this particular orebody. The information that Jim had gathered from his trip was passed on to me along with an engineering report that Kilborn Engineering had made on the same property. The assignment was to take this information and prepare a preliminary report on developing an underground mine so that the operators could then detail a plan to make the transition from open pit to underground mining without losing production during the transition period. Time was of the essence as the operators should have acted sooner in making this change. Had they acted when the Kilborn Report was presented a year before, it would have been easier to make this transition without any loss in production. As it was, they had waited until the eleventh hour. The Kilborn Report was a very good report had the mine been located in a developed nation, but in China it would not fly because the report proposed considerable mechanization. The Chinese want jobs as there is substantial manpower to employ. Also, from what I saw when visiting China a few months later, the Chinese are not high on maintenance, so when the Kilborn Report showed a high level of mechanization and automation, it was bound to fail as it would not be suitable for the Chinese culture. The assignment was to use the Kilborn Report as a guide but change from mechanization to labour intensive operations. Instead of remote-controlled scooptrams working from drawpoints, it was proposed that slusher hoists, working in slusher drifts, be used. In other words, what was proposed were mining methods and equipment used in Canada back in the 1940s and early 1950s. This was an interesting study as I had to think back to what it was like when I first started in mining.

Wright Engineering

Later that summer, (July 1997), Fluor Daniel Wright requested my services for an assignment in China at the same Changma mine that Jim Leader had visited earlier that spring and on which I had written a report on his findings. My travelling mate would be Patrick Lo, a civil engineer from the Vancouver office, who was to be not only a fellow travelling mate but would also act as the interpreter. The sad part of it was, Patrick couldn't speak Mandarin as he was born in Hong Kong and therefore spoke Cantonese, although he could read the Chinese characters. Anyway, that was overcome as there were others whom we met in China who were fluent in both Mandarin and English.

Patrick Lo and I would leave from Vancouver on August 3rd, the day before my 70th birthday. We boarded Canadian Airlines from Vancouver's newly completed terminal and started our 13-hour flight. Something was served on this flight that made me ill, even though we were in Executive Class where one expects the best quality. It wasn't really flying sickness as I have done a lot of flying in my career. It must have been something that was served on board this flight. It wasn't anything I drank because, at the time, I was not drinking alcohol nor tea or coffee due to my high blood pressure. *(My blood pressure has since receded and I'm back drinking.)* It must have been the food. Also, surprise surprise, the service was terrible! *(Is it any wonder that Canadian Airlines went bankrupt!)* The service sure didn't compare with the international flights on such airlines as, Air Canada, KLM, SAS, Finn Air or Lufthansa. The flight attendants were snarly old women who, because of their seniority, probably commanded these prestigious international flights. Anyway, in spite of the poor service, the plane landed in Beijing when my birthday was almost over. That was the shortest August 4th (my 70th birthday) I had ever experienced. In contrast to the Vancouver airport, the Beijing terminal was obsolete and overcrowded but there was construction going on where a new and more modern terminal was underway. *(It was learned since that the new Beijing terminal has been completed and is now in operation.)*

A Chinese national, Sun Yi (Danny) assistant to Vice-President of Pan Asia Mining, met our plane and made all the arrangements for the trip in China. Danny spoke very good English, yet he had never been out of China. He said he learned English at the university in Beijing. *(It should be mentioned that all signs at the airport are written in Chinese characters and in English.)* Danny said that there would be a three-hour wait before catching the plane to Jinan (a city about 400 km south of Beijing in Shandong Province) so to kill time he

Why Mining?

said perhaps we should get something to eat. This we did. In the northern part of China, noodles are the main dish; whereas, in the southern part, rice is served liberally. The meal consisted of noodles and some other dish the name of which escapes my memory. *(Anyway, it wasn't any good or else it would have been remembered. I didn't like the food in China.)* The three-hour wait turned into a six-hour wait and in a crowded airport where there didn't seem to be any airconditioning. The one thing that occupied my interest were the people. There were all kinds of folk from Korea, outer Mongolia, Viet Nam (as they spoke French) and others that I could not place. Everyone seemed to be well dressed and very western in every way. *(It gave one a very good impression of the country.)*

Finally our group boarded Shandong Airlines for the flight to Jinan. Shandong Airlines operates a very good airline: the service was good, the stewardesses were pretty and polite and the plane was well-maintained and clean. *(The food was enjoyable too.)* When arriving in Jinan, a city about the size of Toronto in population but not in area, it was very dark. In China when it's dark, it's really dark as they don't have many street lights. *(It reminded me of Sadonna, Arizona on New Years Eve of 1996: that town didn't believe in street lights either.)* Our driver had to go around a huge traffic pileup which wasted about an hour before he could deliver us to the Guido Hotel where we would spend the night. This hotel was very modern and very clean. The hotel had some English channels on the TV which was rather interesting. I even called home to Cobble Hill B.C. by direct-dial from the room to the surprise of Irene. She shouldn't have been too surprised as when travelling in Kazakhstan, I called her every day. *(We sure live in a technological age as we can communicate easily from the far reaches of the world to home without difficulty.)* One thing that was a surprise, was the wide use of cellular phones in China.

After checking out of the Guido Hotel, there was a two-and-a-half hour drive of about 200 km to the south of Jinan to a place called Meng Yin, the Diamond Capital of China. At the entrance to the town was a large sign in the shape of a cut diamond proclaiming that fact. *(Patrick Lo translated that for me. Although he could not speak Mandarin, he could read the Chinese writing which is common for all Chinese dialects and for Japanese as well.)* Regarding the chauffeur, he took many chances on the 200 km drive to Meng Yin. When driving along a secondary six-lane highway without a median, he was always out into the on-coming lane then he would dodge back into one of the right-

Wright Engineering

hand lanes when an oncoming truck or car was about to head toward us. It was all right on the freeway as there was a median separating the right and left lanes of traffic. I would never drive in China, it's too scary as everyone drives like there's no tomorrow ...honking their horns every few seconds, dodging three-wheeled one-cylinder Beijing tractor trucks that were very much over-loaded. The worst were the horse and ox-drawn carts because they were so slow that it compelled the car drivers to be weaving in and out changing lanes each time a cart was encountered. Also, there are the motor scooters which posed a hazard. It seems all the young ladies rode these things, but what distracted the car drivers were the short skirts these girls wore.

After checking into the Wen He Hotel in Meng Yin, we attended a kick-off meeting. But before discussing the meeting, it would be in order to describe the hotel. The Wen He Hotel was brand spanking new as it had only been opened for about three months. Like all hotels in China, there was a grand entrance and a very large lobby area. The attendants at the desk were very petite and pretty. When Danny Sun told them I was 70 years old, they would not believe him as all Chinese reaching that age are generally wizened up from years of hard work. Our rooms were very modern with western style toilets and showers. *(That says a lot as I encountered the "hole in the floor" type at the Changma mine office in Meng Yin.)* What one has to get used to in this part of China is a lack of hot water. It is turned off at night and is not turned on until late next morning in order to conserve energy. If one is an early riser as I am, the hotel leaves an urn of hot water outside the door for washing and shaving or to make a hot cup of tea as tea is supplied in the room. The Chinese are light on maintenance. It was noticed that some of the carpeting needed stretching and some of the floors and walls were scratched, yet nothing was done about it.

The kick-off meeting, held at one of the meeting rooms in the hotel, was attended by a delegation of the Pan Asia Group headed by Tun Aye Sai, president; Dai Zhen Fei, senior vice president; and Sun Yi (Danny), assistant to vice president of the Pan Asia Group of companies. Also attending were Li Shou Cai, interpreter *(who was gay)*; Dr Anshun Xu, geologist from the Vancouver office; Prof Wang Wen Qian, process consultant and a former UBC assistant professor who did the reverse and returned to China to teach at the Kuming Metallurgical Institute in the southwestern part of China; and David P Taylor, consulting geologist who was a semi-permanent resident in China for the Pan Asia Group. *(As a point of interest, Dave Taylor was a very large fat*

Why Mining?

individual who was in contrast to most Chinese who are thin and wiry. Understandably, he was a curiosity to the general populace.)

Tun Aye Sai started the meeting by outlining the main problem areas in the mine and plant. He said, "...the major problem in the plant was the continued use of grinding mills for size reduction which, it is believed, is the cause of breakage of the larger gem quality stones, thereby decreasing their value." He went on to say that the installation of a DMS unit had been proposed by both Kilborn Engineering and Fluor Daniel Wright (FDW) in the past to replace the grinding and grease table equipment, thereby improving overall recovery and avoiding the degradation of the larger gem grade stones. There were other problem areas such as the secondary crushing area which should be rearranged, and security arrangements need improvement. The problem at the mine is that the open pit will run out of ore before an underground mine can be developed. The WuHan Institute suggested deepening the big pipe section of the pit by 5 metres and the small pipe by 2 to 5 metres which could sustain production for about 18 months and thus prolong the life of the pit. We (FDW) found with the present reserves as well as this added depth in the pit plus some stockpile material at the plant crusher and at the nearby Hogqui mine area, the plant could operate until June 1999 which might be enough time to start underground mining. Our (FDW) task was to do the conceptual design and the WuHan Institute would do the detailed work.

After the meeting there was the usual Chinese lunch which consisted of a large meal of many dishes served around a circular table where eight to ten people sat where the various dishes (*I counted 35 different dishes on one occasion*) were served on a "lazy susan" in the middle. Each person was given a small dish (about six inches in diameter) which one helped themselves to what was wanted from the lazy susan. There was also lots of beer served as well. *(And me not drinking!)* Three things that were not enjoyed was eating with chop sticks; having everyone serve themselves with the same chop sticks that had just come from their mouths; and the Chinese food, period! *(This man doesn't fancy locusts or grasshoppers, eels, and other things that I would rather forget.)* The lunches would last for over an hour and then there would be a short nap. *(The nap part was enjoyable.)*

The next part was a trip out to the mine where our group met the Director Economist of 701 mine, *(that's mine manager)* Mr Zhang Yu Ling and his

Wright Engineering

assistant Li Zeming. Zhang, looked like a young Mao Zedong. *(I didn't dare say anything about his look-alike for fear of either insulting Mao or Zhang himself, so my thoughts were kept to myself.)* After the formal presentation of business cards which is taken very seriously in China, the group *(at this time about 15 persons)* toured the pit and then had an open session with the mine operators and members from the WuHan Institute. Something should be said about presenting and receiving a business card in China. When a business card is received, you nod your head toward the presenter to acknowledge him and then slowly study his card and then carefully (if at a meeting) place it in front of you. At the same time, when presenting your card, one must not toss it at the recipient but present it in a very formal way as if it was a million dollar bill. The Chinese hold dear to this belief in showing respect.

Upon first reaching the mine and after being introduced to its management, the group began the tour of the open pit mine. The ore from the open pit had to be trucked up a long and winding ramp to the top pit edge and from there to the crushing plant about 1.7 km in distance. The open pit was on its last legs both in ore supply as well as the condition of the pit ramp and its equipment. The ramp was very narrow, very poorly graded and generally in very bad shape. The equipment, comprising trucks, drills and shovels was languishing in a deteriorating condition. Fortunately the pit was nearing the end of its life as the equipment would likely not last that much longer in its present condition without extensive maintenance.

Changma 701 Diamond Mine Open Pit

The walls of the pit as seen in the photo were not scaled nor was the ramp graded.

The shovel at the pit bottom was in need of much maintenance.

```
Why Mining?
```

After touring the pit, our group met with the members of WuHan Institute for a session on prolonging the life of the pit and continuing with underground mining. The three members on the WuHan Institute were all academics (Professor Chi Xiuwen and two other representatives who eyed me suspiciously) and convinced me after talking with them for a while that they had no practical experience in mining nor had they worked around an operating mine. Their method of deepening the pit was fine but they proposed to build a wooden trestle from one of the benches down to pit bottom to handle the ore in railcars hauled by a hoist, as access to the ramp would be cut off when deepening the pit. It was difficult to see how a hoist, skip and trestle plus installation cost of that facility could be purchased for 500,000 yuan (about $62,000 US). It was suggested in my report renting a long boom crane fitted with a cactus grab mounted on the bench above as an alternate method to the rather complicated trestle system. This would save the cost of building a trestle and the expensive triple handling of ore from loaders and trucks to skips then to trucks again.

The session on underground mining was a little more exciting or maybe it should be said, heated. The WuHan academics suggested accessing the orebody via two declines of -25° and -30° down to the 110 level, which was 150 metres below surface. In other words, these inclines would be in the order of 300 metres long just to reach the first mining horizon. In order to reach the lowest level which was the -40 level, it would be necessary to drive two dog-leg inclines for another 300 meters. (That's 600 meters altogether.) The intension was to move men, material and ore in 5-car trains through the main decline and use the auxiliary decline for ventilation supply and a second egress. To bring ore up to surface from the very bottom level would require considerable transferring from one hoisting decline to another and then onto another. When questioning them on their method of mucking the decline when it was being driven, they answered ...by mucking machine. *(I have never seen a mucking machine work on a -30° decline so maybe they have something in China from which we can learn???)* The FDW proposal was for a 300-metre vertical main shaft to the then known ore limit with a short 80 metre auxiliary shaft (collared from one of the pit benches) to the first mining level 150 metres below surface. *(Easier to develop and to operate.)* To the academics, the incline was the thing as they said the timing to get into production was shorter *(Which was doubtful),* no truck transportation *(Maybe),* and all water could be pumped up to the decline and taken into the plant *(The same for a shaft).*

It was not until the discussion got into the mining method that one could see they had no understanding of the cost of dilution. The academics suggested mining by sublevel caving this rich ($85 per tonne) orebody with Cavo type loaders hauling to millholes. When it was pointed out that recovery for sublevel caving would be in the order of 85% and dilution could be as high as 35%, they had no understanding of the economic cost whatsoever. They were unable to understand this concept even though graphical examples were drawn to demonstrate the situation. They were only interested in the actual cost of mining a tonne of material whether it was diluted 5% or 35%. For a very rich orebody such as the Changma, I recommended cut and fill mining where high recoveries and low dilution would result.

After the afternoon session, the mine put on a very fancy Chinese dinner where there were many dishes and wine, again served around a lazy-susan table. For some reason, perhaps because I was the oldest, the mine manager saw that I was allowed to serve myself first as all the newly-placed dishes were always set in front of me. *(I appreciated that because I would be the first into the bowl with my chopsticks, I didn't have to worry about hygiene ...catching germs from any of the other diners.)* The only two persons not drinking were Li Shou Cai, our gay interpreter and myself. *(That doesn't mean we were kindred spirits.)* The noise level got louder and louder as the meal progressed with all the wine that was served. The Chinese like their drink. On returning to the hotel, I went to bed but some of the gang continued their partying into the late hours. The Mandarin word for "Cheers" when drinking is "Gambie" and they down the drink in one gulp. The word Gambie was uttered many times that evening. I could hear the celebration very well because my room was only a couple of doors down the hall from the bar. Professor Wang and Dave Taylor felt the affects of too many Gambies the next day as they got into some powerful Chinese "white lightning" that is sold in that part of China.

A Typical Dinner at a Lazy Susan Table

Facing:
    Dr Wang, Patrick Lo,
    The Author, ZangYu Ling
    & Li Shou Cai

```
Why Mining?
```
The following day started off with a tour of the mill plant. The plant buildings consisted of reinforced concrete structures which looked in good shape and covered an area of 150 m by 150 m. For a plant treating about 150 to 250 tonnes daily, the layout was very extensive. The buildings were unheated perhaps because the steam or hot water heaters were not functional. Most of the glass windows were broken and this is the major reason why the plant must shut down during the winter every year. In addition, there seems to be no ventilation system included in the original plant, other than natural circulation through the broken windows. *(The crews deliberately broke the windows during the summer months to get fresh air into the plant.)*

The ore from the pit is trucked to the crusher which was adjacent to the plant. The workmen on the grizzly above the crusher bin wore no lanyards and no safety boots or goggles. There didn't seem to be any regard for safety whatsoever. *(I didn't enquire about their accident frequency, probably because I didn't think they kept records.)* There was a rock breaker but it was very old and needed either replacement or a major overhaul. The primary crusher was a jaw type crusher, and only one of the two cone-type crushers was functional. The whole place suffered from the lack of maintenance.

Changma Diamond Recovery Plant MengYin China 1997

This was one of the more tidy floors in the plant but still there was much debris piled in the far left corner

From Left to Right -
Patrick Lo of Fluor Daniel Wright

Professor Wang Wen Qian
Process Consultant Kunming Metallurgy Research Institute. Prof Wang was on staff at UBC in the '70s

The Author, who celebrated his 70th birthday in Beijing China.

The grinding area was in reasonably good shape and the housekeeping in this area was fairly good. There were six 5-foot diameter x 12-foot long mills installed but only four were operating. In the grease table area there was considerable spillage and leakage in the pipes and launders. Of all the areas visited, the grease table area was probably the messiest as there was considerable spillage of fine particles and grease covering the stairways leading to the floors above. Talking about the floors above, when climbing up to the very top floor to see if it might be a suitable place to position the DSM unit, I was surprised to find a rabbit warren. Apparently, the workmen had boarded up all the openings except the doorway leading into the top floor and were raising rabbits on that floor of the plant for their own personal use. There must have been 50 or more rabbits all scurrying about under the hay that was strewn about the floor. *(Only in China!)*

Wandering around the plant were numerous numbers of goats and geese. The goats kept the grass in check while the geese fertilized the ground so that more grass would grow to feed the goats. One never sees a dog or cat while in China, not that the Chinese don't like these animals: it's because, I'm told, they do, even to the extent of eating them when times were tough during the "Cultural Revolution". But seriously, the Chinese are a very practical people. If an animal doesn't pay its way and it's only use is as a pet, then they don't want to be bothered with that added expense. Only in the cities do you see dogs and they are generally owned by the wealthy folk. What drew my attention while wandering around outside the office building, was the high- pitch buzzing that would change pitch every few seconds. I finally saw that it was a cloud of locusts that would swarm from one part of the garden to another. Having never seen locusts before, it was quite fascinating to see how they would move in unison from one area to another. *(What was not fascinating was to see these same creatures served as a dish to eat at suppertime. None of these were tried, thank you!)*

While at the Changma mine we (Patrick Lo and I) interviewed representatives Messrs Yu Dao Gui and Li Chang of the YanZhou Company, a subsidiary of the China Building Materials Industry General Construction Company *(The Chinese company names seemed to be very long.)* These guys seemed to know their stuff. They gave us typical unit costs and productivity on various items such as shaft sinking, drifting and ramping as well as mobilization and demobilization. They laughed when we told them how the WuHan Institute

Why Mining?

academics proposed to muck a ramp (or decline) round after blasting. The YanZhou method, they said, was by hand mucking just as one would do in shaft sinking... nothing fancy.

XiYu

While in the MengYin area the group visited the XiYu diamond project; otherwise known as the Area 702 exploration project. Our terms of reference were to determine a method where 100,000 tonnes of ore would be mined and sampled from this prospect in order to prepare a feasibility study to bring this property into production. Before visiting the property, the group was given a full description of the geology by Hu Shi Jie of the No 7 Geological Brigade from the Bureau of Geology and Mineral Resources of Shandong Province. *(To people in Canada the word "brigade" sounds very military; whereas we would say geological survey or department.)* Anyway Hu Shi Jie went through the description of the various pipe clusters which amounted to seven clusters and one dyke. This was followed by an update on the geological conclusions and results of the 702 exploration project. When the Chinese explore a property, they really do a very thorough job, much like the Russians did in Kazakhstan. *(Where we in the west, under the capitalist system, must show an economic benefit and try to keep the costs down, they in the east can drill and explore to their heart's content, as what they lose on the apples will be paid for from what they make on the oranges. Everything is thrown in one pot so to speak.)* The No 7 Geological Brigade had determined the reserve and resource calculation by trenching (pits 20m x 40m) on each kimberlite pipe cluster and by diamond drilling at 80m intervals 450m below surface and 40m x 40m intervals above that horizon for a total on 47,567m of drilling. Besides this, 3665m of drifting and cross-cutting were completed on the a horizon 90m below surface. We in the west could never justify the extent of this detailed exploration. While at this meeting the group was given copies of a feasibility study prepared for the MengYin Pan Asia Diamond Development Limited Company assembled by the Suzhou Industrial Design and Research Institute of Non-metal Mine. *(The names of Chinese companies are very long.)* The study was short on mining and technical details but very long on cashflow, market forecast, economical evaluation and sensitivity analysis. *(Give the Chinese a couple of figures, good or bad, and they will work these figures into all kinds of projections.)*

XiPu 702 Project, Shandong Province

Prof Wang, Patrick Lo and the Author examine one of the many Kimberlite pipes that outcropped at the XiPu 702 Project.

The children were from the village adjacent to the area of the outcrops.

The visit to the property was unspectacular other than seeing a number of holes where the No 7 Brigade completed its trenching. Some of these holes were dry where one could go into them and examine the rock while others were full of water. Dave Taylor and I went down a short distance into the decline that was driven from surface to the 90m horizon by the No 7 Brigade. We had flashlights but, as we had only a few matches, we thought it best not to venture down too far for fear of being overcome due to the lack of oxygen. What we could see was that the rock conditions in back and walls of the decline were very competent, thus concluding that the drifts and crosscut driven by the No 7 Brigade would probably be in the same condition. The land on surface, (except where the rock spoil from excavating the drifts and crosscuts, as well as where trenching had taken place), was cultivated into various garden plots which probably sustained the nearby village with food.

When it came to determining a method in which to excavate a 100,000 tonne sample, FDW were at odds with what the WuHan academics proposed. The academics proposed bulldozing the top soil covering the various pipes to one side and drilling and blasting the tops of the pipe clusters. This was a very simple method and would be quite cheap but it had two disadvantages: 1) the surface samples would not be representative of the type of material to be expected at depth because of surface weathering and contamination from the surface soil and 2) it would disrupt the agriculture that is presently conducted in and about the pipe clusters. Later, if the project were to go ahead, arrangements could be made to move or compensate the farmers for the land.

Why Mining?
The FDW (that is, Patrick Lo and I) proposed that sampling should be carried out from underground by taking advantage of the 3665m of drifting and cross-cutting already in existence from previous exploration activities. Access for men and material would be gained from surface to 90m below ground via the existing decline, while the ore and waste would be hoisted to surface via a 90m vertical hoisting raise that could be excavated as part of the preliminary development. The advantage would be that a more representative sample would be obtained and only a minimum amount of farm land would be disrupted at this time. Also as part of this programme, it was proposed that a DSM unit would be set up on the property to extract the diamonds. Then only, if and when the sampling programme proved successful, would one consider full scale removal of the top soil about the pipe clusters for actual mining. *(I think our proposal was much more humane than that of the WuHan academics who wanted to confiscate the land from the villagers immediately.)*

While in the MengYin area we visited the site of the former HuangJie mine to inspect the reported 30,000 tonnes of stockpiled ore that was left behind from previous underground mining. *(After pacing the ground about the stockpile and making some rough calculations, we came to the conclusion that there could be the possibility of 30,000 tonnes of ore, of what grade, one couldn't say.)*

HuiXian
After interviewing the WuHan academics and the YanZhou delegation, our group drove for six hours *(again dangerously)* to the town of ZouPing where we checked into the DaiXi Hotel. Again, like all the hotels in China, this too had a grand entranceway and a very large rotunda. We were told (and later I saw pictures to prove it) that ex-President Carter had stayed at that hotel only three weeks before. They said that the Carters come to ZouPing every three or four years to see firsthand the progress that the common people are making in their housing and in their general welfare as part of their humanitarian activities in which the Carters are involved. *(Ex-president Carter and Mrs Carter are very well respected in that part of China.)*

While in ZouPing, we had a look around and did a bit of shopping. There was a very well-developed shopping area in the center of ZouPing, even a very large department store. Professor Wang, Patrick Lo and Li Sou Cai accompanied me, but I must admit, I was the center of attraction. Following us about the store seemed to be about a dozen Chinese all gawking at this European-looking character which was me. I got a lot of inquisitive stares. *(I guess they had never, or hardly ever, seen an Occidental person before.)* What I was looking for while shopping in ZouPing was an authentic silk Chinese Kimono similar to what one sees the mandarins of old China wear. There was nothing of the sort there in ZouPing as all the clothes sold were of western fashion. They said if I wanted something out of the past, one would have to shop in Beijing where they cater to tourists.

However, the main purpose of the journey to ZouPing was to visit the HuiXian mine located a short distance out of town. FDW were contracted by Consolidated Nu-Media Industries Inc (CNMI) to conduct a site visit and provide a report giving an assessment of the HuiXian copper/gold underground mine property. The report was to be on the prevailing conditions at the property and recommendations for a path forward to be prepared for CNMI's internal use.

We met first with geologists, Zhau Zhiyou, consulting geologist and Yuan Shurong, geologist (formerly from the No 1 Brigade of the Shandong Bureau of Geology and Mineral Resources) who brought us up-to-date on the geology and the reserves of the underground mine. *(I get a big kick out of the use of the word, "Brigade" to describe a geology group or crew.)* At the same time we met the mine manager, Wang Bao Ming. Wang was a very jolly fellow and a heavy chain smoker who had recently recovered from a heart attack but kept on smoking just the same. When he didn't have a cigarette in his mouth, he would have his mouth up to a cellular phone speaking and laughing to someone on the other end.

At this point, something should be said about this mine. The mine is located in ZouPing County, Shandong Province and commenced production in 1993. The mine is presently operating in what is called the No 1 vein of barite-gold-copper in monzonite host rock. There appears to be five similar veins "en echelon" outlined on the surface of the property and the mine is driving cross-cuts south on the 150m level to intersect what is thought to be the No 2 vein. The veins are approximately one meter plus wide, striking roughly east-west and dipping at about $70°$ to the south. *(What could be more ideal?)* Mining started after the No 1 Geological Brigade had explored the top section of the mine about the 185m horizon (above sea-level). Except for the odd remnant,

Why Mining?

most of the ore above this horizon has been mined out. Access to the ore above the 185m horizon was via adits on the 185m and 240m horizons. The mine is currently producing 50 tonnes per day on three shifts with *(get this)* 200 persons on the payroll. This number is down from 400 persons on the payroll one year ago. *(During my first year out of mining school, I worked at the Paradise mine where we mined 60 tons a day on one-shift-per-day for five days a week and milled 50 tons daily on a six-day week basis with only 31 persons, 16 at the mine and 15 at the mill which included the cooks at both camps.)* As one can see the productivity is not that high in China. But to be fair, the 200 persons included the school teachers and doctors as well as the mine employees because in China, industry is responsible for all the day-care, schooling and medical services. In other words, the mine was responsible for all the services about the little village adjacent to the mine where the employees lived. *(Still, the productivity was very low.)*

Author at the HuiXian Shaft

A boy playing at shaft collar

The tour underground was a safetyman's nightmare. First of all, the headframe had no shaft doors guarding against anyone from accidentally falling down the shaft. Little children were playing all around that area. There was quite a gap between the shaft collar and the cage where someone could easily fall between and down the shaft, if one were not careful. Also, there were no changeroom facilities so we had to change into underground clothes in a little shack (that had a dirt floor) next to the hoistroom. The miners had no shower room that could be seen and, like the coal miners in West Virginia, they went home in their dirty clothes and washed (?) there. When arriving back on surface, there were seen three women washing clothes in a little stream that was supplied with water from the underground pumping system. When Patrick Lo and I inspected the tailings pond, there were little boys swimming in a little stream supplied from the outfall of the tailings dam.

Wright Engineering

The mine was accessed via this vertical shaft equipped with a single drum hoist with a cage for men and supplies as well as for hoisting ore in small (0.6^{3}m) railcars from the 155m level to surface, a distance of 45m. The single drum hoist handled two conveyances, a cage and counterweight, where one side of the drum would wind-on, while the other would wind-off. As a consequence, it was not possible to extend the shaft much further than another 30m to the 125m level without changing the hoist for a larger one. The shaft was equipped with rope guides but there were spears at surface and at the 155m level to chair the cage for accommodating the loading and unloading of the railcars on to and off the cage. There were no safety dogs!

Upon reaching the only level (155m horizon) serviced by the shaft, we disembarked the cage and walked the short distance in the shaft cross-cut to the east-west drift on No 1 vein. Instead of drifting parallel to the vein, the operators drifted right on the vein which is OK, if one is prepared to leave some ore at each stope until the level was abandoned. But in the case of these operators, they were not prepared to do this and consequently were limited to one stope per level which will be explained later. No 1 vein was a little wider than one metre and the drift, being only about two metres in width, was supported with a three-piece wooden set and heavily lagged on the back. The walls were open. The timber was something to observe. Most of the caps were cracked and many of the pieces still had some bark on them which told me that they could be rotting beneath the bark. There were no rockbolts or strapping used nor did they seem to be familiar with these.

The mine consisted of three levels (185m, 155m and 125m level horizons) each 30 metres apart. The 185m level was accessed via a portal, the 155m level was accessed by the shaft while the lowest 125m level was accessed via a decline from the 155m level. Servicing the 125m level was very labour intensive and inconvenient as supplies (after being lowered from surface to the 155m level) had to be hand-trammed along the 155m level then lowered down a 30° decline and finally hand-trammed again on the 125m level. The reverse procedure was followed in handling the ore and waste.

Stope development started with a 70° raise driven on dip (within the vein) from the center of the proposed stope (designed to be 40m wide) from the stope floor (155m level horizon - lower level) to the stope crown (185m level horizon - upper level) above. All the ore from blasting this cental stope raise fell onto the lower level tracks at the bottom of the raise and was hand-mucked and hand-trammed from there to the 155m horizon shaft station (the only station). At the shaft station, the ore car was hoisted to surface, dumped and returned to the 155m horizon station for refilling again at the stope raise. Once the stope raise was completed, the miners started silling the stope (at a point about 5m above the track) driving two finger raises on an angle of 45º towards each side limit of the stope. Horizontal breasting, in cuts about 2m high x 2m long and vein width each, started at the central stope raise in both directions towards the stope's side limits. The stope muck was pulled clean, thus necessitating the use of stope props and planking for staging. The miners working on this staging drilling and blasting the breasts were without any safety lanyards. The hangingwall was very crumbly which was dangerous for the men as well as causing considerable dilution to the ore.

The mine at the time of the visit was producing 50 tonnes daily from two stopes (one on the 155m horizon and another on the 125m horizon) because only one stope could be mined on a level at one time. The reason for this was due to no access beyond the active stope as a result of being cut off by the broken stope muck on the track thus preventing further travel on the level. The Shandong Institute, who had done some consulting work for the mine, proposed a change in mining method to cut and fill mining which I think was a good suggestion. They also proposed the driving of a parallel drift so as to enable the tramming past the active stopes. This would have been a good suggestion if we were dealing with a very large mine but not in this case. I suggested that, instead of driving a second drift or the central raise in the vein from the lower level, they first drive a two-round cross-cut into the footwall at track elevation and then drive a knuckle-back raise from the end of this short cross-cut to intersect the vein a few metres above the track drift. In this way, any broken muck would fill up the cross-cut and not spill out onto the track thereby preventing access beyond. When drawing this up for Wang Bao Ming, he was beside himself, he was so excited. He exclaimed, "...western technology, western technology!" *(You see Wang's background was in coal mining and not hardrock mining. Also, what is taken for granted over here, is all very new to them.)*

The ore from the mine was trucked to the crusher in a very old and poorly maintained Russian-built 10-tonne truck. The roadway consisted of two ruts with little or no gravel as a surface. It was a wonder that this truck could navigate such a roadway. The crusher was under-utilized as the mine had difficulty producing more than 70 tonnes per day. Just like conditions found at the grizzly at the Changma crusher, the situation at the HuiXian mine left much to be desired as the workmen worn no lanyards, safety boots or glasses.

The mill was visited to see what could be done to increase its capacity, even though at the time of the visit, they were only using one circuit in the plant. The plant, built in 1992, was originally designed to process 50 tonnes daily but was increased to 150 tonnes per day with the addition of a second line of mills and flotation cells. The full capacity of the mill had never been realised due to the lack of feed from the mine. What could be seen, there was ample space within the present mill building for additional equipment to expand the capacity to 250 tonnes daily even though none of the building was equipped with a crane or hoist beam for lifting heavy equipment or supplies. The main substation (3500kva) located at the mill was more than adequate but the mine (fed by a 400v buried cable) is 850m away resulting in a voltage drop that was unacceptable. The tailings pond was adequate to accommodate any future expansion also.

After the visit, and gathering for discussions back at the hotel in ZouPing, Wang Bao Ming asked for some immediate advice and suggestions as to what they could do to improve the operation. We had already prepared quite a list of immediate things that could be done both for the mine and the mill which are itemized here in order as they came up in the meeting.

- Start the stope raise from a crosscut and then drive a knuckle-back raise to intersect the vein.
- Switch to shrinkage stoping in good areas and modified shrinkage with a delayed pull with wall support in areas of poor ground.
- Reduce the cost of support with the use of rockbolts, strapping and screening.
- Raise the shaft from the 125m to the 155m level to reduce the servicing bottlenecks.
- Improve the roadway between the mine and mill.
- Install an overhead 10kv line to the mine to provide sufficient power for current and future needs.
- The idle processing equipment requires servicing and reconnection to bring the through-put to the 150 tonne per day design capacity.
- Consideration should be given to installing some type of lifting device

Why Mining?

        to assist in servicing the heavy equipment in the mill.
- Maintenance in the plant needs to be improved.
- A copper-gold concentrate was presently being produced, but the mine was only paid for the copper it produced and received nothing for the gold values. *(Some smelter contract!)* We were surprised that the sale of concentrate was not governed by any formal smelter contract as such, and suggested that other potential sources for sale, such as overseas smelters, should be investigated.

Wang Bao Ming took these suggestions very seriously and at one point (as mentioned before) got very excited and enthusiastic regarding the method of starting a raise off a short crosscut. All-in-all he seemed to be very grateful, even though all communication had to be done through our interpreter, Mr Li.

The last night in ZouPing consisted of another formal dinner with delegates of the local party members from the Peoples Republic of China. They were quite a cold and unfriendly lot, not much in the way of smiles, even when they had lots to drink. That didn't stop the rest of the group from whooping it up as the evening wore on. I brought a bottle of Canadian whiskey along for one of the formal toasts that the Chinese like to have when dining out with a group. The Canadian whiskey went over well. I even had some of this myself just to be a good sport as it was our last night in ZouPing. *(I didn't think my high blood pressure would suffer too much by responding to a couple of toasts.)*

The last day in China was a very busy one as we left ZouPing at 6:00 am and drove to Jinan to catch the plane to Beijing. The flight to Beijing was about 90 minutes. It was prearranged that Dai Zhen Fei, whom we met in MengYin, would connect with us at the Beijing Airport and deliver a number of plans and sections of the XiYu project to take back to Canada. He had a surprise for us in that he arranged a trip to the Great Wall of China. As there were six hours to kill between planes this was a very pleasant surprise. Our guide was a little girl by the name of Shelly who had never been out of China but who could speak perfect English all the same.

Shelly had a cab hired for the ride there and back and came with us on our tour of the wall. We took the gondola cable car up to the Great Wall for which I was thankful as it was 30° Celsius with a humidity of 95%. We were wringing wet when we arrived at the Wall. But before describing the Wall, one should say something about the vendors on the way to the Wall. Before getting on the gondola, one has to pass all kinds of hawkers selling everything such as trinkets, T-shirts, paintings, etc. It must be said though that these hawkers were all very polite. They didn't crowd you, as I have been, in Haiti when Irene and I were on one of our Caribbean cruises. The hawkers at the Great Wall all seemed to be able to speak English of a sort and were very funny in their remarks. When refusing to buy something, I said, "I'll be back". This one lady hawker answered, "I'll remember you, you see". There was even a horse at the top of the Wall where one could, for a price, sit on this horse and have one's picture taken. It's hard to figure out the significance of the horse but presumably it was a means of making money.

The Great Wall of China in itself was something to visit. After seeing natural wonders like the Niagara Falls and the Grand Canyon which really thrilled me, the Great Wall was something else as it was built by hundreds of thousands of human beings over many years. It was indeed a marvel and still is. One wondered many times, how many of those persons working on the Great Wall survived the hard work that went into that undertaking? I found that even climbing the stairs to get to the Wall was difficult, but to think that the people who built the Wall had to carry heavy loads as some of the stones in the Wall were fairly large. Of course pictures were taken of ourselves and of the crowds that were on the Wall that day. One could hear many languages spoken ... French, German, English and of course Mandarin. The Great Wall of China is something that one should not miss seeing when visiting the Beijing Area. As the Chinese saying goes, "To be a great man, you must walk the Great Wall of China".

My lasting impression of China is, I would go back there anytime. The people seemed to be very happy and very courteous. The Chinese seemed to be well dressed and many of them had modern transportation in the form of motor scooters and, if one wants to stretch it a bit, the one-cylinder three-wheeled tractor-truck. The super-highways (toll roads) are comparable to our own and many more of these are being built every year. Their secondary highways, which are generally six lanes, are very crowded because they are free. Much of the work is done by hand such as cleaning the roads as I saw for many miles people with straw brooms whisking away dust and dirt off the sides of the roads. I don't recommend to anyone driving in China as the Chinese don't seem to care who is in the way or what is coming towards them. They are

```
Why Mining?
```
constantly blowing on the car's horn. One other thing that didn't appeal to me was the food. I enjoy the North American Chinese food, but what was served at these formal dinners left me wondering at times just what I was eating. But as mentioned before, meeting the many Chinese people on this trip in that part of the world was a pleasant experience.

The Great Wall of China

The Chinese say:

"To be a great man, you must walk the Great Wall of China"

It was 30° Celsius with 95% humidity when our group walked the Great Wall.

Chapter 22
Kilborn Engineering

Potash

I first heard of Kilborn Engineering when attending Ken Dewar's farewell party while working at Steep Rock back in the '50s. Ken, who was the superintendent of underground mining at Steep Rock, was leaving to join Kilborn Engineering in the spring of 1955. By coincidence, Merv Upham was also leaving Steep Rock at about the same time for Blind River. Besides these two, another fellow employee by the name of Paul Lang, who was an engineer-shift boss at Steep Rock, was leaving for Espanola Pulp and Paper at Espanola Ontario. After a few drinks, Ken Dewar and Merv Upham were telling each other how much they enjoyed working with one another ... it was mutual admiration at its best. Paul Lang, who somehow always seemed to say and do the wrong thing at the wrong time, piped up saying, "when you guys are finished sh--ing each other, call on me at Espanola Pulp and Paper, and I'll supply the toilet paper".

It wasn't until later in my career that the name Kilborn Engineering again came up and that was while I was at IMC in Esterhazy. Kilborn Engineering was headed up by Ken Dewar at the time and Merv Upham was the manager at IMC. It is a little known fact that, when IMC first decided to invest in the Saskatchewan Potash Industry, they approached Kilborn Engineering for advice regarding engineering standards and such. At the time, Ken Dewar suggested very strongly that, when they appointed a manager, the person should be a Canadian for political reasons and, at the same time recommended that Merv Upham be given the job. As it happened, Merv Upham was appointed manager, and the close association between IMC and Kilborn began. It was only natural that Kilborn would have the contract to increase the capacity of the IMC potash plant first from 400,000 tons to 1 million tons and then to 1.2 million tons annually. Kilborn Engineering was not only rewarded the 1.2 million ton expansion at K-1 mine but also received the contract for the design of the K-2 plant and later the 2 million ton expansion at K-1 after the 1.2 million ton expansion had been completed. In due respect for both men, Kilborn Engineering was the favoured design group by IMC senior management in Skokie Illinois, the head office of the company. Also, of all the potash plants designed at the time, those designed by Kilborn Engineering were undoubtedly the most modern and the most efficient.

Why Mining?

When moving to Cominco, I steadfastly recommended that Kilborn (after seeing their performance at IMC) should be considered for the design contract of the Cominco Potash plant. A very good friend, Tip Croome, who at the time was business development agent for Stearns-Roger, of course was a bit upset with me when favouring Kilborn over his company, but we still remained friends all the same. It's a laugh when looking back at how Kilborn was chosen over Stearns-Roger Engineering. Tony Banks in the winter of 1964-65 spent four months in Denver Colorado at the Stearns-Roger office preparing a feasibility study for Cominco. He came back to Trail B.C. fully confident that Stearns were the ones who should design and build the Cominco plant. It was not until I started questioning some of the assumptions and conclusions in the Stearns study that Cominco wanted a second opinion. The second opinion was with Kilborn, resulting in Tony Banks being sent to the Kilborn office in Mimico Ontario (Toronto) to oversee the design work on this new study. Again, Tony came back as much in love with Kilborn Engineering as he had been with Stearns-Roger six months before which resulted in Kilborn Engineering being given the design and procurement contract for the Cominco Potash plant.

As part of my duties, after Kilborn was awarded the design contract, I was to attend the various design meetings (every five weeks) in Toronto. At these meetings, I was able to meet people such as Bunny Crocker who was vice-chairman of Kilborn. Bunny was very famous in the mineral processing field where he did considerable work on pebble mills and on semi-autogenous grinding. He was a true gentleman. Dick Roach, who was his right-hand man in processing, specialized in potash and was the oracle in this field. Kilborn Engineering had some very good men amongst their senior people. John Dew for one was a very honest and sincere person, who later became president of Kilborn Engineering. Carl Frietag, young at the time, was Dick Roach's understudy and a very good one too. There was Bill Scott, who was appointed the project engineering manager for the Cominco design. Bill was a bit naive but absolutely honest. His favourite ploy to disarm one when asking a question about some feature in the design was to say, "... now I'm a bit confused..." and he would go on from there.

Later when starting up my own consulting practise in 1986, I visited the Kilborn office in Toronto to see if there was any overload contract work in underground mining. Unfortunately at the time, my old nemesis, Jack Mitchell

Kilborn Engineering

had joined Kilborn and was then the president of their western subsidiary in Vancouver where all mining engineering was being handled. Kilborn had a mining man in the person of Bart Fairbairn who was very capable in handling the underground projects and the fact that Mitchell and I were not on good terms was enough to say that there would be nothing for me from Kilborn Engineering. It wasn't until the untimely demise of Jack Mitchell, from lung cancer a few years later, that Kilborn awarded me an assignment.

Mount Polley

After moving west to Vancouver Island in 1992, I would often travel over to Vancouver to network (making rounds and visiting) with the various mining consulting groups such as Wright Engineers, Simons Metals and Kilborn Engineering. It was the same as starting the networking task all over again just as was done in Toronto in 1986. While networking with Fluor Daniel Wright, and encountering Glenn O'Gorman whom I had met before when he worked back east, we renewed our acquaintance in Vancouver as he now worked for Fluor Daniel Wright. At about the same time, Bart Fairbairn was leaving Kilborn Engineering as he was given a golden hand-shake when SNC Lavalin gained control of Kilborn Engineering. With Bart's retirement, it left an opening for a top mining engineer to assume the position of manager of mining for Kilborn SNC Lavalin in the Vancouver office. Glenn, at the time could see that he would have to wait for John Tulley to leave Fluor Daniel Wright for him to advance higher in that company, so when he was asked if he was interested in joining Kilborn SNC Lavalin as manager of mining, he was receptive and joined them in 1997.

The following spring (1998) mining activity was fairly quiet; in fact, mining was in the doldrums as the price for commodities was very low. Copper was less than half what it was the year before and gold was below $260 US. As a consequence, many mines (especially those mining porphyry copper) served notice that they may have to shut down if the situation persisted. The British Columbia Government Employment Commission was very concerned and commenced a number of studies into how to keep these mines operating during the period of low commodity prices. As a result, Kilborn SNC-Lavalin were selected to investigate ways and means for keeping the Mount Polley mine in Central British Columbia in operation.

Glenn O'Gorman phoned on March 13, 1998 and offered me, along with David

Why Mining?

Price, a Kilborn metallurgical engineer, an assignment to investigate the Mount Polley Mining Corporation's mine at Likely in the Cariboo country of B.C. Never having been to the Cariboo area before, this would be a new experience. So on March 16th, after catching the early ferry at Nanaimo for Vancouver, then proceeding to the Kilborn office and meeting with Glenn who outlined the assignment, I signed a Kilborn contract and confidentiality agreement. From there with David Price, we flew to Williams Lake and, in a rental van, drove to Likely B.C. where we checked into the town's only motel. The next day we drove to the Mount Polley mine and met George Wright the mine manager. George gave the impression of being a fatherly type of individual, perhaps because he was quite a bit older than his staff which he introduced at a specially called meeting. The staff knew Kilborn reps were coming and why and, as a consequence, gave their full cooperation because it was in their interest to do so. Nothing was held back ...we were allowed to see everything from costs to confidential memos regarding the operation.

Mount Polley was an open pit mine composed of three pits, one active (Caribou pit) and two under development (Bell and Springer pits). The mine was expected and was quite capable of producing 20,000 tonnes of ore and 15,000 tonnes of waste daily but the mill at that particular time of this visit was limited to about 15,000 tonnes of ore. The ore was a porphyry copper (0.42% Cu) with some gold (0.84 gm Au). The original thinking was that with both copper and gold, the mine couldn't miss. They thought that if ever the copper price was down, the gold would pay the freight and if the gold price was down, the copper would shoulder the load. But, during these times when both commodity prices were low, the operation was being squeezed, a situation that they didn't consider would happen. In any case, the mining operation was not the problem as there was ample equipment to operate for the short hauls to the crusher. The drilling and blasting were no problem either as the fragmentation was excellent thus requiring no secondary breakage whatsoever. The walls of the pit were very stable also. The drilled and broken reserves were adequate to sustain production at the present rate and for the projected rate of 20,000 tonnes of ore daily.

I was very impressed with the mine superintendent, a relatively young man by the name of Don Parsons. Don was very calm and cool in the way he handled himself. He had the loyalty of his small staff as he ran a very lean and efficient crew which included mine operations, engineering, geology and site

Kilborn Engineering

environmental. He was responsible for not only the mining but also the road maintenance about the site as well as the 12 km access road to the highway. It was noted that while visiting Mount Polley, the mining department supervised the construction of the tailings dams currently being constructed by contractors.

The milling was something else as it seemed to be the problem at Mount Polley mines. There were three main problem areas in the mill besides personnel, namely: crushing plant availability, through-put in the grinding section and gold recovery. The crushing and screen availability was low because of the complications resulting from attempting to generate and supply pebble material for the pebble mills. If the plant did not require pebbles, one could easily eliminate the triple-deck screen at the crusher, six belt conveyors, one magnetic-head pulley, five vibrating feeders, one weightometer, one bin chute and a pebble stockpile. The cost to operate and maintain all this excess equipment in the crushing and screening section by far exceeded the value of the pebbles obtained for the grinding section. It was our recommendation to remove the pebble-producing circuit from the operation.

Mount Polley Mines

Shown are two rod mills in front with a ball mill at the far left. Not shown are the controversial pebble mills that complicated the crushing and screening section of the plant and hampered the grinding of the ore.

Why Mining?

The grinding section had three grinding stages: primary, secondary and tertiary where the load to each stage was distributed by cyclones. Some work had been done on the cyclones in order to distribute the load more evenly but more had to be done. But the real problem was the pebble mills as they could not come up to capacity. What the operator had been doing in the past in order to increase the milling capacity was to add a few steel balls to the pebble mills' charge where Mount Polley was really operating a quasi ball/pebble mill operation. Our recommendation was to strengthen the tertiary mill shells with ribs and charge the mills with balls instead of pebbles. In other words, eliminate the use of pebbles altogether.

The last problem in the milling was gold recovery. This would take some metallurgical testing and some trial and error with the circuits. Even though this was important (gold recovery) the main problem was to get the volume increased in the concentrator so that the operation would show a profit. Mount Polley did employ a metallurgical consultant, one Gary Hawthorne, who could talk the hind leg off an elephant. When having dinner with him one evening at our motel, neither Dave Price nor I were able to get in more than a couple of short sentences each that whole evening.

It didn't take much effort to realize the main problem in the milling operation was personnel, especially the senior management in Vancouver. It was senior management who also set up the pebble mill circuit and brought in the union. The mill general foreman was more concerned (to a point of interfering) with the maintenance problems, according to Jack Zuke, the maintenance superintendent; whereas, he should have been spending much of his effort trying to increase the through-put in the mills as this was where his operating knowledge could have been well spent. The gold recovery problem could use some of this effort but mainly it was the grinding section where a good general foreman could assist the operation. Don Ingram, the mill superintendent was an unassuming soul, who at the time of our visit stated that he was retiring in a month or so and pretty well turned over the milling operation to the general mill foreman. We later found out that Don was mistaken as he had signed a contract when joining the company and had another year to go before he was free to leave unless he wished to forfeit a special retirement benefit which amounted to considerable value.

Kilborn Engineering

When it came to forming a union, the crews had the International Union of Operating Engineers foisted onto them by the Vancouver office who believed "...better a weak union like the Operating Engineers than a militant one like the Steelworkers". Head office did all this without consulting anyone on the property, while those on the property were well prepared to operate non-union if only to be given the chance, as many of those at Mount Polley had come from non-union operations and knew how it was done. Many of the workmen were of the same mind according to both Jack Zuke the maintenance superintendent and Glenn MacDonald, the personnel superintendent. *(It's a shame that the mine management wasn't able to act independently without interference from head office.)*

One of the trying things that the provincial government had saddled this operation with, was a penalty of $180,000 for the "Removal of Young and Immature Timber" when clearing a site for the mine. Yet at the same time, the company had to post a bond of $1.9 million to address site reclamation at the cessation of mining operations. What this is, is double taxation! The Ministry of Forests assesses the operation for immature trees now at the beginning of the project and the Ministry of Employment and Investment requires a bond to pay for the reforestation of the land later at the end of operations. The NDP Government had it both ways. *(Is it any wonder that this mine was feeling the pinch?)* In any case, we were able to point out a number of problem areas which could be addressed. As a result the mine continued to operate for another four years so we must have done some good.

Huckleberry Mine

After completing a study in early April on the Mount Polley mine, Kilborn was given a similar task to perform on the Huckleberry mine in the Omineca area of Northern B.C. Huckleberry Mines Limited (HML) is owned by Imperial Metals Corporation, the same corporation that controls the Mount Polley mine. Again when making this review the staff, both at the mine and at the Vancouver head office, held nothing back and gave Kilborn Engineering everything they asked for and more.

On April 6th, reporting to the Kilborn office in Vancouver, I met Jim Marlow (mining engineer) who would act as project leader and Ron Hall (metallurgical engineer, an independent consultant like myself). The three of us would make up the audit team. Marlow would focus on the geology, mining and

Why Mining?

maintenance, while Hall would concentrate *(no pun intended)* on the metallurgy and tailings disposal, leaving me with management, administration and personnel to research. It was agreed that the following day, I would fly from Victoria and meet these two fellows at the Vancouver Airport and continue on to Smithers, which we did. While on the plane we (Marlow, Hall and myself) were accompanied by Jack Miller, Imperial Metals Corporation vice-president who was travelling to the property along with the mine manager's secretary who also acted as our driver when the plane landed in Smithers. Upon landing in Smithers, she picked up a company van that was left for our transportation, then drove us first to Huston (which is about 70 km southeast of Smithers on Hwy 16) and finally by private road (radio controlled) to the mine, a distance of about 90 km to the south near the top end of Tweedsmuir Park. The reason for the radio control was the narrowness of the road as the large logging trucks took up all of the roadway. Outgoing traffic had the right-of-way, so any incoming traffic had to move off the road into lay-bys to let outgoing traffic proceed. We got to the mine late and were assigned to our rooms in camp which would be home for the next few days.

The camp wasn't that impressive. It was built on a hillside which couldn't be helped as there was little in the way of flat land in the area. We had to climb a number of stairs to get to our billet. The camp consisted of Atco buildings with a covered corridor down the center and the Atco buildings formed on each side of the corridor. There were three rows of these, two rows for single males and the third one for women and married personnel. *(There were a number of married couples who both worked at the mine and therefore shared a double Atco room.)* At the center of each row of Atco buildings were the washrooms and laundry room. If one was unfortunate enough to be assigned a room at the end of the row, then he or she would have a long walk getting to the bathroom. To make it worse, we were also assigned to a mixed gender row and therefore it was necessary to get completely dressed to make our way to use the bathroom facilities.

In the morning after our breakfast, we assembled in the conference room where Emile Brokx, the mine manager, introduced his department heads. At the time of this visit HML operated "Union Free" under seven departments (mine, mill, maintenance, plant, accounting, human relations and environmental). It should be noted that there is both a central maintenance and a plant department reporting to the general manager rather than to operating departments. The maintenance department in this case was concerned with all mechanical maintenance of the mine and mill equipment, while the plant department was involved with all outside construction and electrical and instrumentation. There were only three levels of supervision which made for good communications. Assisting the general manager and, at the same time monitoring the operation, was a Japanese expatriate with the title of executive vice-president. The purpose of having this individual on site was to study Canadian methods of operation and to ensure that the Japanese investment was protected. *(The Japanese companies had loaned HML $60 million.)*

The Surface Plant at Huckleberry Mine

The Open Pit &Crusher at the top and the Mill in the Center.

Emile Brokx the general manager was a likeable fellow, but at the same time he gave one the impression of being a softie. He hired an employee relations man who various personnel consultants cautioned against hiring, but he went ahead and did so just the same. As a result, outside of Amel, no one seemed to care for the employee relations man, not even his own staff whom he had hired personally. All the same, the mine was being operated without a union so presumably an employee relations man doesn't have to be popular to be successful. *(That's a question that could be debated.)*

After the introductions, Bill Dodes, the mine superintendent, started us on a tour of the mine and tailings area. I don't remember too much about Dodes except to say that he seemed to know what had to be done in his area of jurisdiction. The mine was able to supply all the ore that the mill could process. The mining costs were reasonable, particularly in view of the additional support equipment required for dam construction. The mine had to build a number of dams so that all tailings, along with any acid-generating rock from stripping could be submerged underwater to prevent oxidation. The resource and reserve models appeared to be performing adequately since being calculated during the feasibility study. Regarding dilution, this didn't pose a problem either, as the mining personnel were diligent in their concern for grade control.

The mill was toured with Bob Tucker, chief metallurgist. The original mill superintendent was released shortly after startup and his replacement had not assumed this position at the time of the visit. As a result the chief metallurgist from head office was filling in until the new incumbent became familiar with the operation. The tour started at the front end with the gyratory crusher. It appeared that the positioning of the ramp to the crusher was wrong and should have been turned 90º because, at the present orientation, the trucks had to climb an unnecessary grade to dump. The housekeeping in the mill was excellent leaving us with a good impression of the operation. At the time of the visit, the molybdenum section was just being commissioned so there is no comment on that area.

The production plan for the mill was overly aggressive, expecting to achieve design availability, throughput, and recovery almost immediately after commissioning. In practice, design flaws in pumping and piping in the grinding and flotation, coupled with a wide variation in ore hardness, resulted in a copper metal production shortfall in the first five months of approximately 35%. Many of the major issues such as pumping and piping were being addressed. Coarse tuning completed in the circuit had resulted in increasing the throughput and availability above plan. At the time of the visit, the operator was in the process of some fine tuning to increase recovery.

The mill maintenance problems encountered during startup were many. It was necessary to up-grade most of the pumps as many were undersized. Most of the pipelines were undersized also and required changing. The SAG mill had to be relined and the lifter size and gates changed. Hour meters had been installed on all major components in the plant in order to schedule planned maintenance. As for the mine, it always had meters on its equipment. As an example, the 777B truck engines were changed-out after 1800 hours. The Finning Tractor Company overhauls the engines while HML crews do the change-outs. It appeared to us that the maintenance was well handled.

Kilborn Engineering

One of the many difficult problems of doing business in British Columbia is the environmental studies a business must complete before approval is given to commence operations. As an example, HML submitted 1,100 pages of environmental data in nine volumes to various governmental agencies before a licence of operation was issued. Another B.C. governmental department known to punish business is the Ministry of Forests (MF). When constructing the road into HML, the revenue generated from the timber cut along the road right-of-way did not cover the stumpage fee of $5.7 million assessed by MF. HML did, however, receive a $2 million rebate on its stumpage fee for the original work done on the access road. The access road is shared with two logging companies who receive rebates on their stumpage fees when they make expenditures on the road each year. But because HML contemplates no further logging work on its own part (thus no stumpage fees), no rebates are forthcoming on expenditures for its share of the road upgrading. With $500,000 to be spent on road upgrading in 1998, HML would have liked to receive the same consideration as do the logging companies but MF say "no". Normally power companies deliver power to the plant property where metering the power takes place. In the case of HML, B.C. Hydro delivers to the property but meters the load at Huston some 90 km away, thus the mine has to absorb the line loss which, in this case, amounts to 4.8% of the total power or roughly $25,000 per month. Is it any wonder that few mines have started up in B.C. during the decade of 1990-1999.

After touring the site and quizzing all the department heads, our group was driven back to Smithers for the flight to Vancouver and Victoria. The hard part was to complete the report and offer some advice as to how the mine could be kept viable until the commodity prices rebounded. We were able to make a few recommendations such as: 1) increasing the mill throughput by improvements to the crushing circuit, mill piping and pumping; 2) fine-tuning the tank cell-grinding circuit in an effort to increase metal recoveries; 3) blending the ore in the pit at present was hampered by the lack of shovel capacity which should be addressed; 4) planning that considers single handling of till and waste rock for dam construction; 5) reducing inventory and eliminating the amount invested in non-consignment stock; and 6) developing a contingency plan in the event of a failure of the primary loading shovel which could include stockpiling or acquisition of a second shovel.

Some good must have been achieved as the mine continued to operate.

Why Mining?

Epilogue

Mining

Why mining? Why mining indeed! What other endeavour could provide more fun than mining? Mining has everything one would ever want in a career. There is travel, there is money to spend, there is money to be made, but most of all there are people. What more could one ask? I often look at those people of my home town of Cranbrook, *(Yes, I call Cranbrook my home town.)* who have lived there all their lives and wonder what it would be like to have stayed in one place all of one's life? *(Boring!)* While attending a high school reunion I had a conversation with a former high school classmate who, when mentioning that I was now living in Ontario, said, "Dave, I visited London, Ontario once and I wouldn't give you one square inch of British Columbia for all of Ontario." *(How narrow minded can one be?)* Anyway, travel does broaden the mind and being a mining engineer does give one the opportunity to travel and broaden one's horizon. When thinking of the number of places in my working career, I am sure most people outside of mining would find it unbelievable. Mining took me to every province and territory in Canada except Nova Scotia and Prince Edward Island. In the States, mining activities ranged all over the States to New Jersey, New York, Minnesota, Wisconsin, Michigan, West Virginia, Ohio, Pennsylvania, Kentucky, Missouri, New Mexico, Colorado, Wyoming, Idaho, Utah, Montana and Alaska. Overseas, mining activities covered such countries as Sweden, Norway, Finland, Kazakhstan and China.

Mining takes money. When working for Steep Rock Iron Mines, I saw what was needed to develop an orebody and it took a lot of the green stuff. Again in the Potash Industry, in order for the shaft to reach the potash beds some one thousand metres (1000m) below surface, it took a lot of money...anywhere from $12 million in the early 1960s to more than double that in the 21st century. A lot of money was spent on plant and equipment too. To bring the IMC K-1 plant into production in 1962, it cost that company $42 million and then six years later, it cost Cominco $80 million for the same size mine and plant. The Cominco mine did have two shafts while the K-1 mine had only one. But all the same, it was a lot of money that had to be spent to construct and develop those mining operations. A few years later while managing Texasgulf's Kidd Creek No 2 mine, the budget to get the mine into production was $125 million for the construction cost of the mine only. In any league, that's a lot of money!

Epilogue

Regarding money to be made, it was always my belief that the geologists had the best chance of making money. Not so much that they received high salaries as neither geologists nor engineers are compensated extravagantly while working as employees of a mining company. Where probably geologists have it over engineers is that they seem to know who is drilling where and what they have found and then they take a chance on the stock market. Many geologists, either working on their own or just plain playing the market, have made it big time. One exception to working for a company when it paid off, for me that is, was the stock options given to the various department heads at Texasgulf-Kidd Creek. When the Canadian Development Corporation and Elf Aquataine made their hostile bid for Texasgulf (Tg) in 1981, its stock more than quadrupled from $30 to $135. Many at Tg were the beneficiaries of that transaction and have been able to retire in relative comfort as a result.

When travelling in Kazakhstan, I had a look at the passport of one of the geologists in our group and it was full of visa and passport stamps. It made mine look a bit thin in comparison. That same geologist was a very interesting person to know and like most geologists and mining engineers who have worked in a number of places, are very interesting conversationalists. In travelling and as a result of moving around, one meets a large number of various interesting mining people. The people met while working at various mines and while consulting have been phenomenal. When working for various companies, I was fortunate to have worked for some outstanding people. Just to name a few of the outstanding ones, these will include...Henry Doelle, Jack McIntosh, Johnny Parker, Jack Crowhurst (Sheep Creek Mines); Andy Anderson, Herb Cox, Bob Dempsey (Malartic Goldfields); Pop Fotheringham, Wyatt Hegler, Merv Upham, Walt Bannister, Ken Dewer, Harry Mulligan, Ken McRorrie, (Steep Rock Iron Mines); Tom Kierans, (Rockiron); Andy Kyle, Bill Ellis (IMC); Bruce Hurdle, Frank Goodwin, Robin Porter (Cominco); Ray Clarke, Bart Thomson, Merle Marshall, Bruce Gilbert, Dr Chuck Fogarty, Dick Mollison, (Texasgulf - Kidd Creek); and finally Dave McKay, because I did work for myself for the last 16 years. At this time, one should also include some of the various sales personnel, contractors and consultants that were met in my career and who helped me: these include Harold Wright (Wright Engineers); Cam McDonald, Dan Poitevin (Ingersol Rand); Charlie Tompkins (Goodyear and Forano); Gerald Tiley (Westinghouse and G Tiley & Associates), Peter Eastcott (Canadian General Electric); Jim Cook (V B Cook and Associates),

Why Mining?

Alf Norkum (Gorf Construction); Fuzzie MacDonald, Harry Hammerstrom (Joy Manufacturing); Pierre Michaud, Tom Kenrick (Altas Copco) and consultants like Hugh Roberts, Watkin Samuels, Mel Bartley, Eli Robinski, Shosi Serata, Juks Swellnus, Daryl Dudley, Bunny Crocker and many others whose names I have forgotten.

When on the road with Rockiron, I could call into most mines and know many of the people who staffed the organization. This carried on when consulting too. Like myself, many in mining moved around so it was not surprising to see a person from the east pop up at some mining operation in the west and vice versa. It seems no matter where you go in mining you are bound to find someone you know ...you are not without friends. But alas, there comes a time when your old acquaintances seem to disappear. At one time when attending the CIM conventions, I would know many of the delegates who attended. These were fun gatherings and I wouldn't have missed these for one minute. However, once leaving active mining and really retiring, I noticed after a couple of years that fewer of the delegates were known to me as in former years. My generation was retiring too and they were not attending the conventions as in the past. It only takes a few years in retirement to realize that you're not part of the group any more and perhaps forgotten. Oh, but when I look back and think about the many interesting people I had the privilege of meeting while employed in the Mining Industry, remembering those people will remain cherished memories.

Finis

INDEX

Adams, Ralph 162-3, 301
Albright, Clint 147
Allan SK 140, 193
Allan Potash 130,140, 149, 191-5, 197, 209, 211-2, 213,256, 258, 262
Almaty Kazakhastan 285-6, 289-90, 292, 297
Altas Copco 338
Alwinsal Potash 144, 148-9
American mining Congress 178
Amos, Allen 159
Amsden, Mike 194, 242
Amyott, Clarence 165, 168, 170, 172, 175-176
Anderson, Hughie 6
Anderson, Norm 102
Anderson, Andy 47, 53, 337
Annable, Doug 220
Anshun, Dr Xu 307
Aquas Tenidas 300
Armistice Resources 231, 238-9
Armstrong, Merdy 238
Astoria Mine 237-9
Athalmer 1, 27-28, 36
Atikokan ON 69-72, 76, 79, 88, 90-91, 95, 99, 112, 179,198, 205, 299
Atlas Copco 189, 267, 338
Auezov, Kazakhstan 288, 294
Augustin, Paul 214
Auston, John 205, 262
Ayers, George 211, 213, 215, 217
Bakyrchik Kazakhstan 284, 291-2, 294-6
Ball, Bill 62
Banks, Tony 137, 147, 326
Bannister, Walter 69, 72-75, 337
Barabas, Wayne 160
Baron, Chick 74
Barrila, Don 150
Barrington, Ed 28-30
Bartley, Mel 338
Baskin, Dave 192
Battistone, Ross 6
Bechtal 282
Beijing China 305, 307, 317, 322
Belford, Eric 164-5, 179, 191, 196
Belmoral 202, 212, 225-34, 236-9, 248, 250, 262, 276-7, 282
Berthelson, Mike 117
Bertoia, Douglas (Alphonse) 7, 50-51, 69, 242
Bibby, Dr Peter 223
Bizyk, Steve 160
Black, Jim 137
Black Daimond 67, 73
Black, Randle 221
Blair, Bill 2-3
Blairmore 152, 155
Blakeney, Premier Allan 192
Bodas Mine 186
Boliden 185-6
Bond Gold 205, 244-5, 247, 249, 252, 254
Bophey, Glenn 267
Borneman, Elmer 157-8, 165, 251-2
Boyd, Ken 299-300

Boyko, Peter 16
BP Canada 205, 262
Braithwaite, Tom 118, 127
Brandle, Jim 211-2
Bray, Bob 16-17
Braylorne Mine 17,101
Bre-X 245, 281
Breakwater Resources 245
Bridger, Dennis 205
Britannia Mine 16, 36, 102
Brokx, Amel 332-3
Brooks, Lorne 98
Broulan Reef 225-9, 231, 234, 238-9, 242, 257, 276, 281
Brouliard, Ron 159-164
Brown, David 24
Bubnick, Jim 212, 220, 257
Buchar, Chuk 181-2
Budden, Harry 269
Buffalo Ankerite 54, 229-30
Bunker Hill Mine 299
Busch, Ron 122-24, 126
Butler, Kathy 221
Calpine Resources 250, 253-4
Cambrian Engineering 162
Camozzi, Len 3-4
Campbell Trophy 179
Canadian General Electric 172
Canamax Resources 239-40
Canarc Resources 276-8, 281, 299
Canpotex 207
Canterra Oil 201, 204, 207-10, 221-2, 239, 266
Carr, George 115
Carter, ex-President James 316
Casey, Rick 260-1
Castlegar BC 135-6, 138
CDC 166, 190, 199, 201, 207-8, 225, 337
Cedar Isand Shaft 249
Cementation 136-8, 151, 153-4, 165, 176, 257
Changma Mine 304-5, 307, 321
Chapman, Colin 202, 225, 244
Charon, Gerry 160
Chesney, Bud 90
Chi, Professor Xiuwen 310
Childers, Chuck 195
Chimo Mine 237
Chisholm, Ian 285-7, 294
Chizmazia, Tibor 118
CIM 27, 34, 61, 71, 73, 88,114, 132, 156, 178, 205, 225, 254, 338
Clark, Jack 167
Clark, Tex 250
Clarke,Ray 158, 163-4,166, 169, 176, 179,191, 205, 337
Cleveland Cliffs 87
Cluff, Bill 236
Cobble Hill BC 285, 287, 298, 306
Cochrane, Tom 17
Codville, Don 18
Coeur Alaska 282-3

INDEX

Cole, Jack	94	Doran, Rod	159, 176
Coleman Mine	302-3	Dorr Oliver Long	172
Colin, Larry	237-8	Dowe, Tom	159, 168
Cominco	149-51, 153-6, 167, 183, 188, 219-20, 250,254, 325, 336	Drillex	266
		Drillnamic Inc	269
Cominco Potash	326	Dudley, Darle	142
Cominco 2, 17, 20, 26, 102,119, 129,133-7,139-45		Duffield, Hedley	212
Cook, Freddie	47	Dufresne, Norm	262
Cook, Jim	173	Dujardin, Ray	275
Cook VB Company	172-3	Duke, Doug	178, 180-1, 184
Cook, Bill	18	Dumont, Alf	14
Cook, Ron	159	Dumont Mine	231-2, 234-8
Cooke, Bard	276	Dundas ON	139
Coott, Harry	160-1	Duplessis, Premier Maurice 63, 68
Corner Brook NF	262	Durston, Kieth	151-2
Coroccohuayco	274-5	Duval Potash	143,147, 149
Cougher, Harry	299-300	Dyck, Ed	221-23
Coutu, Wilfrid	227	Dynatec	231, 239, 256-8 277, 299-300
Cox, Herb	53, 62, 67, 337	Dynatec Engineering	203, 231, 248, 255, 260
Craigmont Mine	102	East Arm Property	245
Cranbrook BC1, 5, 7,10,19, 25, 49-50, 101, 114, 133- 5,138, 255, 279, 336		East Amphi	245
		Eastman, Jim	18
Crayston, E G	100	Eatock, Bill	257
Creighton Mine	110	Eaton, Cyrus	78
Crocker, Bunny	137, 326, 338	Ecstall	157, 160, 163-4, 167, 176
Croome, Tip (Norm)	326	Edmonstone, Neil	74
Cross, Julian	78	Edwards, Fred 239, 255, 257-8, 260-1, 282, 285, 298, 300
Crossfield AB	213-4, 216-7, 269		
Crouch, Professor Leslie 1		Eldrich-Flavel	229,236
Crowhurst, Jack	28-31, 337	Elf Aquataine	190, 337
Crowley, Dave	262	Elliot Lake ON	110
Curry, Bob	180-1	Ellis, Bill	153, 337
Dai, Zhen Fei	307, 322	Ellison Property	237
Dalton, Ken	239	Emes, Ron	146
Daryl Dudley	338	Engineerings Club	203, 205
Dasovich, John	181	Errington Mine 69, 73, 76, 78-84, 86, 92, 95, 240
David, Wayne	240	Errington, Joe	78
Davies, Don	221	Eskay Creek	250-3
Davies, Kenny	23, 102, 134	Esterhazy 100,112,114,117,120-1,126-28, 133, 139, 146,162, 182-3, 198, 258, 325
Davis, Cliff	276		
Davy-McKee	282-3	Fagan, Jim	302
Day, Alan	196	Fairbairn, Bart	327
de Guzman, Michael	245	Falconbridge 199, 200-01, 209, 225, 269, 302-3
Deason, Bill	301	Fearn, Bill	200
Defelice, Tony	162	Feniuk, Dave	120, 182
Dell, Dick	44	Ferderber Mine	231-8
Demers, Roger	238	Fernz	219-20, 222
Dempsey, Bob	52,54, 337	Fillip, Ned	17,140
Dengler, Bob	255	Finlayson, Dean	12-13
Devine, Premier Grant192-4, 230, 237		Fisher, Frank	271-2
Dew, John	137, 326	Fitzgerald, Fitz	74
Dewer, Ken	74, 76, 137, 325, 337	Flow-Through Shares 225-7, 236, 281
Dewey, Jim	110-11	Fluor Daniel	301-3, 305, 308, 310, 315-7, 327
Dingman, Roger	221		
Dison, Eric	192, 201, 207-8	Fluor Daniel Wright	203, 276, 299
Dobson, Sue	249	Foddal Verk	186
Dodes, Bill	333	Fogarty, Dr Charles	169, 190, 192, 337
Doelle, Henry	27, 34, 42, 337	Foraky	139, 173
Doherty, Jim	176	Forano	337
Donald, Ken	131-32	Fortin, Jim	239
Donaldson, Phil	162	Forward, Professor Frank 12

INDEX

Foss, Paul	221	Harris, Roger	163, 168, 215, 258
Fothergill, Ed	98	Harrison, Paddy	176-7
Fotheringham, Bill	262-3	Hartman, Mark	299-300
Fotheringham, Pop	74, 78, 130, 337	Harty, Fred	271-2
Fournier, Norm	164	Hatherly, Bill	62
Freese, Eugene	221	Haug, Mike	210-11, 213-5, 217-22, 256, 269, 273
Frey, Jack	7, 10, 47, 51, 63, 64, 69, 201		
Friesen, Dave	94	Hawes, Bill	141
Frietag, Carl	137	Hawthorne, Gary	330
Fuller Claim	229-30	Hay, Don	128
Gage, Kieth	161	Hayes, Doug	248
Gagnon, Serge	236-238	Heather, Al	100
Gallagher, Andy	161	Hegler, Wyatt	75-77, 231, 337
Gardner-Denver	44, 95	Hele, Gib	75
Garpensberg Mine	186	Helliwell, John	87-88
Garva, Ron	192	Helsinki, Finland	186, 188-9
Gauthier, Elmer	61	Henningson, Mick	147, 150
Geco Mines	98-99	Henry, John	45-46
General Chemical Corp	259	Herz, Herman C	47
Gibb, Wally	60	Heslop, John	200
Gibney, Bill	17	Highlands, Pappy	62
Gilbert, Bruce	196, 200, 337	Hill, Tony	166
Gilmore, Bill	204, 274, 299-300	Hinz, Ed	219
Giroux, Gary	285-7, 294	Hnatyshan, Leo & Joe	122-24, 126
Golden Patricia Mine	246-8	Hodgson, Stan	17, 134
Gonce, Ed	167	Hogarth	73, 86-87, 90, 92, 204
Good, Don	202	Hogarth, Major-General Don	78
Goodman Company	139, 140, 142-3	Holmlund, Murrary	220
Goodwin, Frank	133-4, 136-8, 142-3, 146, 150, 152-3, 337	Homestake Mining	140, 258-9
		Hope Brook Mine	205, 247, 262-5
Goodyear	337	Hope Princeton Highway	8
Gorf Contracting	174, 226, 337	Hopper, Peter	301
Gould, Ray & Roy	28-30	Horner, Al	16
Graham, Gene	61	Howard, Professor Henry	12
Graham, Jack	98	Howard, John	282-3
GRAND Canal	108	Howe, Gordie	122
Grantholm, Norm	217	Hu, Shi Jie	314
Great Wall of China	322-4	Huckleberry Mine	331
Green, Gar	110-11	Hucklebury Mine	247
Greene, Jack	114	Hughes, Ron	260-1
Gregoire, Yves	238-9	Hughes Bit Company	140
Grenville, Don Sr	157, 159, 179, 183	Hughes, Mike	200
Grenville, Don Jr	161, 183	HuiXian	316-7, 321
Gretzky, Wayne	122, 247	Hurdle, Bruce	133-4, 146, 150, 337
Guse, Allan	266	Husky Oil	210-11, 213-5, 217-8, 269, 273
Haileybury School of Mines	8, 106, 159-60, 164, 201	Huston, Bill	130, 148
		Huxhold, Peter	205, 245
Halet QB	52, 62	Hyder, Alaska	251-2
Hall, Ron	331-2	IMC	100, 112-14, 116, 118-22, 126-7, 129-33, 139, 142, 144-5, 146, 148, 193-5, 213, 258, 301, 325-6, 337
Hall, Corkie	161-2		
Hallanero, Tonelli	186-9		
Hallbauer, Bob	102		
Hallnor Mine	225-7	Imperial Metals Corp	331-2
Hambley, Dave	209	Inco	302-3
Hamilton, Gerry	75, 113, 117, 126-7, 145	Ingersol Rand	17, 61, 95, 337
Hammerstrom, Harry	338	Ingram, Don	330
Hancock, Sid	78-79	Inland Steel	78
Haney, John	205	Invermere BC	1, 27, 36
Harapiak, Steve	192-5, 212, 230-1, 237, 239, 282	Irrancana Ab	223
		Irving, Pappy	9
Hargrave, Bill	107	Isaac, Ernie	64, 68, 113-4, 116-7, 165
Harris, Willaim	10	Jack, Peter	17

INDEX

Jackpine BC 28, 30, 36, 39-40, 54, 63
Jackson, Ed 71
Jarvis-Clark 142, 167
Jean-Charles, Potvin 267
Jeffrey Company 142
Jensen, Henrey 44
Johansen, Maryellen 196
John Bertram Plant 139
Johnnie Mountain Mine 245
Johnson, Stuart 92-94, 119
Johnson, Monty 90-91
Johnstone, Walter 5
Johnstone, Jim 212-3, 239, 248
Jones, Jack 28
Joy Manufacturing 94, 142-3, 338
Jumbo Creek 39-40
Juneau, Alaska 276, 282, 284
Kalium 144,148
Karelse, Peter 276-9
Kazakhstan 284-6, 289-90, 295, 314,306, 336-7
Kearns, Desmond 205, 244, 246, 248
Kaukinen, Brian 192
Kelland, Dave 130, 133, 136, 147, 150
Kennedy, Mary 28
Kennedy, Dave 244-5, 250
Kennedy, Murray 61
Kennington Property 282-4
Kenora Prospectors and Miners Ltd 249
Kenrick, Tom 338
Ketza River Mine 239-41, 244, 247, 264, 277
Kiceluk, Walter 215
Kidd Creek 130,157, 164-5, 166, 170,178-9,183,188, 191-4, 197, 199, 202, 205,209, 225, 240, 242, 244, 251,255, 257, 259, 304, 336
Kidd Creek Potash 166, 191-2, 196, 199, 201, 207
Kierans, Tom 97-106, 108-10, 112, 114-6, 337
Kikanan, Jack 244
Kilborn 75,132,136-8,162, 302, 304, 308 325-8, 331
Kimberley BC 7,15,20-21,25,50,102,103,119, 133-7, 145, 153-4, 254
King of Sweden 243
Kingston ON 199, 202, 204, 205-6, 208, 216, 257, 299, 267-9
Kipnis, Norm 75
Kiruna Sweden 164,185
Kiruna Truck 262-3
KLM 7, 242, 305
Kneen, Tom 17
Knutsen, Chris 183
Korski, Steven 269
Kunkel, Curtis 145
Kvaerner 282, 284, 286, 296, 298
Kvaerner Davy 203
Kvaerner Metals 160
Kyle, Andy 113, 117-8, 120-22, 126-7, 129, 337
La Societe Miniere Louvem Inc 237
Lac Minerals 254
Lacroix, Andre 237
Laforest, Leon 162
Lakehead (Thunder Bay) ON 17,88-89,103
Lalonde, Mike 180-1, 183
Lang, Paul 325
LaPrairie, Leon 203, 229-30, 247
Larocque, Gerry 179
Larsen, Christine 45
Lavigne, Don 225-7
Leader, Jim 299, 302, 304-5
Leask, John 18
Lee, Raymond 213-5
Leo Allarie & Sons 173
Leonard, Kevin 249
Lesage, Premier Jean 104
Levesque, Premier Rene 237
Li, Chang 313
Li, Shou Cai 307, 311, 316, 322
Li, Zeming 308
Lindberg, Bob 112-5, 117, 128-29, 132
Lipic, Bob 266
Little, Douglas 17
Little America 261
LKAB Iron Mine 185
Lo, Patrick 305-6, 313, 316-7
Lockwood, Stewart 277
Lokhorst, Guy 119
Lokken Gruben 186
Longworth, Bill 225-6, 239
Lorimer, Mel 17
Louvem 236-7
Luborksy, Dean 221
Luckey Jim Mine 37-38, 41, 44, 54, 74
Luleå, Sweden 184-5
MacDonald, Glenn 331
MacDonald, Fuzzy 337
MacDougall, Cam 79-80
MacFarland, Jimmy 22
MacIsaac, Joe 165, 177
MacIsaac Mining 165, 174, 175-6, 178
MacIsaac, John 115, 177
MacLaren, Jimmy 22
MacLeod, Jim 245
Macneil, Alex 5-6
MacPhee, Doug 160
MacQuarrie, Roddie 23
Madsen Red Lake Mines 100
Mahood, Doug 191, 197, 201
Makila, Alex 59
Malartic Goldfields (MGF) 47,52,53-62,117,130,234 246, 337
Malartic QB 48, 52,54, 59, 62-65, 69,72,113-4,198, 231,233, 245
Malcohm, Shorty 106
Manitou Mill 237
Mannard, Dr George 190-91
Manning, Doug 160
Mao, Zedong 309

INDEX

Marcon, Dennis 266
Marlow, Jim 331-2
Marshall, Bill 106
Marshall, Merle 98-99, 157-8, 162, 165-6, 168-9, 176, 337
Marti, Christain 237
Martin, Dennis 269
Martin, Angus 62
Mather A & B Mines 87
Matheson, Don 193-4
Matousek, John 158
Matulich, Angelo 159
McColl, Ian 282
McCombe, Bob 240
McCreedy East 301-4
McCreedy, John 110-11, 301
McCroden, Pete 105-6
McDonald, Cam 337
McDonald, Johnnie 20, 22, 24
McDonald, Greg 113, 122, 145, 148-9, 181, 183
McDougall, Angus 17
McEwan, Rupurt 67
McIntosh, Jack 37, 43, 45, 47, 337
McIntosh, Harry 86
McIntyre Mine 86, 105, 226
McKay, Gordon & Elsie 26, 39
McKay, Marguerite 275
McKay, Jimmy 39-40
McKay, Anna Elizabeth 40
McKay, Big Jim 1, 26
McKay, Judy & Bill 1, 2, 3, 4, 8, 10, 26
McKay, Irene (Clark) 140, 158, 164, 168, 185-6, 188, 199, 242, 247, 257, 306
McKee, Maurice 286-7, 298
McMahon, Gordon 219
McManus, Jerome 163
McRae, Duncan 213-6
McRorrie, Ken 69, 78, 93, 204, 337
Meinzinger Bog 203
Melis, Lawrence 239-40
MengYin China 304, 306-7, 314, 322
Merrington, John 257
Merritt, Chip 161
Meta-Probe Inc 267
Metall Mining 274-5
Michaud, Pierre 338
Mick, Art 161
Midan, Ken 274, 299
Middleton, Stuart 219-20
Millar, Jim 18
Miller, Jack 332
Mineral King Mine 38-40
Mines Sullivan Inc 229
Mintronics 267
Mitchell, Jim 212-3
Mitchell, Jack 129-33, 302, 326
Mitchell, Bob 37, 39, 40, 41
Moffat, Bob 71
Mollison, Dick 163, 166-7, 169, 190, 337
Molloy, David 205, 244, 246, 250
Monaghan, Bob 161
Monenco-Agra 219-22
Montanore 255-8, 262
Montfort, Geoff 6
Moore, Edgar 18
Moore, Pat 248
Morash, Jim 61
Morin, Norm 268
Mottahed, Parvis 208
Mount Hope Hydro Inc 271
Mount Polley Mine 327-31
Mulligan, Harry 87-90, 337
Murray, Bob 207-8
Nagai, Eugene 211
NASDAQ 233
National Mine Service 140-41, 143
Nelson BC 19
Nethery, Bryan 274
Netolitzhy, Ron 254
Newton, Cliff 18
Nielson, Eric 221
Nightingale, John 130
Noranada Mines 98, 144, 149, 198, 255
Noranda QB 61
Norkum, Alfie 174, 337
Northern Miner 246, 248
NuFarm 210, 219, 223-4
Olafson, Elmer 240
Orocon Ltd 229, 246
Orotec Inc 258
Outokumpu Finland 188
Ozol, Nelda 13
O'Gorman, Glenn 299, 327-8
O'Halleron, Gerry 61
Padden, Jim 191-2
Palladium Resource Corp 213
Pamo Construction 238
Pamour Mine 106-7
Pan Asia Mining 305, 307, 314
Paradise Mine 27, 30-32, 36-38, 54, 74, 318
Parker, Don 268
Parker, Johnnie 31-32, 36, 337
Parsons, Don 328
Pascalis Nord 237
PCS 149, 191-4, 211, 212-3, 220, 230, 237, 257, 259
Peacock Ltd 227
Pearce, Geoff 246
Pellan, Floyd 211-2
Perry, Alan 157-8, 164
Petersen Mine 87-88
Phipps, Joe 120
Phoenix Belt Scrapers Ltd 268
Pieterse, Karl 177
Pigeon, Ray 162
Pigeon, Gerry 176
Pincock, Allen & Holt 257
Pinski, George 17
Pitts, Randy 261
Pleacock, Mike 37-38, 47
Po, Alexander 229, 245
Podwin, Ken 212-3
Poitevin, Dan 17, 337

INDEX

Polaris Taku	276-8, 280, 285
Popoff, Victor	233, 237-9
Porritt, John	267
Porteous, Dave	140, 158, 164, 181, 220, 253
Porter, Robin	133-5, 254, 337
Potash Company of America	17, 100, 139, 143, 148, 165
Prange, Duncan	209
Price, Ron	181
Price, Herb	258
Price, David	327-8, 330
Prichard, Vic & Dodie	28
Prodecor	214
Produ-Kake Process	268
Prokopchuk, Vic	180
Prugger, Fritz	147, 150-4
Pryor, Rick	163
Pyhasalmi Mine	187-8
Queen Elizabeth II	243
Quick, Martin	194
Ramsay, Bob	221
Ramsay, Jack	179-80, 182-3
Raynor, Rona	15
Reashore, Gil	163
Redpath	247, 255, 283, 299
Reuss, Bob	257-8
Revenue Canada	205, 226-8, 266, 268-9
Richardson, Hap	24
Roach, Dick	137, 326
Roach, Gerry	61
Roberts Pit	204
Roberts, Hugh	338
Robertson, Bob	42-44, 46
Robinski, Eli	338
Robinson, Dr Robbie	47
Rochon, George	155
Rockiron	97-98, 100, 106, 109, 112, 114-5, 137, 157, 301, 337-8
Ross, James Hamilton	239, 284
Ross River YN	239-40, 277
Ross, Barbara (McKay)	239, 283-4
Ross, Mike	229
Rossland BC	1-3, 19
Roth, Ed	139-40
Royal Canadian Mounted Police	1, 36, 114, 120, 145, 206
Royal Oak Resources	265
Ryan Award	179
Sadler, Jim	113, 127, 192, 211
Sadler, Bob	256
Sage, Red	160
Sainas Bill & George	14
Samuel, Watkin	78, 338
Sandvik Company	186
Sarnavka, Alwx	158
Saskatoon SK	17, 100-1, 118, 133, 136, 138, 143, 167, 145-8, 157, 162, 192, 196, 201, 204, 208, 210-11 219, 221, 238, 256, 258, 267
Saskterra	210-11, 213, 222, 238, 242, 257, 259, 266, 273
Sauve, Madame	243
Schauffer, Rudy	191
Scheuring, Joe	255
Schmidt, Hans	213, 217-9
Schmitke, Barry	212-3
Schmitt, Hans	137
Schultz, Bill	139, 142
Schwandt, Art	119
Scobie, Professor Malcolm	18
Scott, Bill	326
Scott, Alex	258
Scott, Angus	118, 127-28
Scott, Nat	205, 244
Scott, Bill	137
Scott, Alex	100
Scott, Angus & Hammish	23
Scoulding, Dave	221
Scragg, Bert	192
Serata, Dr Shoisi	118, 150, 260, 338
Shandong	304-6, 314, 317
Shannon, Bal	17
Shaver, Jack	51
Shaw, Joe	21
Sheel, Paul	271-2
Sheep Creek Gold	27, 29-30, 38, 40, 47, 53, 75, 337
Sherwood Mine	87-88
Shindle, Bill	17
Shively, Harry	147-8
Shoal Lake Property	249
Short, Ken	238
Simons Metals	327
Simons Mining & Metals	274-5
Simunovic, Paul	159
Sinclair, Ken	301
Skrobica, Bob	166
Skyline Property	245
Slack, Malcolm	202-3, 225-6, 230, 237
Slump, Sam	27-8
Smith, John	262
Smith, Hugh	98
Smith, Doogie	22
Solterra	217-9
Sorsa, Seppo	162
Southern Provincial Highway	5, 21
Spehar, Ed	162, 171, 175, 182, 202
Stanley Engineering	221
Stark, Dave	139
Starric Nick	34-35
Stearns-Roger	299, 336
Steep Rock	69, 71-74, 76-81, 83-87, 89-90, 95-96, 92, 97-99, 112-3, 118-9, 121, 130, 137, 154, 145, 148, 204-5, 227, 231, 240, 160, 163, 178-9, 183, 192 299, 220, 222, 263, 302, 325-6, 336-7
Stevens, Earl	222
Stevens, Roy	2-3
Stewart BC	245, 250-1, 253
Stewart, Mike	201, 204, 207-8, 210, 239
Stewart, Johnie	23
Stirling, Doug	237-8
Stockholm Sweden	185-6, 189-90, 242
Stone & Webster Engineering	257, 271

INDEX

Stubb, Ken 166
Sudbury ON 17, 62, 68-70, 98, 103,104, 107, 110 112, 114-5, 168, 177, 266, 269, 301
Sulchem 210, 213-4, 222, 269
SulFer Works 210, 222, 224
Sullivan Mine 7,19,20,22,25,50,74,102-3, 133-4
Sun, Yi (Danny) 305, 307
Sunshine Mine 255, 299-300
Sutherland, Mary 68, 72, 115
Swellnus, Juks 338
Swift Company 140
Sylvite (PCS Rocanville) 144
T-H Arches 99, 106
Tadanac Concetrator 25-26
Tampere Finland 186-7
Tamrock Company 186, 267
Tanaka, Vic 244
Tanguay, Louis 62
Tashme BC 8-9
Taylor, David P 307
Taylor, Stan (Buck) 13
Taylor, Bob 161
Tenajon Property 245
Tereschenko, Ada 286-7, 292
Tessier, Pierre 238-9
Texasgulf Potash 191
Texasgulf 77, 157, 162-3, 166-7, 170-1, 253,173,174 175,177-9,190, 190, 197-8, 157, 162-3, 166-7, 170-1 253, 336
Texasgulf Trona 258-9
Texasgulf-Kidd Creek 61, 99, 107, 160, 163, 167, 173, 301, 337
Thiessen, Larry 44
Thomson, Bart 130, 159, 164-5, 167-8, 176, 179,191, 337
Thomson, Jim 17, 101
Thornton, Sid 28
Thorpe, Anthony 222
Tiley & Assoc 227, 285, 337
Tiley, Peter 285-7, 292
Tiley, Gerald 170-2,175, 177, 337
Timmins ON 17, 54, 57, 61, 64 99, 104-6, 153,186 ,157-8, 162, 164, 167,168-9, 173, 182, 177, 196-8, 225-6, 229, 231,238, 248, 251,258, 253-4, 257
Toby Creek 2, 28, 39-40
Toivanen, Jack 51
Tomco Steel 270
Tompkins, Charlie 337
Torcars 140-42
Totten, Wayne 161
Trail BC 1, 19, 25, 38, 46,133-5,143,145, 155, 326
Trondheim 186
Trotter, Ken 294-6
Trowsdale, George 18
Tucker, Bob 334
Tulley, John 276, 299, 327
Tun, Aye Sai 307-8
Twaddle, Jim 16
U.B.C. 1, 4, 8, 12, 14-15, 31,101-2, 113,119,140 258 307

Upham, Merv 75-6, 86, 100, 112-3, 115, 117,127-31, 325, 337
Upham, Kathryn 115
US Borax 118, 139-40, 144, 148, 150, 259
Ust Kamenogorst 289, 291-4, 297
Vaclavek, Otto 161
Val d'Or 47, 49-52, 59, 61, 63-64, 104-5, 137, 225, 227, 231-4, 237, 250, 262
Van Hees, Ed 227-9
Vanscoy, SK 133, 156
Vedron 203, 229-30, 234, 238, 242, 245
Verreault, Jacques 181
Vicslo, Vaclav 163
Victoria BC 211
Vihantti Mine 187-8
Virginiatown ON 231, 238
Vogt, Anders 250
Vukovich, Ed 159
Vuonos Mine 188
Wagenknecht, Hans 159
Walsh, Al 257
Walton, Terry 225
Wang, Prof Wen Qian 307, 314
Wang, Bao Ming 317, 320-2
Watkinson, Dave 278-9
Waugh, Johnnie 61,130
Wear, Dean 261
Wearing, John 257-8, 271-2
Webber Construction 144-6
Webber, Bob 102
West, Mark 161
Westinghouse 337
Westmin Mine 245, 252
Whelan, Dave 220-1
Whissel, Larry 176
White, Maurice 271-2
White Pine Copper 141
Whitehorse YK 277, 280
Wilkins, Charlie 5
Wilkinson, Jim 260
Williams, Stan 128, 29
Williams, York 106
Williston, Doug 208, 279-80
Williston, Mark 280
Willoughby Property 245
Wilson, Steve 221
Wilson, Michael 205
Wilton, Clarence 60-61
Winters, Steve 151-2
Witte, Peggy 265
Wolfe, Bill 155
World Trade Centre 272-3
Wortman, Dave 299, 300
Wright, Harold 337
Wright, George 328
Wright Engineers 204, 208, 274, 299-300, 327, 337
Wrightbar 234-8
Wrona, Fred 164
WuHan Institute 308-10, 313, 315-6
XiYu Project 314, 322
Yanisiew, Bill 77, 163

INDEX

Yates, Ed	139-43, 147	Zahary, George	118, 127
Yorbeau	237-8	Zawadski, Bill	64
Young, Bob	220	Zhang, Yu Ling	308-9
Young, Roy	159	Zhau, Zhiyou	317
Youngblut, Kieth	179	Zincton BC	37, 41, 44-48, 63, 95
Yu, Dao Gui	313	ZouPing China	316-7, 321-2
Yuan, Shurong	317	Zucchiatti, Louis	74
Yuskiw, Ernie	174, 176-7	Zuke, Jack	330-1